RADAR SIGNAL ANALYSIS AND PROCESSING USING MATLAB®

RADAR SIGNAL ANALYSIS AND PROCESSING USING MATLAB®

Bassem R. Mahafza

deciBel Research Inc.
Huntsville, Alabama, U.S.A.

CRC Press
Taylor & Francis Group
Boca Raton London New York

CRC Press is an imprint of the
Taylor & Francis Group, an informa business

A CHAPMAN & HALL BOOK

Chapman & Hall/CRC
Taylor & Francis Group
6000 Broken Sound Parkway NW, Suite 300
Boca Raton, FL 33487-2742

© 2009 by Taylor & Francis Group, LLC
Chapman & Hall/CRC is an imprint of Taylor & Francis Group, an Informa business

No claim to original U.S. Government works
Printed in the United States of America on acid-free paper
10 9 8 7 6 5 4 3 2 1

International Standard Book Number-13: 978-1-4200-6643-2 (Hardcover)

Library of Congress Cataloging-in-Publication Data

Mahafza, Bassem R.
  Radar signal analysis and processing using MATLAB / Bassem R. Mahafza.
    p. cm.
  "A CRC title."
  Includes bibliographical references and index.
  ISBN 978-1-4200-6643-2 (hardback : alk. paper)
  1. Radar cross sections. 2. Signal processing. 3. Radar targets. 4. MATLAB. I. Title.

TK6575.M267 2008
621.3848--dc22   2008014584

Visit the Taylor & Francis Web site at
http://www.taylorandfrancis.com

and the CRC Press Web site at
http://www.crcpress.com

To my four sons:

Zachary,

Joseph,

Jacob, and

Jordan

Table of Contents

Preface

Chapter 1

Radar Systems - An Overview 1

Chapter 2

Linear Systems and Complex Signal Representation 83

Chapter 8

Pulse Compression   315

Chapter 9

Radar Clutter 353

Chapter 10

Doppler Processing 403

Chapter 11

Adaptive Array Processing 429

Preface

In the year 2000 my book *Radar Systems Analysis and Design Using MAT-LAB*[1]® was published. This book very quickly turned into a bestseller which prompted the publication of its second edition in the year 2005. At the time of its publication, it was based on my years of teaching graduate level courses on radar systems analysis and design including advanced topics in radar signal processing. The motivation behind it was to introduce a college-suitable comprehensive textbook that provides hands-on experience with MATLAB® companion software. Over the years, I have also taught numerous industry courses on the subject of radar systems. Based on my combined teaching experience and real-world work at deciBel Research, Inc., the following conclusion has become very evident to me: There is big appetite and demand for textbooks and reference books that are primarily focused on aspects of radar signals and signal processing. Having arrived at this conclusion, I decided to write this textbook, *Radar Signal Analysis and Processing Using MATLAB*®, which is focused on radar signal analysis and processing.

Unlike other books on the subject, the emphasis is not on signal processing per se, but on signals and signal processing in the context of radar applications. Many good textbooks are already available on signal processing but not on signal processing as it applies to radar applications. This new textbook has many desirable features that include clear and concise presentation of the theory and companion user-friendly MATLAB code. This code is reconfigurable to demonstrate the theory and perform the associated analysis/design trades as well as allow users to vary the inputs in order to better analyze their relevant and unique requirements. This new book should serve as a reference book or as a textbook for a graduate level courses on the subject. It concentrates on the fundamentals and adopts a rigorous mathematical approach of the subject. Many examples and end of chapter problems are included. Finally, a companion Instructor's Manual is also available through the publisher for professors who adopt this book as a text. The Instructor's Manual includes many other problems not listed in the text and their solutions.

---

1. All MATLAB® functions and programs provided in this book were developed using MATLAB R2007b with the Signal Processing Toolbox, on a PC with Windows XP Professional operating system.
® MATLAB® is a registered trademark of the The MathWorks, Inc. For product information, please contact: The MathWorks, Inc., 3 Apple Hill Drive, Natick, MA 01760-2098 USA. Web: *www.mathworks.com*.

Radar Signal Analysis and Processing Using MATLAB® is written so that it can be used as a reference book or as a textbook for two graduate level courses with emphasis on signals and signal processing. Instructors using this book as a text may choose the following chapter breakdown for their curriculum. Chapters 1 through Chapter 7 can be used for the first course, while Chapters 8 through 11 may be used for the second course. Chapter 11 (Target Tracking), Chapter 12 (Synthetic Aperture Radar), and Chapter 13 (Radar Cross Section) from my other book *Radar Systems Analysis and Design Using MATLAB®* may also be used to supplement both courses.

Radar Signal Analysis and Processing Using MATLAB® introduces numerous programs and functions of MATLAB using version R2007a. All MATLAB programs and functions provided in this book can be downloaded from the CRC Press Website. For this purpose and using your favorite Internet browser type in *www.crcpress.com* and hit return. Once you reach the main CRC Press home page, scroll down to the link called *"Electronic Products"* and double click on *"Downloads & Updates,"* then follow the instructions on the screen.

Chapter 1 of this book presents an overview of radar systems operation and design. The approach is to derive the radar range equation and analyze the different radar parameters in the context of this radar equation. The surveillance radar equation is derived. Special topics that affect radar signal processing are presented and analyzed in the context of the radar equation. This includes the effects of system noise, wave propagation, jamming, and target Radar Cross Section (RCS). Chapter 2 introduces a top level review of elements of signal theory that are relevant to radar detection and radar signal processing. It is assumed that the reader has sufficient and adequate background in signals and systems as well as in the Fourier transform and its associated properties.

In Chapter 3 a review of random variables and processes is presented. Instructors using this text may assume that students have already acquired the necessary background as a prerequisite to this course and, thus, may elect to omit this chapter from their syllabus, except for Section 3.6. Chapter 4 is focused on the matched filter. It presents the unique characteristic of the matched filter and develops a general formula for the output of the matched filter that is valid for any waveform. Chapters 5 and 6 analyze the output of the matched filter in the context of the ambiguity function. In Chapter 5 several analog waveforms are analyzed; this includes the single unmodulated pulse, the Linear Frequency Modulation (LFM) pulse, unmodulated pulse train, LFM pulse train, stepped frequency waveforms, and nonlinear FM waveforms. Chapter 6 is concerned with discrete coded waveforms. In this chapter, unmodulated pulse-train codes are analyzed as well as binary codes, polyphase codes, and frequency codes.

Chapter 7 introduces the subject of radar target detection and pulse integration. Swerling models are analyzed in the context of noncoherent integration and the square law detector. The topic of Constant False Alarm Rate (CFAR) is also presented in detail. Chapter 8 introduces the most common techniques in radar signal processing. The matched filter receiver as well as the stretch processor receiver are analyzed. Chapter 9 is concerned with radar clutter. Comprehensive analysis of the subject of clutter is introduced, including the Moving Target Indicator (MTI). Chapter 10 is primarily concerned with radar Doppler processing. Both continuous wave and pulsed radars are considered. Pulse Doppler radars are introduced and analyzed. Chapter 11 is focused on adaptive array processing. For this purpose, a top level overview of phased array antennas is first introduced followed by beamforming and the most common techniques in adaptive array processing.

Bassem R. Mahafza
bmahafza@dbresearch.net
Huntsville, AL
February 2008

Chapter 1   *Radar Systems - An Overview*

This chapter presents an overview of radar systems operation and design. The approach is to introduce few definitions first, followed by detailed derivation of the radar range equation. Different radar parameters are analyzed in the context of the radar equation. The search or surveillance radar equation will also be derived. Where appropriate, a few examples are introduced. Special topics that affect radar signal processing are also presented and analyzed in the context of the radar equation. This includes the effects of system noise, wave propagation, jamming, and target Radar Cross Section (RCS).

1.1. Range Measurements

Consider a radar systems that transmits a periodic sequence, with period T, of square pulses, each of width τ, shown in Fig. 1.1. The period is referred to as the Pulse Repetition Interval (PRI) and the inverse of the PRI is called the Pulse Repetition Frequency (PRF), denoted by f_r. If the peak transmitted power for each pulse is referred to as P_t, then the average transmitted power over one full period is

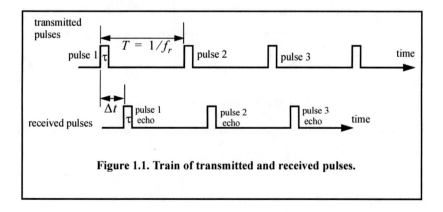

Figure 1.1. Train of transmitted and received pulses.

$$P_{av} = P_t \times \frac{\tau}{T} \tag{1.1}$$

The ratio of the pulse width to the PRI is called transmit duty cycle, denoted by dt. The pulse energy is $E_x = P_t\tau = P_{av}T = P_{av}/f_r$.

The top portion of Fig. 1.1 represents the transmitted sequence of pulses, while the lower portion represents the received radar echoes reflected from a target at some range R. By measuring the two-way time delay, Δt, the radar receiver can determine the range as follows:

$$R = \frac{c\Delta t}{2} \tag{1.2}$$

where: $c = 3 \times 10^8 m/s$ is the speed of light, and the factor 2 is used to account for the round trip (two-way) delay.

The range corresponding to the two-way time delay $\Delta t = T$, where T is the pulse repetition interval is referred to as the radar unambiguous range, R_u. Consider the case shown in Fig. 1.2. Echo 1 represents the radar return from a target at range $R_1 = c\Delta t/2$ due to pulse 1. Echo 2 could be interpreted as the return from the same target due to pulse 2, or it may be the return from a far-away target at range R_2 due to pulse 1 again. That is,

$$R_{2a} = \frac{c\Delta t}{2} \quad or \quad R_{2b} = \frac{c(T+\Delta t)}{2} \tag{1.3}$$

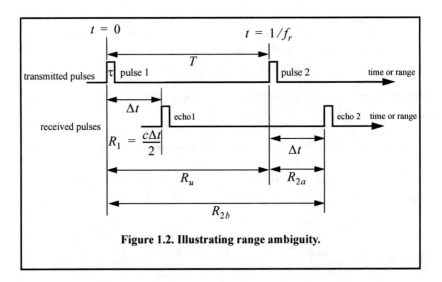

Figure 1.2. Illustrating range ambiguity.

Clearly, range ambiguity is associated with echo 2. Once a pulse is transmitted, the radar must wait a sufficient length of time so that returns from targets at maximum range are back before the next pulse is emitted. It follows that the maximum unambiguous range must correspond to half of the PRI:

$$R_u = c\frac{T}{2} = \frac{c}{2f_r} \tag{1.4}$$

Example:

A certain airborne pulsed radar has peak power $P_t = 10KW$ *and uses two PRFs,* $f_{r1} = 10KHz$ *and* $f_{r2} = 30KHz$. *What are the required pulse widths for each PRF so that the average transmitted power is constant and is equal to* $1500\,Watts$? *Compute the pulse energy in each case.*

Solution:

Since P_{av} *is constant, both PRFs have the same duty cycle,*

$$d_t = \frac{1500}{10 \times 10^3} = 0.15$$

The pulse repetition intervals are

$$T_1 = \frac{1}{10 \times 10^3} = 0.1\,ms$$

$$T_2 = \frac{1}{30 \times 10^3} = 0.0333\,ms$$

It follows that

$$\tau_1 = 0.15 \times T_1 = 15\,\mu s$$

$$\tau_2 = 0.15 \times T_2 = 5\,\mu s$$

$$E_{x1} = P_t\tau_1 = 10 \times 10^3 \times 15 \times 10^{-6} = 0.15 \ Joules$$

$$E_{x2} = P_2\tau_2 = 10 \times 10^3 \times 5 \times 10^{-6} = 0.05 \ Joules$$

1.2. Range Resolution

Range resolution, denoted as ΔR, is a radar metric that describes its ability to detect targets in close proximity to each other as distinct objects. Radar systems are normally designed to operate between a minimum range R_{min} and maximum range R_{max}. The distance between R_{min} and R_{max} along the radar line of sight is divided into M range bins (gates), each of width ΔR,

$$M = \frac{R_{max} - R_{min}}{\Delta R} \tag{1.5}$$

Targets separated by at least ΔR will be completely resolved in range.

In order to derive an exact expression for ΔR, consider two targets located at ranges R_1 and R_2, corresponding to time delays t_1 and t_2, respectively. This is illustrated in Fig. 1.3. Denote the difference between those two ranges as ΔR:

$$\Delta R = R_2 - R_1 = c\frac{(t_2 - t_1)}{2} = c\frac{\delta t}{2} \tag{1.6}$$

The question that needs to be answered is: What is the minimum time, δt, such that target 1 at R_1 and target 2 at R_2 will appear completely resolved in range (different range bins)? In other words, what is the minimum ΔR?

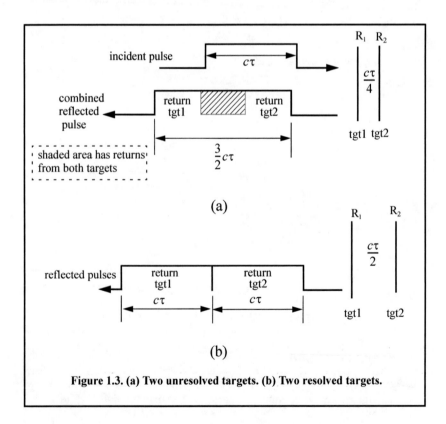

Figure 1.3. (a) Two unresolved targets. (b) Two resolved targets.

First, assume that the two targets are separated by $c\tau/4$, τ is the pulse width. In this case, when the pulse trailing edge strikes target 2, the leading edge would have traveled backward a distance $c\tau$, and the returned pulse would be composed of returns from both targets (i.e., unresolved return), as shown in Fig. 1.3a. If the two targets are at least $c\tau/2$ apart, then as the pulse trailing edge strikes the first target, the leading edge will start to return from target 2, and two distinct returned pulses will be produced, as illustrated by Fig. 1.3b. This means ΔR should be greater or equal to $c\tau/2$. Since the radar bandwidth B is equal to $1/\tau$, then

$$\Delta R = \frac{c\tau}{2} = \frac{c}{2B} \tag{1.7}$$

In general, radar users and designers alike seek to minimize ΔR in order to enhance the radar performance. As suggested by Eq. (1.7), in order to achieve fine range resolution one must minimize the pulse width. This will reduce the average transmitted power and increase the operating bandwidth. Achieving fine range resolution while maintaining adequate average transmitted power can be accomplished by using pulse compression techniques.

Example:

A radar system has an unambiguous range of 100 Km and a bandwidth 0.5 MHz. Compute the required PRF, PRI, ΔR, and τ.

Solution:

$$PRF = \frac{c}{2R_u} = \frac{3 \times 10^8}{2 \times 10^5} = 1500 \ Hz$$

$$PRI = \frac{1}{PRF} = \frac{1}{1500} = 0.6667 \ ms$$

It follows,

$$\Delta R = \frac{c}{2B} = \frac{3 \times 10^8}{2 \times 0.5 \times 10^6} = 300 \ m$$

$$\tau = \frac{2\Delta R}{c} = \frac{2 \times 300}{3 \times 10^8} = 2 \ \mu s$$

1.3. Doppler Frequency

Radars use Doppler frequency to extract target radial velocity (range rate), as well as to distinguish between moving and stationary targets or objects, such as clutter. The Doppler phenomenon describes the shift in the center frequency of an incident waveform due to the target motion with respect to the source of radiation. Depending on the direction of the target's motion, this frequency shift may be positive or negative. A waveform incident on a target has equiphase wavefronts separated by λ, the wavelength. A closing target will cause the reflected equiphase wavefronts to get closer to each other (smaller wavelength). Alternatively, an opening or receding target (moving away from the radar) will cause the reflected equiphase wavefronts to expand (larger wavelength), as illustrated in Fig. 1.4.

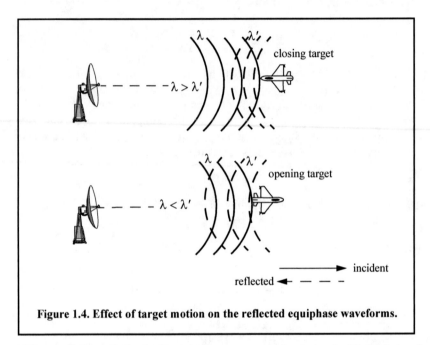

Figure 1.4. Effect of target motion on the reflected equiphase waveforms.

The result formula for the Doppler frequency can be derived with the help of Fig. 1.5. Assume a target closing on the radar with radial velocity (target velocity component along the radar line of sight) v. Let R_0 refer to the range at time t_0 (time reference); then the range to the target at any time t is

$$R(t) = R_0 - vt \qquad (1.8)$$

Assume a radar transmitted signal given by

$$x(t) = A\cos(2\pi f_0 t) \qquad (1.9)$$

where f_0 is the radar operating center frequency. It follows that the signal received by the radar is

$$x_r(t) = x(t - \phi(t)) \qquad (1.10)$$

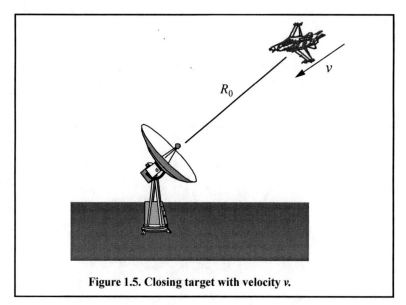

Figure 1.5. Closing target with velocity v.

where

$$\phi(t) = \frac{2}{c}(R_0 - vt) \tag{1.11}$$

Substituting Eq. (1.9) and Eq. (1.11) into Eq. (1.10) and collecting terms yields

$$x_r(t) = A_r \cos\left[2\pi\left(f_0 t - f_0\frac{2R_0}{c} + \frac{2f_0 vt}{c}\right)\right] \tag{1.12}$$

where A_r is a constant. The phase term

$$\psi_0 = 2\pi f_0 \frac{2R_0}{c} \tag{1.13}$$

is used to measure initial target detection range, and the term $2f_0 v/c$ represents a frequency shift due to target velocity (i.e., Doppler frequency shift). The Doppler frequency is given by

$$f_d = \frac{2f_0 v}{c} = \frac{2v}{\lambda} \tag{1.14}$$

where λ is the wavelength given by

$$\lambda = \frac{c}{f_0} \tag{1.15}$$

Note that if the target were going away from the radar (opening or receding target), then

$$f_d = -\frac{2f_0 v}{c} = -\frac{2v}{\lambda} \qquad (1.16)$$

as illustrated in Fig. 1.6.

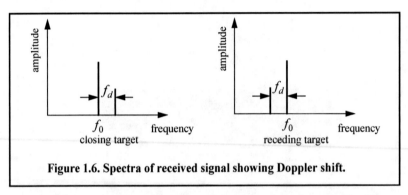

Figure 1.6. Spectra of received signal showing Doppler shift.

In general the target Doppler frequency depends on the target velocity component in the direction of the radar (radial velocity). Figure 1.7 shows three targets all having velocity v. Target 1 has zero Doppler shift; target 2 has maximum Doppler frequency as defined in Eq. (1.15). The amount of Doppler frequency of target 3 is $f_d = 2v\cos\theta/\lambda$, where $v\cos\theta$ is the radial velocity; and θ is the total angle between the radar line of sight and the target.

A more general expression for f_d that accounts for the total angle between the radar and the target is

$$f_d = \frac{2v}{\lambda}\cos\theta \qquad (1.17)$$

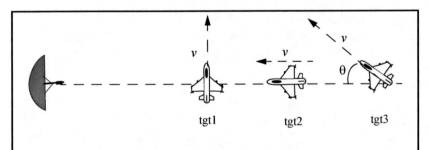

Figure 1.7. Target 1 generates zero Doppler. Target 2 generates maximum Doppler. Target 3 is in between.

and for an opening target is

$$f_d = \frac{-2v}{\lambda}\cos\theta \qquad (1.18)$$

where $\cos\theta = \cos\theta_e \cos\theta_a$. The angles θ_e and θ_a are, respectively, the elevation and azimuth angles; see Fig. 1.8.

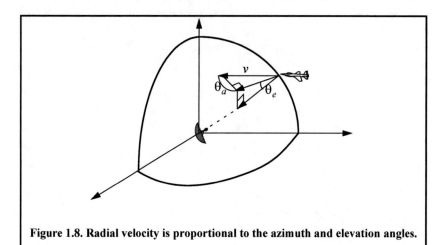

Figure 1.8. Radial velocity is proportional to the azimuth and elevation angles.

Example:

Compute the Doppler frequency measured by the radar shown in the figure below.

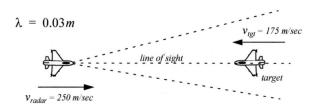

Solution:

The relative radial velocity between the radar and the target is $v_{radar} + v_{tgt}$. *Using Eq. (1.15) yields*

$$f_d = 2\frac{(250 + 175)}{0.03} = 28.3\,KHz$$

Similarly, if the target were opening, the Doppler frequency is

$$f_d = 2\frac{250 - 175}{0.03} = 5KHz$$

1.4. Coherence

A radar is said to be coherent if the phase of any two transmitted pulses is consistent; i.e., there is a continuity in the signal phase from one pulse to the next. One can view coherence as the radar's ability to maintain an integer multiple of wavelengths between the equiphase wavefront from the end of one pulse to the equiphase wavefront at the beginning of the next pulse. Coherency can be achieved by using a STAble Local Oscillator (STALO). A radar is said to be coherent-on-receive or quasi-coherent if it stores in its memory a record of the phases of all transmitted pulses. In this case, the receiver phase reference is normally the phase of the most recently transmitted pulse.

Coherence also refers to the radar's ability to accurately measure (extract) the received signal phase. Since Doppler represents a frequency shift in the received signal, only coherent or coherent-on-receive radars can extract Doppler information. This is because the instantaneous frequency of a signal is proportional to the time derivative of the signal phase.

1.5. The Radar Equation

Consider a radar with an isotropic antenna (one that radiates energy equally in all directions). Since these kinds of antennas have a spherical radiation pattern, we can define the peak power density (power per unit area) at any point in space as

$$P_D = \frac{Peak\ transmitted\ power}{area\ of\ a\ sphere} \qquad \frac{watts}{m^2} \qquad \text{(1.19)}$$

The power density at range R away from the radar (assuming a lossless propagation medium) is

$$P_D = P_t / (4\pi R^2) \qquad \text{(1.20)}$$

where P_t is the peak transmitted power and $4\pi R^2$ is the surface area of a sphere of radius R. Radar systems utilize directional antennas in order to increase the power density in a certain direction. Directional antennas are usually characterized by the antenna gain G and the antenna effective aperture A_e. They are related by

$$G = (4\pi A_e)/\lambda^2 \qquad \text{(1.21)}$$

where λ is the wavelength. The relationship between the antenna's effective aperture A_e and the physical aperture A is

$$A_e = \rho A \qquad (1.22)$$
$$0 \le \rho \le 1$$

ρ is referred to as the aperture efficiency, and good antennas require $\rho \to 1$. In this book we will assume, unless otherwise noted, that A and A_e are the same. We will also assume that antennas have the same gain in the transmitting and receiving modes. In practice, $\rho \approx 0.7$ is widely accepted.

The gain is also related to the antenna's azimuth and elevation beam widths by

$$G = K\frac{4\pi}{\theta_e \theta_a} \qquad (1.23)$$

where $K \le 1$ and depends on the physical aperture shape; the angles θ_e and θ_a are the antenna's elevation and azimuth beam widths, respectively, in radians. When the antenna has a continuous aperture, an excellent approximation of Eq. (1.23) can be written as

$$G \approx \frac{26000}{\theta_e \theta_a} \qquad (1.24)$$

where in this case the azimuth and elevation beam widths are given in degrees.

The power density at a distance R away from a radar using a directive antenna of gain G is then given by

$$P_D = \frac{P_t G}{4\pi R^2} \qquad (1.25)$$

When the radar radiated energy impinges on a target, the induced surface currents on that target radiate electromagnetic energy in all directions. The amount of the radiated energy is proportional to the target size, orientation, physical shape, and material, which are all lumped together in one target-specific parameter called the Radar Cross Section (RCS) denoted by σ.

The radar cross section is defined as the ratio of the power reflected back to the radar to the power density incident on the target,

$$\sigma = \frac{P_r}{P_D} \; m^2 \qquad (1.26)$$

where P_r is the power reflected from the target. The total power delivered to the radar receiver at the back-end of the antenna is

$$P_r = \frac{P_t G \sigma}{(4 \pi R^2)^2} A_e \tag{1.27}$$

Substituting the value of A_e from Eq. (1.21) into Eq. (1.27) yields

$$P_r = \frac{P_t G^2 \lambda^2 \sigma}{(4\pi)^3 R^4} \tag{1.28}$$

Let S_{min} denote the minimum detectable signal power. It follows that the maximum radar range R_{max} is

$$R_{max} = \left(\frac{P_t G^2 \lambda^2 \sigma}{(4\pi)^3 S_{min}} \right)^{1/4} \tag{1.29}$$

Equation (1.29) suggests that in order to double the radar maximum range one must increase the peak transmitted power P_t sixteen times; or equivalently, one must increase the effective aperture four times.

In practical situations the returned signals received by the radar will be corrupted with noise, which introduces unwanted voltages at all radar frequencies. Noise is random in nature and can be described by its Power Spectral Density (PSD) function. The noise power N is a function of the radar operating bandwidth, B. More precisely

$$N = Noise\ PSD \times B \tag{1.30}$$

The receiver input noise power is

$$N_i = kT_0 B \tag{1.31}$$

where $k = 1.38 \times 10^{-23}\ Joule/degree\ Kelvin$ is Boltzmann's constant, and $T_0 = 290$ is the receiver input noise temperature in degrees Kelvin. It is always desirable that the minimum detectable signal (S_{min}) be greater than the noise power. The sensitivity of a radar receiver is normally described by a figure of merit called the noise figure F (see Section 1.9 for details). The noise figure is defined as

$$F = \frac{(SNR)_i}{(SNR)_o} = \frac{S_i/N_i}{S_o/N_o} \tag{1.32}$$

$(SNR)_i$ and $(SNR)_o$ are, respectively, the Signal to Noise Ratios (SNR) at the input and output of the receiver. The input signal power is S_i; and the input noise power immediately at the antenna terminal is N_i. The values S_o and N_o are, respectively, the output signal and noise power.

The receiver effective noise temperature excluding the antenna is (see Section 1.9)

$$T_e = T_0(F-1) \tag{1.33}$$

where F is the receiver noise figure. It follows that the total effective system noise temperature T_s is given by

$$T_s = T_e + T_a = T_0(F-1) + T_a = T_0F - T_0 + T_a \tag{1.34}$$

where T_a is the antenna temperature.

In many radar applications it is desirable to set the antenna temperature T_a to T_0 and thus, Eq. (1.34) is reduced to

$$T_s = T_0F \tag{1.35}$$

Using Eq. (1.35) and Eq. (1.31) in Eq. (1.32) yields

$$S_i = kT_0BF(SNR)_o \tag{1.36}$$

The minimum detectable signal power can be written as

$$S_{min} = kT_0BF(SNR)_{o_{min}} \tag{1.37}$$

The radar detection threshold is set equal to the minimum output SNR, $(SNR)_{o_{min}}$. Substituting Eq. (1.37) in Eq. (1.29) gives

$$R_{max} = \left(\frac{P_tG^2\lambda^2\sigma}{(4\pi)^3kT_0BF(SNR)_{o_{min}}} \right)^{1/4} \tag{1.38}$$

or equivalently,

$$(SNR)_{o_{min}} = \frac{P_tG^2\lambda^2\sigma}{(4\pi)^3kT_0BFR_{max}^4} \tag{1.39}$$

In general, radar losses denoted as L reduce the overall SNR, and hence

$$(SNR)_o = \frac{P_tG^2\lambda^2\sigma}{(4\pi)^3kT_0BFLR^4} \tag{1.40}$$

Equivalently, Eq. (1.40) can be rewritten using Eq. (1.35) as

$$(SNR)_o = \frac{P_tG^2\lambda^2\sigma}{(4\pi)^3kT_sBLR^4} \tag{1.41}$$

In this book, the antenna temperature is assumed to be negligible; therefore, Eq. (1.40) will be dominantly used as the Radar Equation.

Example:

Given a certain C-band radar with the following parameters: Peak power $P_t = 1.5 MW$, operating frequency $f_0 = 5.6 GHz$, antenna gain $G = 45 dB$, effective temperature $T_0 = 290 K$, noise figure $F = 3 dB$, pulse width $\tau = 0.2 \mu sec$. The radar threshold is $(SNR)_{min} = 20 dB$. Assume target cross section $\sigma = 0.1 m^2$. Compute the maximum range.

Solution:

The radar bandwidth is

$$B = \frac{1}{\tau} = \frac{1}{0.2 \times 10^{-6}} = 5 MHz$$

The wavelength is

$$\lambda = \frac{c}{f_0} = \frac{3 \times 10^8}{5.6 \times 10^9} = 0.054 m$$

From Eq. (1.40) we have

$$(R^4)_{dB} = (P_t + G^2 + \lambda^2 + \sigma - (4\pi)^3 - kT_0B - F - (SNR)_{o_{min}})_{dB}$$

where, before summing, the dB calculations are carried out for each of the individual parameters on the right-hand side. We can now construct the following table with all parameters computed in dB:

P_t	λ^2	G^2	kT_0B	$(4\pi)^3$	F	$(SNR)_{o_{min}}$	σ
61.761	−25.421	90 dB	−136.987	32.976	3 dB	20 dB	−10

It follows that

$$R^4 = 61.761 + 90 - 25.352 - 10 - 32.976 + 136.987 - 3 - 20 = 197.420 dB$$

$$R^4 = 10^{(197.420/10)} = 55.208 \times 10^{18} m^4$$

$$R = \sqrt[4]{55.208 \times 10^{18}} = 86.199 Km$$

Thus, the maximum detection range is 86.2 Km.

Figure 1.9 shows plots of the SNR versus detection range for the following parameters: Peak power $P_t = 1.5MW$, operating frequency $f_0 = 5.6GHz$, antenna gain $G = 45dB$, radar losses $L = 6dB$, and noise figure $F = 3dB$. The radar bandwidth is $B = 5MHz$. The radar minimum and maximum detection ranges are $R_{min} = 25Km$ and $R_{max} = 165Km$. This figure can be reproduced using the following MATLAB code which utilizes MATLAB function *"radar_eq.m."*

```
close all;
clear all
pt = 1.5e+6; % peak power in Watts
freq = 5.6e+9; % radar operating frequency in Hz
g = 45.0; % antenna gain in dB
sigma = 0.1; % radar cross section in m squared
b = 5.0e+6; % radar operating bandwidth in Hz
nf = 3.0; %  noise figure in dB
loss = 6.0; % radar losses in dB
range = linspace(25e3,165e3,1000);
snr = radar_eq(pt, freq, g, sigma, b, nf, loss, range);
rangekm = range ./ 1000;
plot(rangekm,snr,'linewidth',1.5)
grid;
xlabel ('Detection range in Km');
ylabel ('SNR in dB');
```

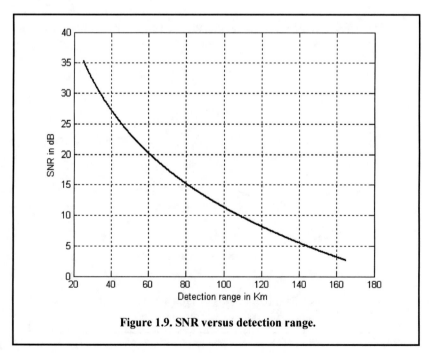

Figure 1.9. SNR versus detection range.

1.6. *Surveillance Radar Equation*

The first task a certain radar system has to accomplish is to continuously scan a specified volume in space searching for targets of interest. Once detection is established, target information such as range, angular position, and possibly target velocity are extracted by the radar signal and data processors. Depending on the radar design and antenna, different search patterns can be adopted.

Search volumes are normally specified by a search solid angle Ω in steradians, as illustrated in Fig. 1.10. Define the radar search volume extent for both azimuth and elevation as Θ_A and Θ_E. Consequently, the search volume is computed as

$$\Omega = (\Theta_A \Theta_E)/(57.296)^2 \ steradians \tag{1.42}$$

where both Θ_A and Θ_E are given in degrees. The radar antenna $3dB$ beamwidth can be expressed in terms of its azimuth and elevation beam widths θ_a and θ_e, respectively. It follows that the antenna solid angle coverage is $\theta_a \theta_e$ and, thus, the number of antenna beam positions n_B required to cover a solid angle Ω is

$$n_B = \frac{\Omega}{\theta_a \theta_e} \tag{1.43}$$

In order to develop the search radar equation, start with Eq. (140). Using the relations $\tau = 1/B$ and $P_t = P_{av}T/\tau$, where T is the PRI and τ is the pulse width, yields

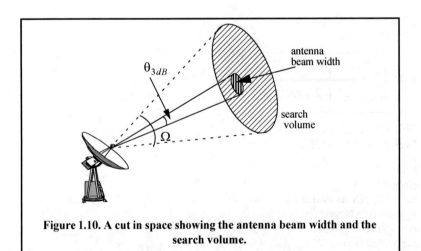

Figure 1.10. A cut in space showing the antenna beam width and the search volume.

$$SNR = \frac{T}{\tau} \frac{P_{av}G^2\lambda^2\sigma\tau}{(4\pi)^3 kT_0 FLR^4} \tag{1.44}$$

Define the time it takes the radar to scan a volume defined by the solid angle Ω as the scan time T_{sc}. The time on target can then be expressed in terms of T_{sc} as

$$T_i = \frac{T_{sc}}{n_B} = \frac{T_{sc}}{\Omega}\theta_a\theta_e \tag{1.45}$$

Assume that during a single scan only one pulse per beam per PRI illuminates the target. It follows that $T_i = T$ and, thus, Eq. (1.44) can be written as

$$SNR = \frac{P_{av}G^2\lambda^2\sigma}{(4\pi)^3 kT_0 FLR^4} \frac{T_{sc}}{\Omega}\theta_a\theta_e \tag{1.46}$$

Substituting Eq. (1.21) and Eq. (1.45) into Eq. (1.46) and collecting terms yield the search radar equation (based on a single pulse per beam per PRI) as

$$SNR = \frac{P_{av}A_e\sigma}{4\pi kT_0 FLR^4} \frac{T_{sc}}{\Omega} \tag{1.47}$$

The quantity $P_{av}A$ in Eq. (1.47) is known as the power aperture product. In practice, the power aperture product (PAP) is widely used to categorize the radar's ability to fulfill its search mission. Normally, a power aperture product is computed to meet a predetermined SNR and radar cross section for a given search volume defined by Ω.

Figure 1.11 shows a plot of the PAP versus detection range. using the following parameters:

σ	T_{sc}	$\theta_e = \theta_a$	R	$F+L$	SNR
$0.1\ m^2$	$2.5\,sec$	$2°$	$250Km$	$13dB$	$15dB$

This figure can be reproduced using the following MATLAB code which utilizes the MATLAB function *"power_aperture.m."*

```
close all;
clear all;
tsc = 2.5; % scan time is 2.5 seconds
sigma = 0.1; % radar cross section in m squared
te = 900.0; % effective noise temperature in Kelvin
snr = 15; % desired SNR in dB
nf = 6.0; % noise figure in dB
```

loss = 7.0; % radar losses in dB
az_angle = 2; % search volume azimuth extent in degrees
el_angle = 2; % search volume elevation extent in degrees
range = linspace(20e3,250e3,1000);
pap = power_aperture(snr,tsc,sigma/10,range,nf,loss,az_angle,el_angle);
rangekm   = range ./ 1000;
plot(rangekm,pap,'linewidth',1.5)
grid
xlabel ('Detection range in Km');
ylabel ('Power aperture product in dB');

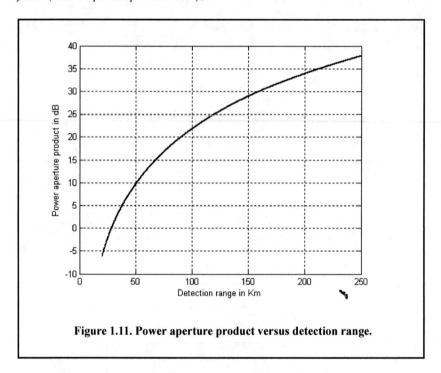

Figure 1.11. Power aperture product versus detection range.

Example:

Compute the power aperture product corresponding to the radar that has the following parameters: Scan time $T_{sc} = 2s$, noise figure $F = 8dB$, losses $L = 6dB$, search volume $\Omega = 7.4$ steradians, range of interest $R = 75Km$, and required SNR $20dB$. Assume that $\sigma = 3.162m^2$.

Solution:

Note that $\Omega = 7.4$ steradians corresponds to a search sector that is three fourths of a hemisphere. Thus, we conclude that $\Theta_a = 180°$ and $\Theta_e = 135°$. Using the MATLAB function "power_aperture.m" with the following syntax:

$$PAP = power\_aperture(20, 2, 3.162, 75e3, 8, 6, 180, 135)$$

one computes the power aperture product as 36.2 dB.

Example:

Compute the power aperture product for an X-band radar with the following parameters: Signal-to-noise ratio $SNR = 15dB$; losses $L = 8dB$; search volume $\Omega = 2°$; scan time $T_{sc} = 2.5s$; noise figure $F = 5dB$. Assume a $-10dBsm$ target cross section, and range $R = 250Km$. Also, compute the peak transmitted power corresponding to 30% duty factor if the antenna gain is 45 dB. Assume a circular aperture.

Solution:

The angular coverage is $2°$ in both azimuth and elevation. It follows that the solid angle coverage is

$$\Omega = \frac{2 \times 2}{(57.23)^2} = -29.132dB$$

Note that the factor $360/2\pi = 57.23$ converts degrees into steradians. When the aperture is circular Eq. (1.47) is reduced to (details are left as an exercise)

$$(SNR)_{dB} = (P_{av} + A + \sigma + T_{sc} - 16 - R^4 - kT_0 - L - F - \Omega)_{dB}$$

σ	T_{sc}	16	R^4	kT_0
-10	3.979	12.041	215.918	-203.977

It follows that

$$15 = P_{av} + A - 10 + 3.979 - 12.041 - 215.918 + 203.977 - 5 - 8 + 29.133$$

Then the power aperture product is

$$P_{av} + A = 38.716dB$$

Now, assume the radar wavelength to be $\lambda = 0.03m$, then

$$A = \frac{G\lambda^2}{4\pi} = 3.550dB$$

$$P_{av} = -A + 38.716 = 35.166dB$$

$$P_{av} = 10^{3.5166} = 3285.489W$$

$$P_t = \frac{P_{av}}{d_t} = \frac{3285.489}{0.3} = 10.9512\,KW$$

1.7. Radar Cross Section

Electromagnetic waves are normally diffracted or scattered in all directions when incident on a target. These scattered waves are broken down into two parts. The first part is made of waves that have the same polarization as the receiving antenna. The other portion of the scattered waves will have a different polarization to which the receiving antenna does not respond. The two polarizations are orthogonal and are referred to as the Principal Polarization (PP) and Orthogonal Polarization (OP), respectively. The intensity of the *back-scattered* energy that has the same polarization as the radar's receiving antenna is used to define the target RCS. When a target is illuminated by RF energy, it acts like a virtual antenna and will have near and far scattered fields. Waves reflected and measured in the near field are, in general, spherical. Alternatively, in the far field the wavefronts are decomposed into a linear combination of plane waves. Assume the power density of a wave incident on a target located at range R away from the radar is P_{Di}, as illustrated in Fig. 1.12. The amount of reflected power from the target is

$$P_r = \sigma P_{Di} \tag{1.48}$$

where σ denotes the target cross section. Define P_{Dr} as the power density of the scattered waves at the receiving antenna. It follows that

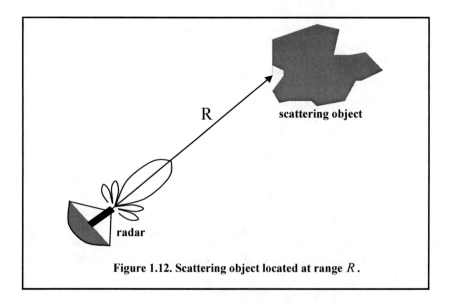

R

scattering object

radar

Figure 1.12. Scattering object located at range R.

$$P_{Dr} = P_r / (4\pi R^2) \tag{1.49}$$

Equating Eqs. (1.48) and (1.49) yields

$$\sigma = 4\pi R^2 \left(\frac{P_{Dr}}{P_{Di}} \right) \tag{1.50}$$

and in order to ensure that the radar receiving antenna is in the far field (i.e., scattered waves received by the antenna are planar), Eq. (1.50) is modified to

$$\sigma = 4\pi R^2 \lim_{R \to \infty} \left(\frac{P_{Dr}}{P_{Di}} \right) \tag{1.51}$$

The RCS defined by Eq. (1.51) is often referred to as either the monostatic RCS, the backscattered RCS, or simply target RCS.

The backscattered RCS is measured from all waves scattered in the direction of the radar and has the same polarization as the receiving antenna. It represents a portion of the total scattered target RCS σ_t, where $\sigma_t > \sigma$. Assuming a spherical coordinate system defined by (ρ, θ, φ), then at range ρ the target scattered cross section is a function of (θ, φ). Let the angles (θ_i, φ_i) define the direction of propagation of the incident waves. Also, let the angles (θ_s, φ_s) define the direction of propagation of the scattered waves. The special case, when $\theta_s = \theta_i$ and $\varphi_s = \varphi_i$, defines the monostatic RCS. The RCS measured by the radar at angles $\theta_s \neq \theta_i$ and $\varphi_s \neq \varphi_i$ is called the bistatic RCS.

The total target scattered RCS is given by

$$\sigma_t = \frac{1}{4\pi} \int_{\varphi_s = 0}^{2\pi} \int_{\theta_s = 0}^{\pi} \sigma(\theta_s, \varphi_s) \sin\theta_s \; d\theta \; d\varphi_s \tag{1.52}$$

The amount of backscattered waves from a target is proportional to the ratio of the target extent (size) to the wavelength, λ, of the incident waves. In fact, a radar will not be able to detect targets much smaller than its operating wavelength. The frequency region, where the target extent and the wavelength are comparable, is referred to as the Rayleigh region. Alternatively, the frequency region where the target extent is much larger than the radar operating wavelength is referred to as the optical region.

1.7.1. RCS Dependency on Aspect Angle and Frequency

Radar cross section fluctuates as a function of radar aspect angle and frequency. For the purpose of illustration, isotropic point scatterers are considered. Consider the geometry shown in Fig. 1.13. In this case, two unity ($1m^2$)

isotropic scatterers are aligned and placed along the radar line of sight (zero aspect angle) at a far field range R. The spacing between the two scatterers is 1 meter. The radar aspect angle is then changed from zero to 180 degrees, and the composite RCS of the two scatterers measured by the radar is computed.

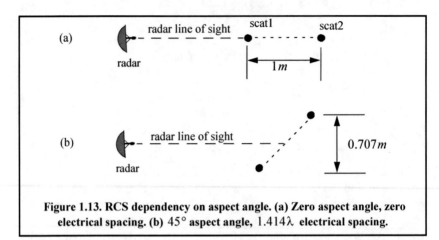

Figure 1.13. RCS dependency on aspect angle. (a) Zero aspect angle, zero electrical spacing. (b) $45°$ aspect angle, 1.414λ electrical spacing.

This composite RCS consists of the superposition of the two individual radar cross sections. At zero aspect angle, the composite RCS is $2m^2$. Taking scatterer-1 as a phase reference, when the aspect angle is varied, the composite RCS is modified by the phase that corresponds to the electrical spacing between the two scatterers. For example, at aspect angle $10°$, the electrical spacing between the two scatterers is

$$elec\text{--}spacing = \frac{2 \times (1.0 \times \cos(10°))}{\lambda} \tag{1.53}$$

λ is the radar operating wavelength.

Figure 1.14 shows the composite RCS corresponding to this experiment. This plot can be reproduced using the MATLAB code listed below. As clearly indicated by Fig. 1.14, RCS is dependent on the radar aspect angle; thus, knowledge of this constructive and destructive interference between the individual scatterers can be very critical when a radar tries to extract the RCS of complex or maneuvering targets. This is true for two reasons. First, the aspect angle may be continuously changing. Second, complex target RCS can be viewed to be made up from contributions of many individual scattering points distributed on the target surface. These scattering points are often called scattering centers. Many approximate RCS prediction methods generate a set of scattering centers that define the backscattering characteristics of such complex targets. The figures can be reproduced using the following MATLAB program.

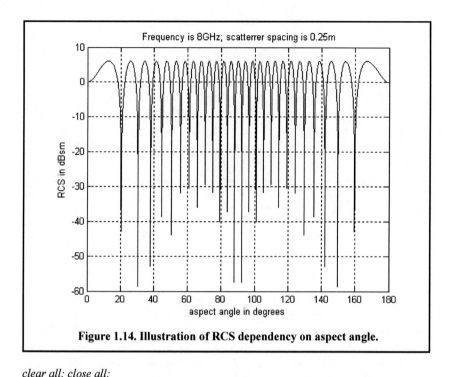

Figure 1.14. Illustration of RCS dependency on aspect angle.

```
clear all; close all;
% This program produces Fig. 1.14. This code demonstrates the effect of aspect angle
% on RCS. The radar is observing two unity point scatterers separated by scat_spacing.
% Initially the two scatterers are aligned with radar line of sight. The aspect angle is
% changed from 0 degrees to 180 degrees and the equivalent RCS is computed.
% The RCS as measured by the radar versus aspect angle is then plotted.
scat_spacing = 0.25; % 0.25 meter scatterers spacing
freq = 8e9; % operating frequency
eps = 0.00001;
wavelength = 3.0e+8 / freq;
% Compute aspect angle vector
aspect_degrees = linspace(0, 180, 500);
aspect_radians = (pi/180) .* aspect_degrees;
% Compute electrical scatterer spacing vector in wavelength units
elec_spacing = (2.0 * scat_spacing / wavelength) .* cos(aspect_radians);
% Compute RCS (rcs = RCS_scat1 + RCS_scat2)
% Scat1 is taken as phase reference point
rcs = abs(1.0 + cos((2.0 * pi) .* elec_spacing) + i * sin((2.0 * pi) .* elec_spacing));
rcs = rcs + eps;
rcs = 20.0*log10(rcs); % RCS in dBsm
% Plot RCS versus aspect angle
figure (1);
plot(aspect_degrees,rcs);
```

grid; xlabel('aspect angle in degrees'); ylabel('RCS in dBsm');
title(' Frequency is 8GHz; scatterer spacing is 0.25m');

Next, to demonstrate RCS dependency on frequency, consider the experiment shown in Fig. 1.15. In this case, two far field unity isotropic scatterers are aligned with radar line of sight, and the composite RCS is measured by the radar as the frequency is varied from 8 GHz to 12.5 GHz (X-band). Figs. 1.16 and 1.17 show the composite RCS versus frequency for scatterer spacing of 0.25 and 0.75 meters. The figures can be reproduced using the following MATLAB function.

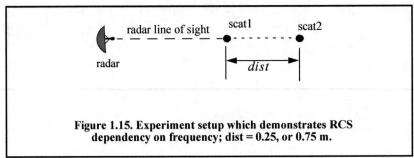

Figure 1.15. Experiment setup which demonstrates RCS dependency on frequency; dist = 0.25, or 0.75 m.

clear all; close all;
% This program demonstrates the dependency of RCS on wavelength
% The radar line of sight is aligned with the two scatterers
% A plot of RCS variation versus frequency if produced
eps = 0.0001;
scat_spacing = 0.25;
freql = 8e9;
frequ = 12.5e9;
freq = linspace(freql,frequ,500);
wavelength = 3.0e+8 ./ freq;
% Compute electrical scatterer spacing vector in wavelength units
*elec_spacing = 2.0 * scat_spacing ./ wavelength;*
% Compute RCS (RCS = RCS_scat1 + RCS_scat2)
*rcs = abs (1 + cos((2.0 * pi) .* elec_spacing) ...*
*   + i * sin((2.0 * pi) .* elec_spacing));*
rcs = rcs + eps;
*rcs = 20.0*log10(rcs); % RCS ins dBsm*
% Plot RCS versus frequency
figure (1);
plot(freq./1e9,rcs);
grid;
xlabel('Frequency in GHz');
ylabel('RCS in dBsm');
% title(' X=Band; scatterer spacing is 0.25 m'); % Fig. 1.16
% title(' X=Band; scatterer spacing is 0.75 m'); % Fig. 1.17

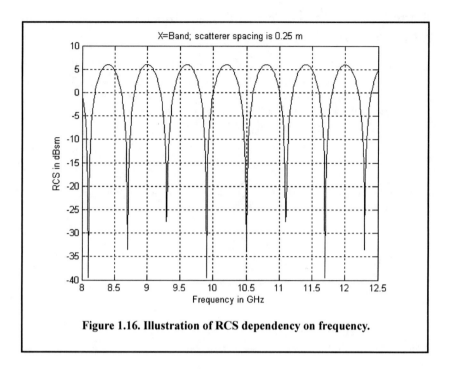

Figure 1.16. Illustration of RCS dependency on frequency.

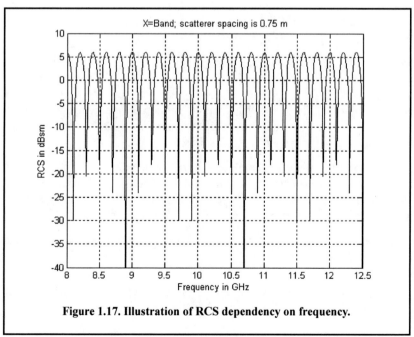

Figure 1.17. Illustration of RCS dependency on frequency.

1.7.2. RCS Dependency on Polarization

Normalized Electric Field

In most radar simulations, it is desirable to obtain the complex-valued electric field scattered by the target at the radar. In such cases, it is useful to use a quantity called the normalized electric field. It is assumed that the incident electric field has a magnitude of unity and is phase centered at a point at the target (usually the center of gravity). More precisely,

$$E_i = e^{jk(\vec{r_i} \cdot \vec{r})} \tag{1.54}$$

where $\vec{r_i}$ is the direction of incidence and $\vec{r}$ a location at the target, each with respect to the phase center. The normalized scattered field is then given by

$$E_s = \sigma E_i \tag{1.55}$$

The quantity E_s is independent of radar and target location. It may be combined with an incident magnitude and phase.

Polarization

The x and y electric field components for a wave traveling along the positive z direction are given by

$$E_x = E_1 \sin(\omega t - kz) \tag{1.56}$$

$$E_y = E_2 \sin(\omega t - kz + \delta) \tag{1.57}$$

where $k = 2\pi/\lambda$, ω is the wave frequency, the angle δ is the time phase angle at which E_y leads E_x, and finally, E_1 and E_2 are, respectively, the wave amplitudes along the x and y directions. When two or more electromagnetic waves combine, their electric fields are integrated vectorially at each point in space for any specified time. In general, the combined vector traces an ellipse when observed in the x-y plane. This is illustrated in Fig. 1.18.

The ratio of the major to the minor axes of the polarization ellipse is called the Axial Ratio (AR). When AR is unity, the polarization ellipse becomes a circle, and the resultant wave is then called circularly polarized. Alternatively, when $E_1 = 0$ and $AR = \infty$, the wave becomes linearly polarized.

Equations (1.56) and (1.57) can be combined to give the instantaneous total electric field,

$$\vec{E} = \hat{a}_x E_1 \sin(\omega t - kz) + \hat{a}_y E_2 \sin(\omega t - kz + \delta) \tag{1.58}$$

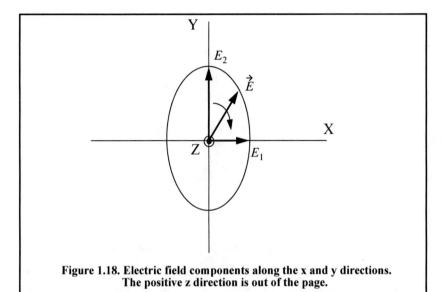

**Figure 1.18. Electric field components along the x and y directions.
The positive z direction is out of the page.**

where $\hat{a}_x$ and $\hat{a}_y$ are unit vectors along the x and y directions, respectively. At $z = 0$, $E_x = E_1 \sin(\omega t)$ and $E_y = E_2 \sin(\omega t + \delta)$, then by replacing $\sin(\omega t)$ by the ratio E_x/E_1 and by using trigonometry properties Eq. (1.58) can be rewritten as

$$\frac{E_x^2}{E_1^2} - \frac{2E_xE_y\cos\delta}{E_1E_2} + \frac{E_y^2}{E_2^2} = (\sin\delta)^2 \qquad (1.59)$$

which has no dependency on ωt.

In the most general case, the polarization ellipse may have any orientation, as illustrated in Fig. 1.19. The angle ξ is called the tilt angle of the ellipse. In this case, AR is given by

$$AR = \frac{OA}{OB} \qquad (1 \le AR \le \infty) \qquad (1.60)$$

When $E_1 = 0$, the wave is said to be linearly polarized in the y direction, while if $E_2 = 0$, the wave is said to be linearly polarized in the x direction. Polarization can also be linear at an angle of $45°$ when $E_1 = E_2$ and $\xi = 45°$. When $E_1 = E_2$ and $\delta = 90°$, the wave is said to be Left Circularly Polarized (LCP), while if $\delta = -90°$ the wave is said to Right Circularly Polarized (RCP). It is a common notation to call the linear polarizations along the x and y directions by the names horizontal and vertical polarizations, respectively.

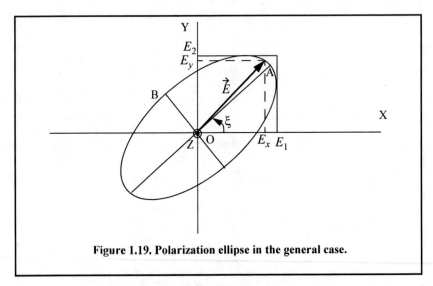

Figure 1.19. Polarization ellipse in the general case.

In general, an arbitrarily polarized electric field may be written as the sum of two circularly polarized fields. More precisely,

$$\vec{E} = \vec{E_R} + \vec{E_L} \tag{1.61}$$

where $\vec{E_R}$ and $\vec{E_L}$ are the RCP and LCP fields, respectively. Similarly, the RCP and LCP waves can be written as

$$\vec{E_R} = \vec{E_V} + j\vec{E_H} \tag{1.62}$$

$$\vec{E_L} = \vec{E_V} - j\vec{E_H} \tag{1.63}$$

where $\vec{E_V}$ and $\vec{E_H}$ are the fields with vertical and horizontal polarizations, respectively. Combining Eqs. (1.62) and (1.63) yields

$$E_R = \frac{E_H - jE_V}{\sqrt{2}} \tag{1.64}$$

$$E_L = \frac{E_H + jE_V}{\sqrt{2}} \tag{1.65}$$

Using matrix notation, Eqs. (1.64) and (1.65) can be rewritten as

$$\begin{bmatrix} E_R \\ E_L \end{bmatrix} = \frac{1}{\sqrt{2}} \begin{bmatrix} 1 & -j \\ 1 & j \end{bmatrix} \begin{bmatrix} E_H \\ E_V \end{bmatrix} = [T] \begin{bmatrix} E_H \\ E_V \end{bmatrix} \tag{1.66}$$

$$\begin{bmatrix} E_H \\ E_V \end{bmatrix} = \frac{1}{\sqrt{2}} \begin{bmatrix} 1 & 1 \\ j & -j \end{bmatrix} \begin{bmatrix} E_R \\ E_L \end{bmatrix} = [T]^{-1} \begin{bmatrix} E_H \\ E_V \end{bmatrix} \qquad (1.67)$$

For many targets the scattered waves will have different polarization than the incident waves. This phenomenon is known as depolarization or cross-polarization. However, perfect reflectors reflect waves in such a fashion that an incident wave with horizontal polarization remains horizontal, and an incident wave with vertical polarization remains vertical but is phase shifted $180°$. Additionally, an incident wave that is RCP becomes LCP when reflected, and a wave that is LCP becomes RCP after reflection from a perfect reflector. Therefore, when a radar uses LCP waves for transmission, the receiving antenna needs to be RCP polarized in order to capture the PP RCS, and LCP to measure the OP RCS.

Target Scattering Matrix

Target backscattered RCS is commonly described by a matrix known as the scattering matrix and is denoted by $[S]$. When an arbitrarily linearly polarized wave is incident on a target, the backscattered field is then given by

$$\begin{bmatrix} E_1^s \\ E_2^s \end{bmatrix} = [S] \begin{bmatrix} E_1^i \\ E_2^i \end{bmatrix} = \begin{bmatrix} s_{11} & s_{12} \\ s_{21} & s_{22} \end{bmatrix} \begin{bmatrix} E_1^i \\ E_2^i \end{bmatrix} \qquad (1.68)$$

The superscripts i and s denote incident and scattered fields. The quantities s_{ij} are in general complex and the subscripts 1 and 2 represent any combination of orthogonal polarizations. More precisely, $1 = H, R$, and $2 = V, L$. From Eq. (1.50), the backscattered RCS is related to the scattering matrix components by the following relation:

$$\begin{bmatrix} \sigma_{11} & \sigma_{12} \\ \sigma_{21} & \sigma_{22} \end{bmatrix} = 4\pi R^2 \begin{bmatrix} |s_{11}|^2 & |s_{12}|^2 \\ |s_{21}|^2 & |s_{22}|^2 \end{bmatrix} \qquad (1.69)$$

It follows that once a scattering matrix is specified, the target backscattered RCS can be computed for any combination of transmitting and receiving polarizations. The reader is advised to see Ruck et al. (1970) for ways to calculate the scattering matrix $[S]$. Rewriting Eq. (1.69) in terms of the different possible orthogonal polarizations yields

$$\begin{bmatrix} E_H^s \\ E_V^s \end{bmatrix} = \begin{bmatrix} s_{HH} & s_{HV} \\ s_{VH} & s_{VV} \end{bmatrix} \begin{bmatrix} E_H^i \\ E_V^i \end{bmatrix} \qquad (1.70)$$

$$\begin{bmatrix} E_R^s \\ E_L^s \end{bmatrix} = \begin{bmatrix} s_{RR} & s_{RL} \\ s_{LR} & s_{LL} \end{bmatrix} \begin{bmatrix} E_R^i \\ E_L^i \end{bmatrix} \tag{1.71}$$

By using the transformation matrix $[T]$ in Eq. (1.66), the circular scattering elements can be computed from the linear scattering elements

$$\begin{bmatrix} s_{RR} & s_{RL} \\ s_{LR} & s_{LL} \end{bmatrix} = [T] \begin{bmatrix} s_{HH} & s_{HV} \\ s_{VH} & s_{VV} \end{bmatrix} \begin{bmatrix} 1 & 0 \\ 0 & -1 \end{bmatrix} [T]^{-1} \tag{1.72}$$

and the individual components are

$$s_{RR} = \frac{-s_{VV} + s_{HH} - j(s_{HV} + s_{VH})}{2} \tag{1.73}$$

$$s_{RL} = \frac{s_{VV} + s_{HH} + j(s_{HV} - s_{VH})}{2} \tag{1.74}$$

$$s_{LR} = \frac{s_{VV} + s_{HH} - j(s_{HV} - s_{VH})}{2} \tag{1.75}$$

$$s_{LL} = \frac{-s_{VV} + s_{HH} + j(s_{HV} + s_{VH})}{2} \tag{1.76}$$

Similarly, the linear scattering elements are given by

$$\begin{bmatrix} s_{HH} & s_{HV} \\ s_{VH} & s_{VV} \end{bmatrix} = [T]^{-1} \begin{bmatrix} s_{RR} & s_{RL} \\ s_{LR} & s_{LL} \end{bmatrix} \begin{bmatrix} 1 & 0 \\ 0 & -1 \end{bmatrix} [T] \tag{1.77}$$

and the individual components are

$$s_{HH} = \frac{-s_{RR} + s_{RL} + s_{LR} - s_{LL}}{2} \tag{1.78}$$

$$s_{VH} = \frac{j(s_{RR} - s_{LR} + s_{RL} - s_{LL})}{2} \tag{1.79}$$

$$s_{HV} = \frac{-j(s_{RR} + s_{LR} - s_{RL} - s_{LL})}{2} \tag{1.80}$$

$$s_{VV} = \frac{s_{RR} + s_{LL} + js_{RL} + s_{LR}}{2} \tag{1.81}$$

1.8. Radar Equation with Jamming

Any deliberate electronic effort intended to disturb normal radar operation is usually referred to as an Electronic Countermeasure (ECM). This may also include chaff, radar decoys, radar RCS alterations (e.g., radio frequency absorbing materials), and of course, radar jamming.

Jammers can be categorized into two general types: (1) barrage jammers and (2) deceptive jammers (repeaters). When strong jamming is present, detection capability is determined by receiver signal-to-noise plus interference ratio rather than SNR. In fact, in most cases, detection is established based on the signal-to-interference ratio alone.

Barrage jammers attempt to increase the noise level across the entire radar operating bandwidth. Consequently, this lowers the receiver SNR and, in turn, makes it difficult to detect the desired targets. This is the reason barrage jammers are often called maskers (since they mask the target returns). Barrage jammers can be deployed in the main beam or in the sidelobes of the radar antenna. If a barrage jammer is located in the radar main beam, it can take advantage of the antenna maximum gain to amplify the broadcasted noise signal. Alternatively, sidelobe barrage jammers must either use more power or operate at a much shorter range than main-beam jammers. Main-beam barrage jammers can either be deployed on-board the attacking vehicle or act as an escort to the target. Sidelobe jammers are often deployed to interfere with a specific radar, and since they do not stay close to the target, they have a wide variety of stand-off deployment options.

Repeater jammers carry receiving devices on board in order to analyze the radar's transmission and then send back false target-like signals in order to confuse the radar. There are two common types of repeater jammers: spot noise repeaters and deceptive repeaters. The spot noise repeater measures the transmitted radar signal bandwidth and then jams only a specific range of frequencies. The deceptive repeater sends back altered signals that make the target appear in some false position (ghosts). These ghosts may appear at different ranges or angles than the actual target. Furthermore, there may be several ghosts created by a single jammer. By not having to jam the entire radar bandwidth, repeater jammers are able to make more efficient use of their jamming power. Radar frequency agility may be the only way possible to defeat spot noise repeaters.

In general a jammer can be identified by its effective operating bandwidth B_J and by its Effective Radiated Power (ERP), which is proportional to the jammer transmitter power P_J. More precisely,

$$ERP = P_J G_J / L_J \tag{1.82}$$

where G_J is the jammer antenna gain and L_J is the total jammer loss. The effect of a jammer on a radar is measured by the Signal-to-Jammer ratio (S/J).

Consider a radar system whose detection range R in the absence of jamming is governed by

$$SNR = \frac{P_t G^2 \lambda^2 \sigma}{(4\pi)^3 k T_s B_r L R^4} \qquad (1.83)$$

The term Range Reduction Factor (RRF) refers to the reduction in the radar detection range due to jamming. More precisely, in the presence of jamming the effective radar detection range is

$$R_{dj} = R \times RRF \qquad (1.84)$$

In order to compute RRF, consider a radar characterized by Eq. (1.83) and a barrage jammer whose output power spectral density is J_o (i.e., Gaussian-like). Then the amount of jammer power in the radar receiver is

$$J = k T_J B_r \qquad (1.85)$$

where T_J is the jammer effective temperature. It follows that the total jammer plus noise power in the radar receiver is given by

$$N_i + J = k T_s B_r + k T_J B_r \qquad (1.86)$$

In this case, the radar detection range is now limited by the receiver signal-to-noise plus interference ratio rather than SNR. More precisely,

$$\left(\frac{S}{J+N}\right) = \frac{P_t G^2 \lambda^2 \sigma}{(4\pi)^3 k (T_s + T_J) B_r L R^4} \qquad (1.87)$$

The amount of reduction in the signal-to-noise plus interference ratio because of the jammer effect can be computed from the difference between Eqs. (1.83) and (1.87). It is expressed (in dB) by

$$\Upsilon = 10.0 \times \log\left(1 + \frac{T_J}{T_s}\right) \qquad (1.88)$$

Consequently, the RRF is

$$RRF = 10^{\frac{-\Upsilon}{40}} \qquad (1.89)$$

Figures 1.20 a and b show typical value for the RRF versus the radar wavelength and detection range using the following parameters

Symbol	Value
te	500 kelvin
pj	500 KW
gj	3 dB
g	45 dB
freq	10 GHz
bj	10 MHZ
rangej	750 Km
lossj	1 dB

This figure can be reproduced using the following MATLAB code

```
clear all;
close all;
te = 730.0; % radar effective temp in Kelvin
pj  = 15; % jammer peak power in W
gj = 3.0; % jammer antenna gain in dB
g = 40.0; % radar antenna gain
freq = 10.0e+9; % radar operating frequency in Hz
bj  = 1.0e+6; % radar operating bandwidth in Hz
rangej = 400.0; % radar to jammer range in Km
lossj = 1.0; % jammer losses in dB
c = 3.0e+8;
k = 1.38e-23;
lambda = c / freq;
gj_10 = 10^( gj/10);
g_10 = 10^( g/10);
lossj_10 = 10^(lossj/10);
index = 0;
for wavelength = .01:.001:1
   index = index +1;
   jamer_temp = (pj * gj_10 * g_10 *wavelength^2) / ...
     (4.0^2 * pi^2 * k * bj * lossj_10 * (rangej * 1000.0)^2);
   delta = 10.0 * log10(1.0 + (jamer_temp / te));
   rrf(index) = 10^(-delta /40.0);
end
w = 0.01:.001:1;
figure (1)
semilogx(w,rrf,'k')
grid
xlabel ('Wavelength in meters')
ylabel ('Range reduction factor')
index = 0;
```

```
for ran =rangej*.3:10:rangej*2
   index = index + 1;
   jamer_temp = (pj * gj_10 * g_10 *lambda^2) / ...
      (4.0^2 * pi^2 * k * bj * lossj_10 * (ran * 1000.0)^2);
   delta = 10.0 * log10(1.0 + (jamer_temp / te));
   rrf1(index) = 10^(-delta /40.0);
end
figure(2)
ranvar = rangej*.3:10:rangej*2 ;
plot(ranvar,rrf1,'k')
grid
xlabel ('Radar to jammer range in Km')
ylabel ('Range reduction factor')
index = 0;
for pjvar = pj*.01:100:pj*2
   index = index + 1;
   jamer_temp = (pjvar * gj_10 * g_10 *lambda^2) / ...
      (4.0^2 * pi^2 * k * bj * lossj_10 * (rangej * 1000.0)^2);
   delta = 10.0 * log10(1.0 + (jamer_temp / te));
   rrf2(index) = 10^(-delta /40.0);
end
```

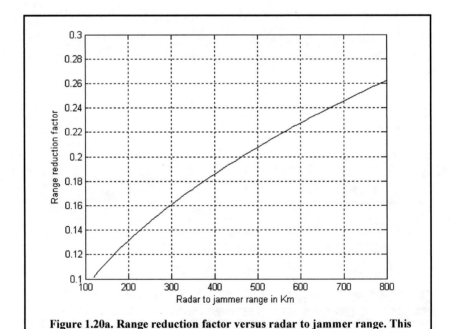

Figure 1.20a. Range reduction factor versus radar to jammer range. This plot was generated using the function *"range_red_factor.m."*

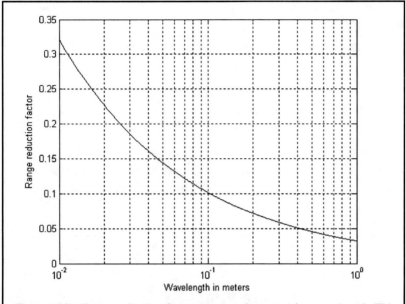

Figure 1.20b. Range reduction factor versus radar operating wavelength. This plot was generated using the function *"range_red_factor.m."*

1.9. Noise Figure

Any signal other than the target returns in the radar receiver is considered to be noise. This includes interfering signals from outside the radar and thermal noise generated within the receiver itself. Thermal noise (thermal agitation of electrons) and shot noise (variation in carrier density of a semiconductor) are the two main internal noise sources within a radar receiver.

The power spectral density of thermal noise is given by

$$S_n(\omega) = \frac{|\omega|h}{\pi\left[\exp\left(\frac{|\omega|h}{2\pi kT}\right) - 1\right]} \qquad (1.90)$$

where $|\omega|$ is the absolute value of the frequency in radians per second, T is the temperature of the conducting medium in degrees Kelvin, k is Boltzman's constant, and h is Plank's constant ($h = 6.625 \times 10^{-34}$ *Joules*). When the condition $|\omega| \ll 2\pi kT/h$ is true, it can be shown that Eq. (1.90) is approximated by

$$S_n(\omega) \approx 2kT \qquad (1.91)$$

This approximation is widely accepted, since, in practice, radar systems operate at frequencies less than $100GHz$; and, for example, if $T = 290K$, then $2\pi kT/h \approx 6000GHz$.

The mean-square noise voltage (noise power) generated across a 1 *ohm* resistance is then

$$\langle n^2 \rangle = \frac{1}{2\pi} \int\limits_{-2\pi B}^{2\pi B} 2kT \ d\omega = 4kTB \tag{1.92}$$

where B is the system bandwidth. Any electrical system containing thermal noise and having input resistance R_{in} can be replaced by an equivalent noiseless system with a series combination of a noise equivalent voltage source and a noiseless input resistor R_{in} added at its input. This is illustrated in Fig. 1.21.

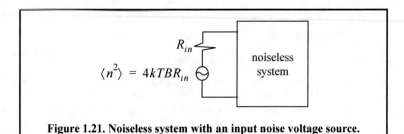

Figure 1.21. Noiseless system with an input noise voltage source.

The amount of noise power that can physically be extracted from $\langle n^2 \rangle$ is one fourth the value computed in Eq. (1.92). Consider a noisy system with power gain A_P, as shown in Fig. 1.22. The noise figure is defined by

$$F_{dB} = 10 \ \log \frac{total \ \ noise \ \ power \ \ out}{noise \ \ power \ \ out \ \ due \ \ to \ \ R_{in} \ \ alone} \tag{1.93}$$

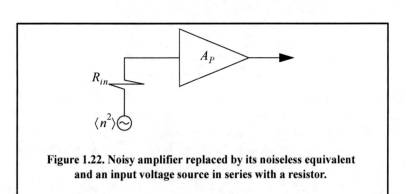

Figure 1.22. Noisy amplifier replaced by its noiseless equivalent and an input voltage source in series with a resistor.

More precisely,

$$F_{dB} = 10 \ \log \frac{N_o}{N_i \ A_p} \tag{1.94}$$

where N_o and N_i are, respectively, the noise power at the output and input of the system.

If we define the input and output signal power by S_i and S_o, respectively, then the power gain is

$$A_p = \frac{S_o}{S_i} \tag{1.95}$$

It follows that

$$F_{dB} = 10\log\left(\frac{S_i/Ni}{S_o/N_o}\right) = \left(\frac{S_i}{N_i}\right)_{dB} - \left(\frac{S_o}{N_o}\right)_{dB} \tag{1.96}$$

where

$$\left(\frac{S_i}{N_i}\right)_{dB} > \left(\frac{S_o}{N_o}\right)_{dB} \tag{1.97}$$

Thus, the noise figure is the loss in the signal-to-noise ratio due to the added thermal noise of the amplifier $((SNR)_o = (SNR)_i - F \ in \ dB)$.

One can also express the noise figure in terms of the system's effective temperature T_e. Consider the amplifier shown in Fig. 1.22, and let its effective temperature be T_e. Assume the input noise temperature is T_0. Thus, the input noise power is

$$N_i = kT_0B \tag{1.98}$$

and the output noise power is

$$N_o = kT_0B \ A_p + kT_eB \ A_p \tag{1.99}$$

where the first term on the right-hand side of Eq. (1.99) corresponds to the input noise, and the latter term is due to thermal noise generated inside the system. It follows that the noise figure can be expressed as

$$F = \frac{(SNR)_i}{(SNR)_o} = \frac{S_i}{kT_0B} kBA_p \frac{T_0 + T_e}{S_o} = 1 + \frac{T_e}{T_0} \tag{1.100}$$

Equivalently, we can write

$$T_e = (F-1)T_0 \tag{1.101}$$

Example:

An amplifier has a 4dB noise figure; the bandwidth is $B = 500$ KHz. Calculate the input signal power that yields a unity SNR at the output. Assume $T_0 = 290K$ and an input resistance of one ohm.

Solution:

The input noise power is

$$kT_0B = 1.38 \times 10^{-23} \times 290 \times 500 \times 10^3 = 2.0 \times 10^{-15} W$$

Assuming a voltage signal, then the input noise mean squared voltage is

$$\langle n_i^2 \rangle = kT_0B = 2.0 \times 10^{-15} \ v^2$$

$$F = 10^{0.4} = 2.51$$

From the noise figure definition we get

$$\frac{S_i}{N_i} = F\left(\frac{S_o}{N_o}\right) = F$$

and

$$\langle s_i^2 \rangle = F\langle n_i^2 \rangle = 2.51 \times 2.0 \times 10^{-15} = 5.02 \times 10^{-15} \ v^2$$

Finally,

$$\sqrt{\langle s_i^2 \rangle} = 70.852nv$$

Consider a cascaded system as in Fig. 1.23. Network 1 is defined by noise figure F_1, power gain G_1, bandwidth B, and temperature T_{e1}. Similarly, network 2 is defined by F_2, G_2, B, and T_{e2}. Assume the input noise has temperature T_0.

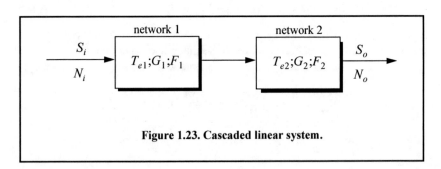

Figure 1.23. Cascaded linear system.

The output signal power is

$$S_o = S_i G_1 G_2 \tag{1.102}$$

The input and output noise powers are, respectively, given by

$$N_i = kT_0 B \tag{1.103}$$

$$N_o = kT_0 B G_1 G_2 + kT_{e1} B G_1 G_2 + kT_{e2} B G_2 \tag{1.104}$$

where the three terms on the right-hand side of Eq. (1.104), respectively, correspond to the input noise power, thermal noise generated inside network 1, and thermal noise generated inside network 2.

Now if we use the relation $T_e = (F - 1)T_0$ along with Eq. (1.02), we can express the overall output noise power as

$$N_o = F_1 N_i G_1 G_2 + (F_2 - 1) N_i G_2 \tag{1.105}$$

It follows that the overall noise figure for the cascaded system is

$$F = \frac{(S_i/N_i)}{(S_o/N_o)} = F_1 + \frac{F_2 - 1}{G_1} \tag{1.106}$$

In general, for an n-stage system we get

$$F = F_1 + \frac{F_2 - 1}{G_1} + \frac{F_3 - 1}{G_1 G_2} + \dots + \frac{F_n - 1}{G_1 G_2 G_3 \cdot \cdot \cdot G_{n-1}} \tag{1.107}$$

Also, the n-stage system effective temperatures can be computed as

$$T_e = T_{e1} + \frac{T_{e2}}{G_1} + \frac{T_{e3}}{G_1 G_2} + \dots + \frac{T_{en}}{G_1 G_2 G_3 \cdot \cdot \cdot G_{n-1}} \tag{1.108}$$

As suggested by Eq. (1.107) and Eq. (1.108), the overall noise figure is mainly dominated by the first stage. Thus, radar receivers employ low noise power amplifiers in the first stage in order to minimize the overall receiver noise figure. However, for radar systems that are built for low RCS operations every stage should be included in the analysis.

Example:

A radar receiver consists of an antenna with cable loss $L = 1dB = F_1$, *an RF amplifier with* $F_2 = 6dB$, *and gain* $G_2 = 20dB$, *followed by a mixer whose noise figure is* $F_3 = 10dB$ *and conversion loss* $L = 8dB$, *and finally, an integrated circuit IF amplifier with* $F_4 = 6dB$ *and gain* $G_4 = 60dB$. *Find the overall noise figure.*

Solution:

From Eq. (1.107) we have

$$F = F_1 + \frac{F_2 - 1}{G_1} + \frac{F_3 - 1}{G_1 G_2} + \frac{F_4 - 1}{G_1 G_2 G_3}$$

G_1	G_2	G_3	G_4	F_1	F_2	F_3	F_4
$-1 dB$	$20 dB$	$-8 dB$	$60 dB$	$1 dB$	$6 dB$	$10 dB$	$6 dB$
0.7943	100	0.1585	10^6	1.2589	3.9811	10	3.9811

It follows that

$$F = 1.2589 + \frac{3.9811 - 1}{0.7943} + \frac{10 - 1}{100 \times 0.7943} + \frac{3.9811 - 1}{0.158 \times 100 \times 0.7943} = 5.3629$$

$$F = 10 \log(5.3628) = 7.294 dB$$

1.10. Effects of the Earth's Surface on the Radar Equation

So far, in developing the radar equation it was implicitly assumed that the radar electromagnetic waves travel as if they were in free space. Furthermore, all analysis presented did not account for the effects of the earth's atmosphere nor the effects of the earth's surface. Despite the fact that *"free space analysis"* may be adequate to provide a general understanding of radar systems, it is only an approximation. In order to accurately predict radar performance, we must modify free space analysis to include the effects of the earth and its atmosphere. Radar clutter is not considered to be part of this analysis. This is true because clutter is almost always assumed to be a distributed target that can be dealt with by the radar signal processor separately. Clutter is the subject of discussion in a later chapter of this book.

In this chapter, the effects of the earth's atmosphere are considered first. Then, the effect of the earth' surface on the radar equation is analyzed. The earth's surface impact on the radar equation manifests itself by introducing an additional power term in the radar equation. This term is called the *pattern propagation factor* and is denoted by symbol F. The propagation factor, can actually introduce constructive as well as distructive interference in the SNR depending on the radar frequency and the geometry under consideration. In general, the pattern propagation factor is defined by

$$F = |E / E_0| \tag{1.109a}$$

where E is the electric field in the medium and E_0 is the free space electric field. In this case the radar equation is now given by

$$(SNR)_o = \frac{P_t G^2 \lambda^2 \sigma}{(4\pi)^3 k T_0 BFLR^4} F^4 \qquad \text{(1.109b)}$$

1.10.1. Earth's Atmosphere

The earth's atmosphere is composed of several layers, as illustrated in Fig. 1.24. The first layer, which extends in altitude to about 20 Km, is known as the troposphere. Electromagnetic waves refract (bend downward) as they travel in the troposphere. The troposphere refractive effect is related to its dielectric constant, which is a function of pressure, temperature, water vapor, and gaseous content. Additionally, due to gases and water vapor in the atmosphere, radar energy suffers a loss. This loss is known as the atmospheric attenuation. Atmospheric attenuation increases significantly in the presence of rain, fog, dust, and clouds.

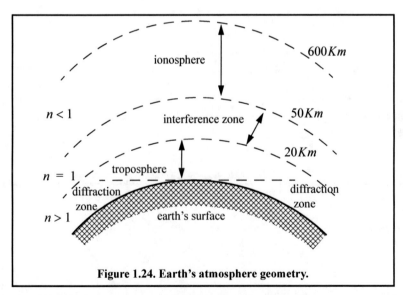

Figure 1.24. Earth's atmosphere geometry.

The region above the troposphere (altitude from 20 to 50 Km) behaves like free space, and thus little refraction occurs in this region. This region is known as the interference zone. The ionosphere extends from about 50 Km to about 600 Km. It has very low gas density compared to the troposphere. It contains a significant amount of ionized free electrons. The ionization is primarily caused by the sun's ultraviolet and X-rays. This presence of free electrons in the ionosphere affects electromagnetic wave propagation in different ways. These effects include refraction, absorption, noise emission, and polarization rotation. The degree of degradation depends heavily on the frequency of the incident waves. For example, frequencies lower than about 4 to 6 MHz are completely reflected from the lower region of the ionosphere. Frequencies higher than 30 MHz may penetrate the ionosphere with some level of attenuation. In general, as the frequency is increased the ionosphere's effects become less prominent. The region below the horizon, close to the earth's surface, is called the diffraction region. *Diffraction* is a term used to describe the bending of radar waves around physical objects. Two types of diffraction are common. They are knife edge and cylinder edge diffraction.

In order to effectively study the effects of the atmosphere on the propagation of radar waves, it is necessary to have accurate knowledge of the height-variation of the index of refracting in the troposphere and the ionosphere. The index of refraction is a function of the geographic location on the earth, weather, time of day or night, and the season of the year. Therefore, analyzing the atmospheric propagation effects under all parametric conditions is an overwhelming task. Typically, this problem is simplified by analyzing atmospheric models that are representative of an average of atmospheric conditions.

1.10.2. Refraction

In free space, electromagnetic waves travel in straight lines. However, in the presence of the earth's atmosphere, they bend (refract), as illustrated in Fig. 1.25. *Refraction* is a term used to describe the deviation of radar wave propagation from a straight line. The deviation from straight line propagation is caused by the variation of the index of refraction. The index of refraction is defined as

$$n = c/v \tag{1.110}$$

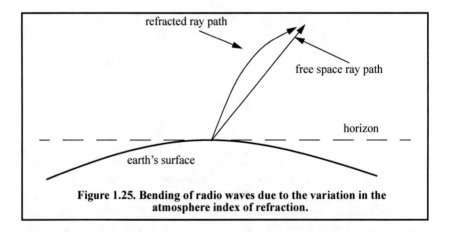

Figure 1.25. Bending of radio waves due to the variation in the atmosphere index of refraction.

where c is the velocity of electromagnetic waves in free space and v is the wave group velocity in the medium. Close to the earth's surface the index of refraction is almost unity; however, with increasing altitude the index of refraction decreases gradually.

The discussion presented in this chapter assumes a well-mixed atmosphere, where the index of refraction decreases in a smooth monotonic fashion with height. The rate of change of the earth's index of refraction n with altitude h is normally referred to as the refractivity gradient, dn/dh. As a result of the negative rate of change in dn/dh, electromagnetic waves travel at slightly higher velocities in the upper troposphere than in the lower part. As a result of this, waves traveling horizontally in the troposphere gradually bend downward. In general, since the rate of change in the refractivity index is very slight, waves do not curve downward appreciably unless they travel very long distances through the troposphere.

Refraction affects radar waves in two different ways depending on height. For targets that have altitudes typically above 100 meters, the effect of refraction is illustrated in Fig. 1.26. In this case, refraction imposes limitations on the radar's capability to measure target position. Refraction introduces an error in measuring the elevation angle. In a well mixed atmosphere, the refractivity gradient close to the earth's surface is almost constant. However, temperature changes and humidity lapses close to the earth's surface may cause serious changes in the refractivity profile. When the refractivity index becomes large enough, electromagnetic waves bend around the curve of the earth beyond the expected curvature due to earth surface. This phenomenon is called ducting and is illustrated in Fig. 1.27. Ducting can be extensive over the sea surface during a hot summer.

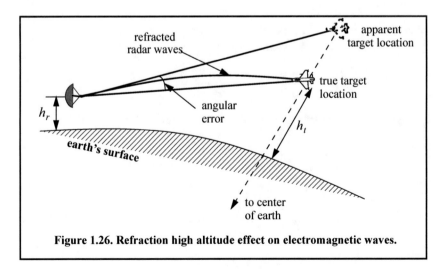

Figure 1.26. Refraction high altitude effect on electromagnetic waves.

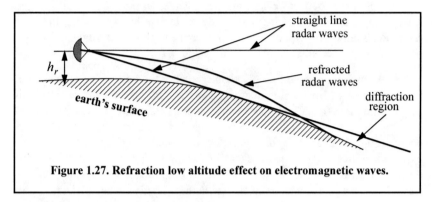

Figure 1.27. Refraction low altitude effect on electromagnetic waves.

Stratified Atmospheric Refraction Model

An approximation method for calculating the range measurement errors and the time-delay errors experienced by radar waves due to refraction is presented. This method is referred to as the "stratified atmospheric model" and is capable of producing very accurate theoretical estimates of the propagation errors. The basic assumption for this approach is that the atmosphere is stratified into M spherical layers, each of thickness $\{h_m;\ m = 0, 1, ..., M\}$, and a constant refractive index $\{n_m;\ m = 0, 1, ..., M\}$, as illustrated in Fig. 1.28. In this figure, β_0 is the apparent elevation angle and β_{0M} is the true elevation angle. The free space path is denoted by R_{0M}, while the refracted path is composed of $\{R_0, R_1, R_2, ..., R_M\}$. From the figure,

$$r_{(m+1)} = r_0 + \sum_{i=0}^{m} h_i \qquad ;\ m = 0, 1, ..., M \qquad (1.111)$$

where r_0 is the actual radius of the earth and is equal to 6375 Km. Using the law of sines, the angle of incidence α_0 is given by

$$\frac{\sin\alpha_0}{r_0} = \frac{\sin(180 + \beta_0)}{r_1} \qquad (1.112)$$

Using Snell's law for spherically symmetrical surfaces, the mth angle, β_m, that the ray makes with the horizon in layer m is given by

$$n_{(m-1)}r_{(m-1)}\cos\beta_{(m-1)} = n_m r_m \cos\beta_m \qquad (1.113)$$

Consequently,

$$\beta_m = \text{acos}\left[\frac{n_{(m-1)}r_{(m-1)}}{n_m r_m}\cos\beta_{(m-1)}\right] \qquad (1.114)$$

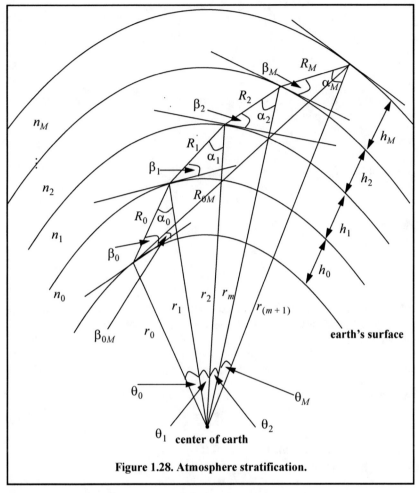

Figure 1.28. Atmosphere stratification.

From Eq. (1.112) one can write the general expression for the angle of incidence. More precisely,

$$\alpha_m = \text{asin}\left[\frac{r_m}{r_{(m+1)}} \cos\beta_m\right] \qquad (1.115)$$

Applying the law of sines of the direct path R_{0M} yields

$$\beta_{0M} = \text{acos}\left\{\frac{r_{(M+1)}}{R_{0M}} \sin\left[\sum_{i=0}^{M} \theta_i\right]\right\} \qquad (1.116)$$

where

$$R_{0M}^2 = r_0^2 + r_{(m+1)}^2 - 2r_0 r_{(m+1)} \cos\left[\sum_{i=0}^{M} \theta_i\right] \tag{1.117}$$

$$\theta_i = \frac{\pi}{2} - \beta_i - \alpha_i \tag{1.118}$$

The refraction angle error is measured as the difference between the apparent and true elevation angles. Thus, it is given by

$$\Delta\beta_M = \beta_0 - \beta_{0M} \tag{1.119}$$

Note that for $M = 0$,

$$R_{00} = R_0 \ \ and \ \ \Delta\beta_M = 0 \tag{1.120}$$

Furthermore, when $\beta_0 = 90°$,

$$R_{0M} = \sum_{i=0}^{M} h_i \tag{1.121}$$

Now, in order to determine the time-delay error due to refraction, refer again to Fig. 1.28. The time it takes an electromagnetic wave to travel through a given layer, $\{R_i; \ i = 0, 1, ..., M\}$, is defined as $\{t_i; \ i = 0, 1, ..., M\}$ where

$$t_i = R_i / \varphi_i \tag{1.122}$$

and where φ_i is the phase velocity in the *ith* layer and is defined by

$$\varphi_i = c / n_i \tag{1.123}$$

It follows that the total time of travel of the refracted wave in a stratified atmosphere is given by

$$t_T = \frac{1}{c} \sum_{i=0}^{M} n_i R_i \tag{1.124}$$

The free space travel time of an unrefracted wave is denoted by t_{0M},

$$t_{0M} = R_{0M} / c \tag{1.125}$$

Therefore, the range error ΔR that results from refraction is

$$\Delta R = \sum_{i=0}^{M} n_i R_i - R_{0M} \tag{1.126}$$

By using the law of cosines one computes R_i as

$$R_i^2 = r_i^2 + r_{(i+1)}^2 - 2r_i r_{(i+1)} \cos\theta_i \qquad (1.127)$$

The results stated in Eqs. (1.125) and (1.26) are valid only in the troposphere. In the ionosphere, which is a dispersive medium, the index of refraction is also a function of frequency. In this case, the group velocity must be used when estimating the range errors of radar measurements. Thus, the total time of travel in the medium is now given by

$$t_T = \frac{1}{c} \sum_{i=0}^{M} \frac{R_i}{n_i} \qquad (1.128)$$

Finally, the range error in the ionosphere is

$$\Delta R = \sum_{i=0}^{M} \frac{R_i}{n_i} - R_{0M} \qquad (1.129)$$

1.10.3. Four-Third Earth Model

An effective and fairly accurate technique for dealing with refraction is to replace the actual earth with an imaginary earth whose radius is $r_e = kr_0$, where $r_0 = 6375 Km$ is the actual earth radius, and k is

$$k = \frac{1}{1 + r_0(dn/dh)} \qquad (1.130)$$

When the refractivity gradient is assumed to be constant with altitude and is equal to 39×10^{-9} per meter, then $k = 4/3$. Using an effective earth radius $r_e = (4/3)r_0$ produces what is known as the *"four-third earth model."* In general, choosing

$$r_e = r_0(1 + 6.37 \times 10^{-3}(dn/dh)) \qquad (1.131)$$

produces a propagation model where waves travel in straight lines. Selecting the correct value for k depends heavily on the region's meteorological conditions. At low altitudes (typically less than 10 Km) when using the 4/3 earth model, one can assume that radar waves (beams) travel in straight lines and do not refract. This is illustrated in Fig. 1.29.

1.10.4. Ground Reflection

When radar waves are reflected from the earth's surface, they suffer a loss in amplitude and a change in phase. Three factors that contribute to these changes they are the smooth surface reflection coefficient, the divergence factor due to earth's curvature, and the surface roughness.

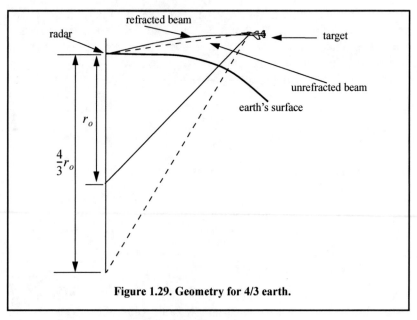

Figure 1.29. Geometry for 4/3 earth.

Smooth Surface Reflection Coefficient

The smooth surface reflection coefficient depends on the frequency, the surface dielectric coefficient, and the radar grazing angle. The vertical polarization and the horizontal polarization reflection coefficients are

$$\Gamma_v = \frac{\varepsilon \sin \psi_g - \sqrt{\varepsilon - (\cos \psi_g)^2}}{\varepsilon \sin \psi_g + \sqrt{\varepsilon - (\cos \psi_g)^2}} \qquad (1.132)$$

$$\Gamma_h = \frac{\sin \psi_g - \sqrt{\varepsilon - (\cos \psi_g)^2}}{\sin \psi_g + \sqrt{\varepsilon - (\cos \psi_g)^2}} \qquad (1.133)$$

where ψ_g is the grazing angle (incident angle) and ε is the complex dielectric constant of the surface, and are given by

$$\varepsilon = \varepsilon' - j\varepsilon'' = \varepsilon' - j60\lambda\sigma \qquad (1.134)$$

where λ is the wavelength and σ the medium conductivity in mhos/meter. Typical values of ε' and ε'' can be found tabulated in the literature.

Note that when $\psi_g = 90°$, we get

$$\Gamma_h = \frac{1-\sqrt{\varepsilon}}{1+\sqrt{\varepsilon}} = -\frac{\varepsilon - \sqrt{\varepsilon}}{\varepsilon + \sqrt{\varepsilon}} = -\Gamma_v \qquad (1.135)$$

while when the grazing angle is very small ($\psi_g \approx 0$), we have

$$\Gamma_h = -1 = \Gamma_v \qquad (1.136)$$

Tables 1.1 and 1.2 show some typical values for the electromagnetic properties of soil and sea water. Figure 1.30 shows the corresponding magnitude plots for Γ_h and Γ_v, while Fig. 1.31 shows the phase plots for seawater at $28°C$ where $\varepsilon' = 65$ and $\varepsilon'' = 30.7$ at X-band. The plots shown in these figures show the general typical behavior of the reflection coefficient.

Table 1.1. Electromagnetic properties of soil.

Frequency GHz	Moisture content by volume							
	0.3%		10%		20%		30%	
	ε'	ε''	ε'	ε''	ε'	ε''	ε'	ε''
0.3	2.9	0.071	6.0	0.45	10.5	0.75	16.7	1.2
3.0	2.9	0.027	6.0	0.40	10.5	1.1	16.7	2.0
8.0	2.8	0.032	5.8	0.87	10.3	2.5	15.3	4.1
14.0	2.8	0.350	5.6	1.14	9.4	3.7	12.6	6.3
24	2.6	0.030	4.9	1.15	7.7	4.8	9.6	8.5

Table 1.2. Electromagnetic properties of sea water.

Frequency GHz	Temperature					
	$T = 0°C$		$T = 10°C$		$T = 20°C$	
	ε'	ε''	ε'	ε''	ε'	ε''
0.1	77.8	522	75.6	684	72.5	864
1.0	77.0	59.4	75.2	73.8	72.3	90.0
2.0	74.0	41.4	74.0	45.0	71.6	50.4
3.0	71.0	38.4	72.1	38.4	70.5	40.2
4.0	66.5	39.6	69.5	36.9	69.1	36.0
6.0	56.5	42.0	63.2	39.0	65.4	36.0
8.0	47.0	42.8	56.2	40.5	60.8	36.0

Observation of Fig. 1.30 indicates the following conclusions: (1) The magnitude of the reflection coefficient with horizontal polarization is equal to unity at very small grazing angles and it decreases monotonically as the angle is increased. (2) The magnitude of the vertical polarization has a well-defined minimum. The angle that corresponds to this condition is called Brewster's polarization angle. For this reason, airborne radars in the look-down mode utilize mainly vertical polarization to significantly reduce the terrain bounce reflections. (3) For horizontal polarization the phase is almost π; however, for vertical polarization the phase changes to zero around the Brewster's angle. (4) For very small angles (less than $2°$) both $|\Gamma_h|$ and $|\Gamma_v|$ are nearly one; $\angle\Gamma_h$ and $\angle\Gamma_v$ are nearly π. Thus, little difference in the propagation of horizontally or vertically polarized waves exists at low grazing angles. Figure 1.30 can be reproduced using the following MATLAB code.

```
close all; clear all
psi = 0.01:0.05:90;
[rh,rv] = ref_coef (psi, 65,30.7);
gamamodv = abs(rv); gamamodh = abs(rh); subplot(2,1,1)
plot(psi,gamamodv,'k',psi,gamamodh,'k -.','linewidth',1.5); grid
legend ('Vertical Polarization', 'Horizontal Polarization')
title('Reflection coefficient - magnitude')
pv = -angle(rv); ph = angle(rh); subplot(2,1,2)
plot(psi,pv,'k',psi,ph,'k -.','linewidth',1.5); grid
legend ('Vertical Polarizatio', 'Horizontal Polarization')
title('Reflection coefficient - phase'); xlabel('Grazing angle in degrees');
```

Figures 1.31 and 1.32 show the magnitudes of the horizontal and vertical reflection coefficients as a function of grazing angle for four soils at 8 GHz.

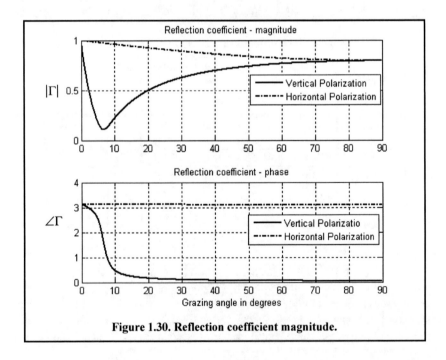

Figure 1.30. Reflection coefficient magnitude.

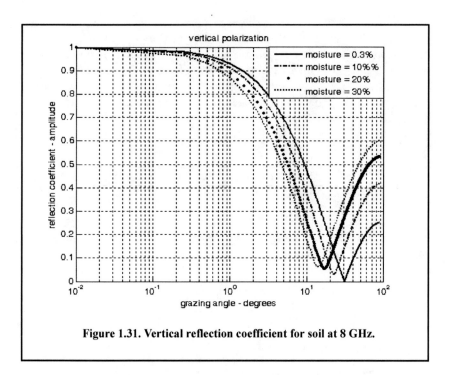

Figure 1.31. Vertical reflection coefficient for soil at 8 GHz.

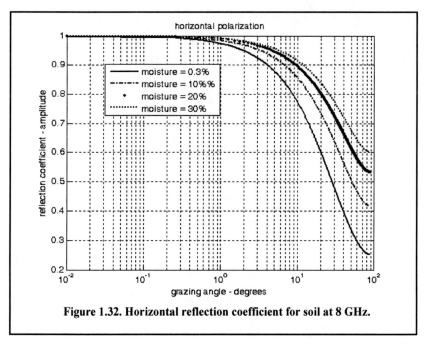

Figure 1.32. Horizontal reflection coefficient for soil at 8 GHz.

Divergence

The overall reflection coefficient is also affected by the round earth divergence factor, D. When an electromagnetic wave is incident on a round earth surface, the reflected wave diverges because of the earth's curvature. This is illustrated in Fig. 1.33. Due to divergence the reflected energy is defocused, and the radar power density is reduced. The divergence factor can be derived using geometrical considerations.

The divergence factor can be expressed as

$$D = \sqrt{\frac{r_e \ r \ \sin\psi_g}{[(2r_1r_2/\cos\psi_g) + r_e r \sin\psi_g](1 + h_r/r_e)(1 + h_t/r_e)}} \qquad (1.137)$$

where all the parameters in Eq. (1.137) are defined in Fig. 1.34. Since the grazing ψ_g is always small when the divergence D is very large, the following approximation is adequate in most radar cases of interest:

$$D \approx \frac{1}{\sqrt{1 + \frac{4r_1r_2}{r_e r \sin 2\psi_g}}} \qquad (1.138)$$

Rough Surface Reflection

In addition to divergence, surface roughness also affects the reflection coefficient. Surface roughness is given by

$$S_r = e^{-2\left(\frac{2\pi h_{rms} \sin\psi_g}{\lambda}\right)^2} \qquad (1.139)$$

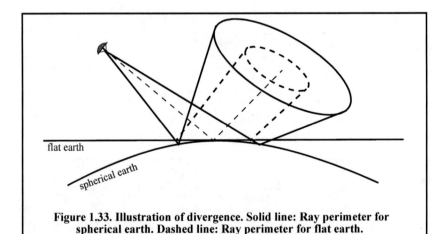

Figure 1.33. Illustration of divergence. Solid line: Ray perimeter for spherical earth. Dashed line: Ray perimeter for flat earth.

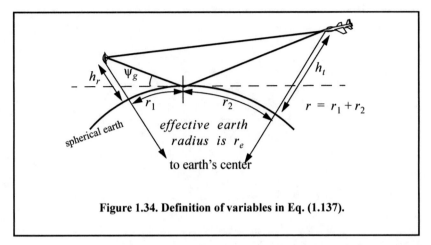

Figure 1.34. Definition of variables in Eq. (1.137).

where h_{rms} is the root mean square (rms) surface height irregularity. Another form for the rough surface reflection coefficient that is more consistent with experimental results is given by

$$S_r = e^{-z}I_0(z) \qquad (1.140)$$

$$z = 2\left(\frac{2\pi h_{rms}\sin\psi_g}{\lambda}\right)^2 \qquad (1.141)$$

where I_0 is the modified Bessel function of order zero.

Total Reflection Coefficient

In general, rays reflected from rough surfaces undergo changes in phase and amplitude, which results in the diffused (noncoherent) portion of the reflected signal. Combining the effects of smooth surface reflection coefficient, divergence, and the rough surface reflection coefficient, one express the total reflection coefficient Γ_t as

$$\Gamma_t = \Gamma_{(h,v)}DS_r \qquad (1.142)$$

$\Gamma_{(h,v)}$ is the horizontal or vertical smooth surface reflection coefficient, D is divergence, and S_r is the rough surface reflection coefficient.

1.10.5. The Pattern Propagation Factor - Flat Earth

Consider the geometry shown in Fig. 1.35. The radar is located at height h_r. The target is at range R, and is located at a height h_t. The grazing angle is ψ_g. The radar energy emanating from its antenna will reach the target via two paths: the "direct path" AB and the "indirect path" ACB.

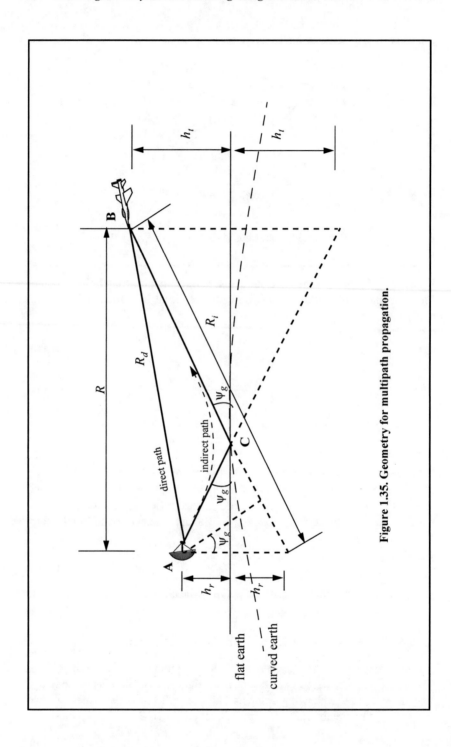

Figure 1.35. Geometry for multipath propagation.

The lengths of the paths AB and ACB are normally very close to one another and thus, the difference between the two paths is very small. Denote the direct path as R_d, the indirect path as R_i, and the difference as $\Delta R = R_i - R_d$. It follows that the phase difference between the two paths is given by

$$\Delta \Phi = \frac{2\pi \Delta R}{\lambda} \qquad (1.143)$$

where λ is the radar wavelength.

The indirect signal amplitude arriving at the target is less than the signal amplitude arriving via the direct path. This is because the antenna gain in the direction of the indirect path is less than that along the direct path, and because the signal reflected from the earth's surface at point C is modified in amplitude and phase in accordance to the earth's reflection coefficient, Γ. The earth reflection coefficient is given by

$$\Gamma = \rho e^{j\varphi} \qquad (1.144)$$

where ρ is less than unity and φ describes the phase shift induced on the indirect path signal due to surface roughness.

The direct signal (in volts) arriving at the target via the direct path can be written as

$$E_d = e^{j\omega_0 t} e^{j\frac{2\pi}{\lambda}R_d} \qquad (1.145)$$

where the time harmonic term $\exp(j\omega_0 t)$ represents the signal's time dependency, and the exponential term $\exp(j(2\pi/\lambda)R_d)$ represents the signal spatial phase. The indirect signal at the target is

$$E_i = \rho e^{j\varphi} e^{j\omega_0 t} e^{j\frac{2\pi}{\lambda}R_i} \qquad (1.146)$$

where $\rho \exp(j\varphi)$ is the surface reflection coefficient. Therefore, the overall signal arriving at the target is

$$E = E_d + E_i = e^{j\omega_0 t} e^{j\frac{2\pi}{\lambda}R_d}\left(1 + \rho e^{j\left(\varphi + \frac{2\pi}{\lambda}(R_i - R_d)\right)}\right) \qquad (1.147)$$

Due to reflections from the earth's surface, the overall signal strength is then modified at the target by the ratio of the signal strength in the presence of earth to the signal strength at the target in free space. From Eq. (1.147) the modulus of this ratio is the propagation factor is

$$F = \left| \frac{E_d}{E_d + E_i} \right| = \left| 1 + \rho e^{j\varphi} e^{j\Delta\Phi} \right| \tag{1.148}$$

which can be rewritten as

$$F = \left| 1 + \rho e^{j\alpha} \right| \tag{1.149}$$

where $\alpha = \Delta\Phi + \varphi$. Using Euler's identity ($e^{j\alpha} = \cos\alpha + j\sin\alpha$), Eq. (1.149) can be written as

$$F = \sqrt{1 + \rho^2 + 2\rho\cos\alpha} \tag{1.150}$$

It follows that the signal power at the target is modified by the factor F^2. By using reciprocity, the signal power at the radar is computed by multiplying the radar equation by the factor F^4. In the following two sections we will develop exact expressions for the propagation factor for flat and curved earth.

In order to calculate the propagation factor defined in Eq. (1.150), consider the geometry of Fig. 1.35; the direct and indirect paths are computed as

$$R_d = \sqrt{R^2 + (h_t - h_r)^2} \tag{1.151}$$

$$R_i = \sqrt{R^2 + (h_t + h_r)^2} \tag{1.152}$$

which can be approximated using the truncated binomial series expansion as

$$R_d \approx R + \frac{(h_t - h_r)^2}{2R} \tag{1.153}$$

$$R_i \approx R + \frac{(h_t + h_r)^2}{2R} \tag{1.154}$$

This approximation is valid for low grazing angles, where $R \gg h_t, h_r$. It follows that

$$\Delta R = R_i - R_d \approx \frac{2h_t h_r}{R} \tag{1.155}$$

Substituting Eq. (1.155) into Eq. (1.143) yields the phase difference due to multipath propagation between the two signals (direct and indirect) arriving at the target. More precisely,

$$\Delta\Phi = \frac{2\pi}{\lambda} \Delta R \approx \frac{4\pi h_t h_r}{\lambda R} \tag{1.156}$$

As a special case, assume smooth surface with reflection coefficient $\Gamma = -1$. This assumption means that waves reflected from the surface suffer no amplitude loss, and that the induced surface phase shift is equal to $180°$. It follows that

$$F^2 = 2 - 2\cos\Delta\Phi = 4(\sin(\Delta\Phi/2))^2 \tag{1.157}$$

Substituting Eq. (1.156) into Eq. (1.157) yields

$$F^2 = 4\left(\sin\frac{2\pi h_t h_r}{\lambda R}\right)^2 \tag{1.158}$$

By using reciprocity, the expression for the propagation factor at the radar is then given by

$$F^4 = 16\left(\sin\frac{2\pi h_t h_r}{\lambda R}\right)^4 \tag{1.159}$$

Finally, the signal power at the radar is computed by multiplying the radar equation by the factor F^4:

$$P_r = \frac{P_t G^2 \lambda^2 \sigma}{(4\pi)^3 R^4} \; 16\left(\sin\frac{2\pi h_t h_r}{\lambda R}\right)^4 \tag{1.160}$$

Since the sine function varies between 0 and 1, the signal power will then vary between 0 and 16. Therefore, the fourth power relation between signal power and the target range results in varying the target range from 0 to twice the actual range in free space. In addition to that, the field strength at the radar will now have holes that correspond to the nulls of the propagation factor.

The nulls of the propagation factor occur when the sine is equal to zero. More precisely,

$$\frac{2h_r h_t}{\lambda R} = n \tag{1.161}$$

where $n = \{0, 1, 2, \ldots\}$. The maxima occur at

$$\frac{4h_r h_t}{\lambda R} = n + 1 \tag{1.162}$$

The target heights that produce nulls in the propagation factor are $\{h_t = n(\lambda R/2h_r); n = 0, 1, 2, \ldots\}$, and the peaks are produced from target heights $\{h_t = n(\lambda R/4h_r); n = 1, 2, \ldots\}$. Therefore, due to the presence of surface reflections, the antenna elevation coverage is transformed into a lobed pattern structure.

For small angles, Eq. (1.160) can be approximated by

$$P_r \approx \frac{4\pi P_t G^2 \sigma}{\lambda^2 R^8} (h_t h_r)^4 \tag{1.163}$$

Thus, the received signal power varies as the eighth power of the range instead of the fourth power. Also, the factor $G\lambda$ is now replaced by G/λ.

1.10.6. The Pattern Propagation Factor - Spherical Earth

In order to model the effects of multipath propagation on radar performance more accurately, we need to remove the flat earth condition and account for the earth's curvature. When considering round earth, electromagnetic waves travel in curved paths because of the atmospheric refraction. And as mentioned earlier, the most commonly used approach to mitigating the effects of atmospheric refraction is to replace the actual earth by an imaginary earth such that electromagnetic waves travel in straight lines. The fictitious effective earth radius is

$$r_e = kr_0 \tag{1.164}$$

where k is a constant and r_0 is the actual earth radius. Using the geometry in Fig. 1.36, the direct and indirect path difference is

$$\Delta R = R_1 + R_2 - R_d \tag{1.165}$$

The propagation factor is computed by using ΔR from Eq. (1.150). To compute $(R_1, R_2, \text{and } R_d)$, the following cubic equation must first be solved for r_1:

$$2r_1^3 - 3rr_1^2 + (r^2 - 2r_e(h_r + h_t))r_1 + 2r_e h_t r = 0 \tag{1.166}$$

The solution is

$$r_1 = \frac{r}{2} - p\sin\frac{\xi}{3} \tag{1.167}$$

where

$$p = \frac{2}{\sqrt{3}}\sqrt{r_e(h_t + h_r) + \frac{r^2}{4}} \tag{1.168}$$

$$\xi = \operatorname{asin}\left(\frac{2r_e r(h_t - h_r)}{p^3}\right) \tag{1.169}$$

Next, we solve for R_1, R_2, and R_d. From Fig. 1.36 (assume flat 4/3 earth and use small angle approximation),

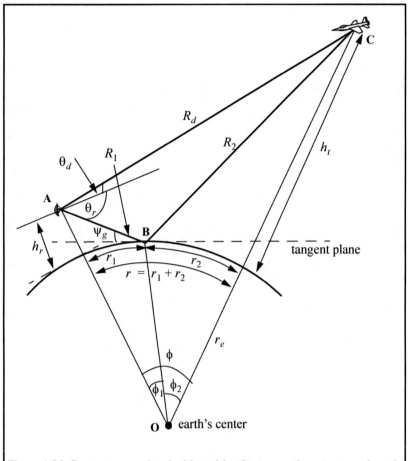

Figure 1.36. Geometry associated with multipath propagation over round earth.

$$\phi_1 = r_1/r_e; \quad \phi_2 = r_2/r_e \qquad (1.170)$$

$$\phi = r/r_e \qquad (1.171)$$

Using the law of cosines to the triangles ABO and BOC yields

$$R_1 = \sqrt{r_e^2 + (r_e + h_r)^2 - 2r_e(r_e + h_r)\cos\phi_1} \qquad (1.172)$$

$$R_2 = \sqrt{r_e^2 + (r_e + h_t)^2 - 2r_e(r_e + h_t)\cos\phi_2} \qquad (1.173)$$

Eqs. (1.172) and (1.173) can be written in the following simpler forms:

$$R_1 = \sqrt{h_r^2 + 4r_e(r_e + h_r)(\sin(\phi_1/2))^2}$$ **(1.174)**

$$R_2 = \sqrt{h_t^2 + 4r_e(r_e + h_t)(\sin(\phi_2/2))^2}$$ **(1.175)**

Using the law of cosines on the triangle AOC yields

$$R_d = \sqrt{(h_r - h_t)^2 + 4(r_e + h_t)(r_e + h_r)\left(\sin\left(\frac{\phi_1 + \phi_2}{2}\right)\right)^2}$$ **(1.176)**

Additionally

$$r = r_e \mathrm{acos}\left(\sqrt{\frac{(r_e + h_r)^2 + (r_e + h_t)^2 - R_d^2}{2(r_e + h_r)(r_e + h_t)}}\right)$$ **(1.177)**

Substituting Eqs. (1.174) through (1.176) directly into Eq. (1.165) may not be conducive to numerical accuracy. A more suitable form for the computation of ΔR is then derived. The detailed derivation is in Blake (1980). The results are listed below. For better numerical accuracy use the following expression to compute ΔR:

$$\Delta R = \frac{4R_1 R_2 (\sin\psi_g)^2}{R_1 + R_2 + R_d}$$ **(1.178)**

where

$$\psi_g = \mathrm{asin}\left(\frac{2r_e h_r + h_r^2 - R_1^2}{2r_e R_1}\right) \approx \mathrm{asin}\left(\frac{h_r}{R_1} - \frac{R_1}{2r_e}\right)$$ **(1.179)**

MATLAB Program "multipath.m"

The MATLAB program *"multipath.m"* calculates the two-way propagation factor using the 4/3 earth model for spherical earth. It assumes a known free space radar-to-target range. It can be easily modified to assume a known true spherical earth ground range between the radar and the target. Additionally, this program generates three types of plots. They are (1) the propagation factor as a function of range, (2) the free space relative signal level versus range, and (3) the relative signal level with multipath effects included. This program uses the equations presented in the previous few sections and includes the effects of the total surface reflection coefficient Γ_t. Finally, it can also be easily modified to plot the propagation factor versus target height at a fixed target range.

Using this program, Fig. 1.37 presents a plot for the propagation factor loss versus range using $f = 3\,GHz$, $h_r = 30.48m$, and $h_t = 60.96m$. In this example, vertical polarization is assumed. Divergence effects are not included; neither is the reflection coefficient. More precisely in this example $D = \Gamma_t = 1$ is assumed.

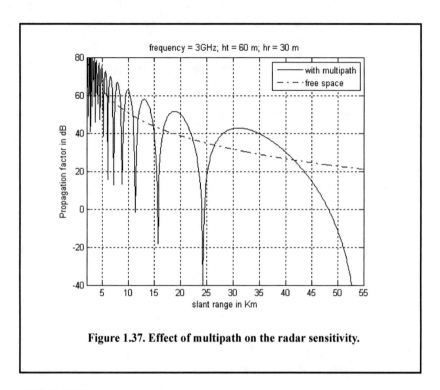

Figure 1.37. Effect of multipath on the radar sensitivity.

1.10.7. Diffraction

The analysis that led to creating the multipath model described in the previous section applies only to ground reflections from the intermediate region, as illustrated in Fig. 1.38. The effects of ground reflection below the radar horizon is governed by another physical phenomenon referred to as diffraction. The diffraction model requires calculations of the Airy function and its roots. For this purpose, the numerical approximation presented in Shatz and Polychronopoulos[1] is adopted. This numerical algorithm, described by Shatz and Polychronopoulos, is very accurate and its implementation using MATLAB is straight forward.

---

1. Shatz, M. P., and Polychronopoulos, G. H., *An Algorithm for Evaluation of Radar Propagation in the Spherical Earth Diffraction Region.* IEEE Transactions on Antenna and Propagation, Vol. 38, August 1990, pp. 1249-1252.

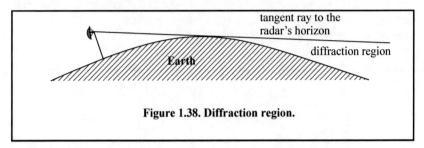

Figure 1.38. Diffraction region.

Define the following parameters,

$$x = \frac{R}{r_0} \ , \quad y = \frac{h_r}{h_0} \ , \quad t = \frac{h_t}{h_0} \tag{1.180}$$

where h_r is the radar altitude, h_t is target altitude, R is range to the target, h_0 and r_0 are normalizing factors given by

$$h_0 = \frac{1}{2}\left(\frac{r_e\lambda^2}{\pi^2}\right)^{1/3} \tag{1.181}$$

$$r_0 = \left(\frac{r_e^2\lambda}{\pi}\right)^{1/3} \tag{1.182}$$

λ is the wavelength and r_e is the effective earth radius. Let $A_i(u)$ denote the Airy function defined by

$$A_i(u) = \frac{1}{\pi}\int\limits_0^\infty \cos\left(\frac{q^3}{3} + uq\right) \, dq \tag{1.183}$$

The general expression for the propagation factor in the diffraction region is equal to

$$F = 2\sqrt{\pi x}\sum_{n=1}^\infty f_n(y)f_n(t)\exp[(e^{j\pi/6})a_n x] \tag{1.184}$$

where (x, y, t) are defined in Eq. (1.180) and

$$f_n(u) = \frac{A_i(a_n + ue^{j\pi/3})}{e^{j\pi/3}A_i{}'(a_n)} \tag{1.185}$$

where a_n is the nth root of the Airy function and A_i' is the first derivative of the Airy function. Shatz and Polychronopoulos showed that Eq. (1.184) can be approximated by

$$F = 2\sqrt{\pi x} \sum_{n=1}^{\infty} \frac{\widehat{A_i}(a_n + ye^{j\pi/3})}{e^{j\pi/3}A_i'(a_n)} \frac{\widehat{A_i}(a_n + te^{j\pi/3})}{e^{j\pi/3}A_i'(a_n)} \tag{1.186}$$

$$\exp\left[\frac{1}{2}(\sqrt{3}+j)a_n x - \frac{2}{3}(a_n + ye^{j\pi/3})^{3/2} - \frac{2}{3}(a_n + te^{j\pi/3})^{3/2}\right]$$

where

$$\widehat{A_i}(u) = A_i(u)e^{j\frac{2}{3}u^{3/2}} \tag{1.187}$$

Shatz and Polychronopoulos showed that sum in Eq. (1.186) represents accurate computation of the propagation factor within the diffraction region. In this book, a MATLAB program called *"diffraction.m"* was written by this author to implement Eq. (1.86) where the sum is terminated at $n \le 1500$ for accurate computation. For this purpose, another MATLAB function called *"airyzol.m"* was used to compute the roots of Airy function and the roots of its first derivative. Figure 1.39 (after Shatz) shows a typical output generated by this program for $h_t = 1000m$, $h_r = 8000m$, and $frequency = 167MHz$.

This figure can be reproduced using the following MATLAB code.

```
% Figure 1.39 or Figure 1.40
clc
clear all
close all
freq =167e6;
hr = 8000;
ht = 1000;
R = linspace(400e3,600e3,200); % range in Km
nt =1500; % number of point used in calculating infinite series
F = diffraction(freq, hr, ht, R, nt);
figure(1)
plot(R/1000,10*log10(abs(F).^2),'k','linewidth',1)
grid
xlabel('Range in Km')
ylabel('One way propagation factor in dB')
title('frequency = 167MHz; hr = 8000 m; ht = 1000m')
```

Figure 1.40 is similar to Fig. 1.39 except in this case the following parameters are used: $h_t = 3000m$, $h_r = 200m$, and $frequency = 428MHz$. Figure 1.41 shows a plot for the propagation factor using the same parameters in Fig.

1.40; however, in this figure, both intermediate and diffraction regions are shown.

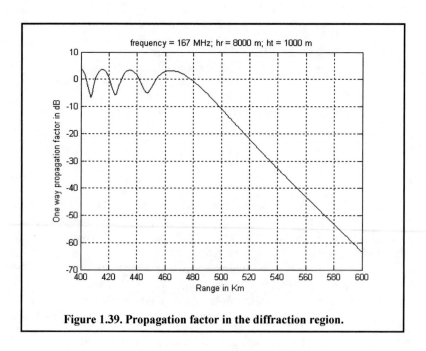

Figure 1.39. Propagation factor in the diffraction region.

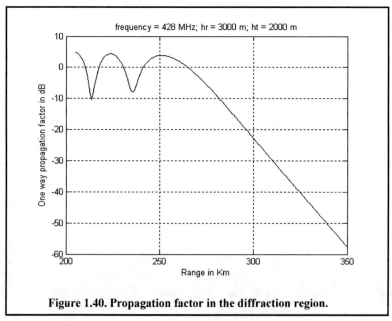

Figure 1.40. Propagation factor in the diffraction region.

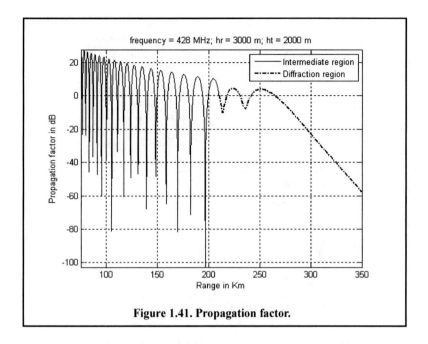

Figure 1.41. Propagation factor.

1.11. Atmospheric Attenuation

Electromagnetic waves travel in free space without suffering any energy loss. Alternatively, due to gases and water vapor in the atmosphere, radar energy suffers a loss. This loss is known as atmospheric attenuation. Atmospheric attenuation increases significantly in the presence of rain, fog, dust, and clouds. Most of the lost radar energy is normally absorbed by gases and water vapor and transformed into heat, while a small portion of this lost energy is used in molecular transformation of the atmosphere particles.

The two-way atmospheric attenuation over a range R can be expressed as

$$L_{atmosphere} = e^{-2\alpha R} \qquad \textbf{(1.188)}$$

where α is the one-way attenuation coefficient. Water vapor attenuation peaks at about $22.3\,GHz$, while attenuation due to oxygen peaks at between 60 and $118\,GHz$. Atmospheric attenuation is severe for frequencies higher than $35\,GHz$. This is the reason ground-based radars rarely use frequencies higher than $35\,GHz$. Atmospheric attenuation is a function range, frequency, and elevation angle. Figure 1.42 shows a typical two-way atmospheric attenuation plot versus range at $3\,GHz$, with the elevation angle as a parameter.

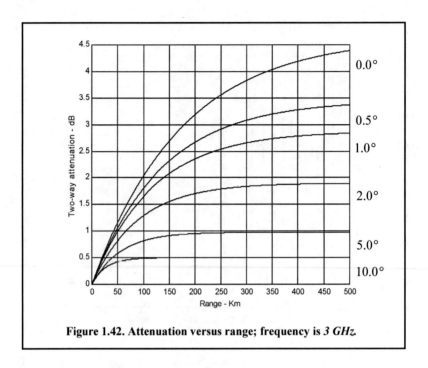

Figure 1.42. Attenuation versus range; frequency is *3 GHz*.

1.12. MATLAB Program Listings

This section presents listings for all the MATLAB programs used to produce all of the MATLAB-generated figures in this chapter. They are listed in the same order they appear in the text.

1.12.1. MATLAB Function "range_resolution.m"

The MATLAB function *"range_resolution.m"* calculates range resolution; its syntax is as follows:

$$[delta\_R] = range\_resolution(var, indicator)$$

where

Symbol	Description	Units	Status
var, indicator	*bandwidth, "hz"*	*Hz, none*	*inputs*
var, indicator	*pulse width, "s"*	*seconds, none*	*inputs*
delta_R	*range resolution*	*meters*	*output*

MATLAB Function *"range_resolution.m"* **Listing**

function [delta_R] = range_resolution(bandwidth, indicator)
% This function computes radar range resolution in meters
% the bandwidth must be in Hz ==> indicator = Hz.
% Bandwidth may be equal to (1/pulse width)==> indicator = seconds
c = 3.e+8; % speed of light
if(indicator == 'hz')
  delta_R = c / 2.0 / bandwidth;
else
  delta_R = c * bandwidth / 2.0;
end
return

1.12.2. MATLAB Function "radar_eq.m"

The function *"radar_eq.m"* implements Eq. (1.40); its syntax is as follows:

$$[snr] = radar\_eq \ (pt, freq, g, sigma, b, nf, loss, range)$$

where

Symbol	Description	Units	Status
pt	peak power	Watts	input
freq	radar center frequency	Hz	input
g	antenna gain	dB	input
sigma	target cross section	m^2	input
b	bandwidth	Hz	input
nf	noise figure	dB	input
loss	radar losses	dB	input
range	target range (can be either a single value or a vector)	meters	input
snr	SNR (single value or a vector, depending on the input range)	dB	output

MATLAB Function *"radar_eq.m"* **Listing**

function [snr] = radar_eq(pt, freq, g, sigma, b, nf, loss, range)
% This program implements Eq. (1.40)
c = 3.0e+8; % speed of light
lambda = c / freq; % wavelength
p_peak = 10*log10(pt); % convert peak power to dB
lambda_sqdb = 10*log10(lambda^2); % compute wavelength square in dB
sigmadb = 10*log10(sigma); % convert sigma to dB
four_pi_cub = 10*log10((4.0 * pi)^3); % (4pi)^3 in dB

```
k_db = 10*log10(1.38e-23); % Boltzmann's constant in dB
to_db = 10*log10(290); % noise temp. in dB
b_db = 10*log10(b); % bandwidth in dB
range_pwr4_db = 10*log10(range.^4); % vector of target range^4 in dB
% Implement Equation (1.63)
num = p_peak + g + g + lambda_sqdb + sigmadb;
den = four_pi_cub + k_db + to_db + b_db + nf + loss + range_pwr4_db;
snr = num - den;
return
```

1.12.3. MATLAB Function "power_aperture.m"

The function *"power_aperture.m"* implements the search radar equation given in Eq. (1.47); its syntax is as follows:

PAP = power_aperture (snr, tsc, sigma, range, nf, loss, az_angle, el_angle)

where

Symbol	Description	Units	Status
snr	*sensitivity snr*	*dB*	*input*
tsc	*scan time*	*seconds*	*input*
sigma	*target cross section*	m^2	*input*
range	*target range*	*meters*	*input*
nf	*noise figure*	*dB*	*input*
loss	*radar losses*	*dB*	*input*
az_angle	*search volume azimuth extent*	*degrees*	*input*
el_angle	*search volume elevation extent*	*degrees*	*input*
PAP	*power aperture product*	*dB*	*output*

MATLAB Function *"power_aperture.m"* **Listing**

```
function PAP = power_aperture(snr,tsc,sigma,range,nf,loss,az_angle,el_angle)
% This program implements Eq. (1.47)
Tsc = 10*log10(tsc); % convert Tsc into dB
Sigma = 10*log10(sigma); % convert sigma to dB
four_pi = 10*log10(4.0 * pi); % (4pi) in dB
k_db = 10*log10(1.38e-23); % Boltzmann's constant in dB
To = 10*log10(290); % noise temp. in dB
range_pwr4_db = 10*log10(range.^4); % target range^4 in dB
omega = (az_angle/57.296) * (el_angle / 57.296);
% compute search volume in steraradians
Omega = 10*log10(omega); % search volume in dB
% implement Eq. (1.79)
PAP = snr + four_pi + k_db + To + nf + loss + range_pwr4_db + Omega ...
```

- Sigma - Tsc;
return

1.12.4. MATLAB Function "range_red_factor.m"

The function *"range_red_factor.m"* implements Eqs. (1.88) and (1.89). This function generates plots of RRF versus (1) the radar operating frequency, (2) radar to jammer range, and (3) jammer power. Its syntax is as follows:

$$[RRF] = range\_red\_factor\ (te,\ pj,\ gj,\ g,\ freq,\ bj,\ rangej,\ lossj)$$

where

Symbol	Description	Units	Status
te	radar effective temperature	K	input
pj	jammer peak power	W	input
gj	jammer antenna gain	dB	input
g	radar antenna gain on jammer	dB	input
freq	radar operating frequency	Hz	input
bj	jammer bandwidth	Hz	input
rangej	radar to jammer range	Km	input
lossj	jammer losses	dB	input

MATLAB Function *"range_red_factor.m"* **Listing**

```
function RRF = range_red_factor (ts, pj, gj, g, freq, bj, rangej, lossj)
% This function computes the range reduction factor and produces
% plots of RRF versus wavelength, radar to jammer range, and jammer power
c = 3.0e+8;
k = 1.38e-23;
lambda = c / freq;
gj_10 = 10^( gj/10);
g_10 = 10^( g/10);
lossj_10 = 10^(lossj/10);
index = 0;
for wavelength = .01:.001:1
  index = index +1;
  jamer_temp = (pj * gj_10 * g_10 *wavelength^2) /...
    (4.0^2 * pi^2 * k * bj * lossj_10 * (rangej * 1000.0)^2);
  delta = 10.0 * log10(1.0 + (jamer_temp / ts));
  rrf(index) = 10^(-delta /40.0);
end
w = 0.01:.001:1;
figure (1)
semilogx(w,rrf,'k')
```

```
grid
xlabel ('Wavelength in meters')
ylabel ('Range reduction factor')
index = 0;
for ran =rangej*.3:10:rangej*2
  index = index + 1;
  jamer_temp = (pj * gj_10 * g_10 *lambda^2) / ...
     (4.0^2 * pi^2 * k * bj * lossj_10 * (ran * 1000.0)^2);
  delta = 10.0 * log10(1.0 + (jamer_temp / ts));
  rrf1(index) = 10^(-delta /40.0);
end
figure(2)
ranvar = rangej*.3:10:rangej*2 ;
plot(ranvar,rrf1,'k')
grid
xlabel ('Radar to jammer range in Km')
ylabel ('Range reduction factor')
index = 0;
for pjvar = pj*.01:100:pj*2
  index = index + 1;
  jamer_temp = (pjvar * gj_10 * g_10 *lambda^2) / ...
     (4.0^2 * pi^2 * k * bj * lossj_10 * (rangej * 1000.0)^2);
  delta = 10.0 * log10(1.0 + (jamer_temp / ts));
  rrf2(index) = 10^(-delta /40.0);
end
figure(3)
pjvar = pj*.01:100:pj*2;
plot(pjvar,rrf2,'k')
grid
xlabel ('Jammer peak power in Watts')
ylabel ('Range reduction factor')
```

1.12.5. MATLAB Function *"ref_coef.m"*

The function *"ref_coef.m"* calculates the horizontal and vertical magnitude and phase response of the reflection coefficient. The syntax is as follows

$$[rh,rv] = ref\_coef \, (psi, \, epsp, \, epspp)$$

where

Symbol	Description	Status
psi	grazing angle in degrees (can be a vector or a scalar)	input
epsp	ε'	input

Symbol	Description	Status
epspp	ε''	*input*
rh	*horizontal reflection coefficient complex vector*	*output*
rv	*vertical reflection coefficient complex vector*	*output*

MATLAB Function *"ref_coef.m"* Listing

```
function [rh,rv] = ref_coef (psi, epsp, epspp)
eps = epsp - i .* epspp;
psirad = psi.*(pi./180.);
arg1 = eps - (cos(psirad).^2);
arg2 = sqrt(arg1);
arg3 = sin(psirad);
arg4 = eps.*arg3;
rv = (arg4-arg2)./(arg4+arg2);
rh = (arg3-arg2)./(arg3+arg2);
```

1.12.6. MATLAB Function "divergence.m"

The MATLAB function *"divergence.m"* calculates the divergence. The syntax is as follows:

$$D = divergence\ (psi,\ r1,\ r2,\ hr,\ ht)$$

where

Symbol	Description	Status
psi	*grazing angle in degrees (can be vector or scalar)*	*input*
r1	*ground range between radar and specular point in Km*	*input*
r2	*ground range between specular point and target in Km*	*input*
hr	*radar height in meters*	*input*
ht	*target height in meters*	*input*
D	*divergence*	*output*

MATLAB Function *"divergence.m"* Listing

```
function [D] = divergence(psi, r1, r2, hr, ht)
% calculates divergence
%     inputs %%%%%%%%%%%%%%%%%%%%%%%%%%%
% r1  ground range between radar and specular point in Km
% r2  ground range between specular point and target in Km
% psi grazing angle in degrees
%     parameters %%%%%%%%%%%%%%%%%%%%%%%
% re  4/3 earth radius 4/3 * 6375 Km
```

```
%   r = r1 + r2
psi = psi .* pi ./180; % psi in radians
re = (4/3) * 6375e3;
r = r1 + r2;
arg1 = re .* r . * sin(psi) ;
arg2 = ((2 .* r1 .* r2 ./ cos(psi)) + re .* r. * sin(psi)) .* (1+hr./re) .* (1+ht./re);
D = sqrt(arg1 ./ arg2);
return
```

1.12.7. MATLAB Function "surf_rough.m"

The MATLAB function *"surf_rough.m"* calculates the surface roughness reflection coefficient. The syntax is as follows:

$$Sr = surf\_rough\ (hrms, freq, psi)$$

where

Symbol	Description	Status
hrms	surface rms roughness value in meters	input
freq	frequency in Hz	input
psi	grazing angle in degrees	input
Sr	surface roughness coefficient	output

MATLAB Function *"surf_rough.m"* **Listing**

```
function Sr = surf_rough(hrms, freq, psi)
clight = 3e8;
psi = psi .* pi ./ 180; % angle in radians
lambda = clight / freq; % wavelength
g = (2.* pi .* hrms .* sin(psi) ./ lambda).^2;
Sr = exp(-2 .* g);
return
```

1.12.8. MATLAB Program "multipath.m"

```
% This program calculates and plots the propagation factor versus
% target range with a fixed target height.
% The free space radar-to-target range is assumed to be known.
clear all;
close all;
eps = 0.01;
%%%%%%%%%%%%%%%% input %%%%%%%%%%%%%%%%%%%%%%
ro = 6375e3; % earth radius
re = ro * 4 /3; % 4/3 earth radius
freq = 3000e6; % frequency
```

```
lambda = 3.0e8 / freq; % wavelength
hr = 30.48; % radar height in meters
ht = 2 .* hr; % target height in meters
Rd1 = linspace(2e3, 55e3, 500); % slant range 3 to 55 Km 500 points
%%%%%%%%%%%%%%%%%%%%%%%%%%%%%%%%%%%%%%%%%%%%%%%
% determine whether the traget is beyond the radar's line of sight
range_to_horizon = sqrt(2*re) * (sqrt(ht) + sqrt(hr)); % range to horizon
index = find(Rd1 > range_to_horizon);
if isempty(index);
   Rd = Rd1;
else
   Rd = Rd1(1:index(1)-1);
   fprintf('****** WARNING ****** \n')
   fprintf('Maximum range is beyond radar line of sight. \n')
   fprintf('Target is in diffraction region \n')
   fprintf('****** WARNING ****** \n')
end
%%%%%%%%%%%%%%%%%%%%%%%%%%%%%%%%%%%%%%%%%%%%%%%%%%
val1 = Rd.^2 - (ht -hr).^2;
val2 = 4 .* (re + hr) .* (re + ht);
r = 2 .* re .* asin(sqrt(val1 ./ val2));
phi = r ./ re;
p = sqrt(re .* (ht + hr) + (r.^2 ./4)) .* 2 ./ sqrt(3);
exci = asin((2 .* re .* r .* (ht - hr) ./ p.^3));
r1 = (r ./ 2) - p .* sin(exci ./3);
phi1 = r1 ./ re;
r2 = r - r1;
phi2 = r2 ./ re;
R1 = sqrt( re.^2 + (re + hr).^2 - 2 .* re .* (re + hr) .* cos(phi1));
R2 = sqrt( re.^2 + (re + ht).^2 - 2 .* re .* (re + ht) .* cos(phi2));
psi = asin((2 .* re .* hr + hr^2 - R1.^2) ./ (2 .* re .* R1));
deltaR = R1 + R2 - Rd;
%%%%%%%%%%%%%%% input surface roughness %%%%%%%%%%%%%%%%%%%
hrms = 1;
psi = psi .* 180 ./ pi;
[Sr] = surf_rough(hrms, freq, psi);
%%%%%%%%%%%%%%%%%%% input divergence %%%%%%%%%%%%%%%%%%%%%
[D] = divergence(psi, r1, r2, hr, ht);
%%%%%%%%%%%%%%%%%% input smooth earth ref. coefficient %%%%%%%%%%%%%%
epsp = 50;
epspp = 15;
[rh,rv] = ref_coef (psi, epsp, epspp);
D = 1;
 Sr =1;
gamav = abs(rv);
phv = angle(rv);
gamah = abs(rh);
phh = angle (rh);
```

```
gamav =1;
phv = pi;
Gamma_mod = gamav .* D .* Sr;
Gamma_phase = phv; %
rho = Gamma_mod;
delta_phi = 2 .* pi .* deltaR ./ lambda;
alpha = delta_phi + phv;
F = sqrt( 1 + rho.^2 + 2 .* rho .* cos( alpha));
Ro = 185.2e3; % reference range in Km
F_free = 40 .* log10(Ro ./ Rd);
F_dbr = 40 .* log10( F .* Ro ./ Rd);
F_db = 40 .* log10( eps + F );
figure(1)
plot(Rd./1000, F_db,'k','linewidth',1)
grid
xlabel('slant range in Km')
ylabel('propagation factor in dB')
axis tight
axis([2 55 -60 20])
figure(2)
plot(Rd./1000, F_dbr,'k',Rd./1000, F_free,'k-.','linewidth',1)
grid
xlabel('slant range in Km')
ylabel('Propagation factor in dB')
axis tight
axis([2 55 -40 80])
legend('with multipath','free space')
title('frequency = 3GHz; ht = 60 m; hr = 30 m')
```

1.12.9. MATLAB Program "diffraction.m"

This function utilizes Shatz's model to calculate the propagation factor in the diffraction region. It utilizes the MATLAB function *"airy.m"* which is part of the Signal Processing Toolbox. Its syntax is as follows

$$F = diffraction(freq, hr, ht, R, nt);$$

where

% Generalized spherical earth propagation factor calculations

Symbol	Description	Status
freq	radar operating frequency	Hz
hr	radar height	meters
ht	target height	meters

Symbol	Description	Status
R	*range over which to calculate the propagation factor*	*Km*
nt	*number of data point is the series given in Eq. (1.186)*	*none*
F	*propagation factor in diffraction region*	*dB*

MATLAB Program *"diffraction.m"* Listing

```
function F = diffraction(freq, hr, ht,R,nt);
%   Generalized spherical earth propagation factor calculations
%   After Shatz: Michael P. Shatz, and George H. Polychronopoulos, An
%   Algorithm for Elevation of Radar Propagation in the Spherical Earth
%   Diffraction Region. IEEE Transactions on Antenna and Propagation,
%   VOL. 38, NO.8, August 1990.
format long
re = 6373e3 * (4/3); % 4/3 earth radius in Km
[an] = airyzo1(nt);% calculate the roots of the Airy function
c = 3.0e8; % speed of light
lambda = c/freq; % wavelength
r0 = (re*re*lambda / pi)^(1/3);
h0 = 0.5 * (re*lambda*lambda/pi/pi)^(1/3);
y = hr / h0;
z = ht / h0;
%%%%%%%%%%%%%%
par1 = exp(sqrt(-1)*pi/3);
pary1 = ((2/3).*(an + y .* par1).^(1.5));
   pary = exp(pary1);
   parz1 = ((2/3).*(an + z .* par1).^(1.5));
   parz = exp(parz1);
   f1n = airy(an + y * par1) .* airy(an + z * par1) .* pary .*parz ;
   f1d = par1 .* par1 .* airy(1,an) .* airy(1,an);
   f1 = f1n ./ f1d;
   index = find(f1<1e6);
%%%%%%%%%%%%%
F = zeros(1,size(R,2));
for range = 1:size(R,2)
   x(range) = R(range)/r0;
   f2 = exp(0.5 .* (sqrt(3) +sqrt(-1)) .*an.*x(range) - pary1 -parz1);
   victor = f1(index) .* f2(index);
   fsum = sum(victor);
   F(range) = 2 .*sqrt(pi.*x(range)) .* fsum;
end
```

1.12.10. MATLAB Program "airyzo1.m"

The function "airyzo1.m" was developed to compute the roots of the Airy function. Its syntax is as follows:

$$[an] = airyzo1(nt)$$

where the input *nt* is the number of required roots, and the output *[an]* is the roots (zeros) vector.

MATLAB Program *"airyzo1.m"* **Listing**

```
function [an] = airyzo1(nt)
%   This program is a modified version of a function obtained from
%   free internet source www.mathworks.com/matlabcentral/fileexchange/
%   modified by B. Mahafza (bmahafza@dbresearch.net) in 2005
%   ==============================
%   Purpose: This program computes the first nt zeros of Airy
%   functions Ai(x)
%   Input : nt  --- Total number of zeros
%   Output: an ---   first nt roots for Ai(x)
format long
an = zeros(1,nt);
xb = zeros(1,nt);
ii = linspace(1,nt,nt);
u = 3.0.*pi.*(4.0.*ii-1)./8.0;
u1 = 1./(u.*u);
rt0 = -(u.*u).^(1.0./3.0).*((((-15.5902.*u1+.929844).* ...
u1-.138889).*u1+.10416667).*u1+1.0);
rt = 1.0e100;
while(abs((rt-rt0)./rt)> 1.e-12);
x = rt0;
ai = airy(0,x);
ad = airy(1,x);
rt=rt0-ai./ad;
if(abs((rt-rt0)./rt)> 1.e-12);
rt0 = rt;
end;
end;
an(ii)= rt;
end
```

1.12.11. MATLAB Program "fig_31_32.m"

```
% This program produces Figs. 1.31 and 1.32
close all
clear all
```

```
psi = 0.01:0.25:90;
epsp = [2.8];
epspp = [0.032];% 0.87 2.5 4.1];
[rh1,rv1] = ref_coef(psi, epsp,epspp);
gamamodv1 = abs(rv1);
gamamodh1 = abs(rh1);
epsp = [5.8] ;
epspp = [0.87];
[rh2,rv2] = ref_coef(psi, epsp,epspp);
gamamodv2 = abs(rv2);
gamamodh2 = abs(rh2);
epsp = [10.3];
epspp = [2.5];
[rh3,rv3] = ref_coef(psi, epsp,epspp);
gamamodv3 = abs(rv3);
gamamodh3 = abs(rh3);
epsp = [15.3];
epspp = [4.1];
[rh4,rv4] = ref_coef(psi, epsp,epspp);
gamamodv4 = abs(rv4);
gamamodh4 = abs(rh4);
figure(1)
semilogx(psi,gamamodh1,'k',psi,gamamodh2,'k-. ', ...
psi,gamamodh3,'k. ',psi,gamamodh4,'k: ','linewidth',1.5);
grid
xlabel('grazing angle - degrees');
ylabel('reflection coefficient - amplitude')
legend('moisture = 0.3%','moisture = 10%%','moisture = 20%','moisture = 30%')
title('horizontal polarization')
% legend ('Vertical Polarization','Horizontal Polarization')
% pv = -angle(rv);
% ph = angle(rh);
% figure(2)
% plot(psi,pv,'k',psi,ph,'k -.');
% grid
% xlabel('grazing angle - degrees');
% ylabel('reflection coefficient - pahse')
% legend ('Vertical Polarization','Horizontal Polarization')
```

Problems

1.1. (a) Calculate the maximum unambiguous range for a pulsed radar with PRF of $200 Hz$ and $750 Hz$. (b) What are the corresponding PRIs?

1.2. For the same radar in Problem 1.1, assume a duty cycle of 30% and peak power of $5 KW$. Compute the average power and the amount of radiated energy during the first $20 ms$.

1.3. A certain pulsed radar uses pulse width $\tau = 1\mu s$. Compute the corresponding range resolution.

1.4. An X-band radar uses PRF of $3KHz$. Compute the unambiguous range and the required bandwidth so that the range resolution is $30m$. What is the duty cycle?

1.5. Compute the Doppler shift associated with a closing target with velocity 100, 200, and 350 meters per second. In each case compute the time dilation factor. Assume that $\lambda = 0.3m$.

1.6. In reference to Fig. 1.8, compute the Doppler frequency for

$v = 150m/s$, $\theta_a = 30°$, and $\theta_e = 15°$. Assume that $\lambda = 0.1m$.

1.7. (a) Develop an expression for the minimum PRF of a pulsed radar; (b) compute $f_{r_{min}}$ for a closing target whose velocity is $400m/s$. (c) What is the unambiguous range? Assume that $\lambda = 0.2m$.

1.8. An L-band pulsed radar is designed to have an unambiguous range of $100Km$ and range resolution $\Delta R \leq 100m$. The maximum resolvable Doppler frequency corresponds to $v_{target} \leq 350m/sec$. Compute the maximum required pulse width, the PRF, and the average transmitted power if $P_t = 500W$.

1.9. Compute the aperture size for an X-band antenna at $f_0 = 9GHz$. Assume antenna gain $G = 10, 20, 30$ dB.

1.10. An L-band radar (1500 MHz) uses an antenna whose gain is $G = 30dB$. Compute the aperture size. If the radar duty cycle is $d_t = 0.2$ and the average power is $25KW$, compute the power density at range $R = 50Km$.

1.11. For the radar described in Problem 1.9, assume the minimum detectable signal is $5dBm$. Compute the radar maximum range for $\sigma = 1.0, 10.0, 20.0m^2$.

1.12. Consider an L-band radar with the following specifications: operating frequency $f_0 = 1500MHz$, bandwidth $B = 5MHz$, and antenna gain $G = 5000$. Compute the peak power, the pulse width, and the minimum detectable signal for this radar. Assume target RCS $\sigma = 10m^2$, the single pulse SNR is $15.4dB$, noise figure $F = 5dB$, temperature $T_0 = 290K$, and maximum range $R_{max} = 150Km$.

1.13. Consider a low PRF C-band radar operating at $f_0 = 5000MHz$. The antenna has a circular aperture with radius $2m$. The peak power is $P_t = 1MW$ and the pulse width is $\tau = 2\mu s$. The PRF is $f_r = 250Hz$, and the effective temperature is $T_0 = 600K$. Assume radar losses $L = 15dB$ and target RCS $\sigma = 10m^2$. (a) Calculate the radar's unambiguous range; (b) calculate the range R_0 that corresponds to $SNR = 0dB$; (c) calculate the SNR at $R = 0.75R_0$.

1.14. Repeat the second example in Section 1.6 with $\Omega = 4°$, $\sigma = 1m^2$, and $R = 400Km$.

1.15. The atmospheric attenuation can be included in the radar equation as another loss term. Consider an X-band radar whose detection range at $20Km$ includes a $0.25dB/Km$ atmospheric loss. Calculate the corresponding detection range with no atmospheric attenuation.

1.16. Let the maximum unambiguous range for a low PRF radar be R_{max}. (a) Calculate the SNR at $(1/2)R_{max}$ and $(3/4)R_{max}$. (b) If a target with $\sigma = 10m^2$ exists at $R = (1/2)R_{max}$, what should the target RCS be at $R = (3/4)R_{max}$ so that the radar has the same signal strength from both targets.

1.17. A Millie-Meter Wave (MMW) radar has the following specifications: operating frequency $f_0 = 94GHz$, PRF $f_r = 15KHz$, pulse width $\tau = 0.05ms$, peak power $P_t = 10W$, noise figure $F = 5dB$, circular antenna with diameter $D = 0.254m$, antenna gain $G = 30dB$, target RCS $\sigma = 1m^2$, system losses $L = 8dB$, radar scan time $T_{sc} = 3s$, radar angular coverage $200°$, and atmospheric attenuation $3dB/Km$. Compute the following: (a) wavelength λ, (b) range resolution ΔR, (c) bandwidth B, (d) the SNR as a function of range, (e) the range for which $SNR = 15dB$, (f) antenna beam width, (g) antenna scan rate, (h) time on target, (i) the effective maximum range when atmospheric attenuation is considered.

1.18. A radar with antenna gain G is subject to a repeater jammer whose antenna gain is G_J. The repeater illuminates the radar with three fourths of the incident power on the jammer. (a) Find an expression for the ratio between the power received by the jammer and the power received by the radar. (b) What is this ratio when $G = G_J = 200$ and $R/\lambda = 10^5$?

1.19. A radar has the following parameters: peak power $P_t = 65KW$, total losses $L = 5dB$, operating frequency $f_0 = 8GHz$, PRF $f_r = 4KHz$, duty cycle $d_t = 0.3$, circular antenna with diameter $D = 1m$, effective aperture is 0.7 of physical aperture, noise figure $F = 8dB$. (a) Derive the various parameters needed in the radar equation. (b) What is the unambiguous range? (c) Plot the SNR versus range (1 Km to the radar unambiguous range) for a 5dBsm target, and (d) if the minimum SNR required for detection is 14 dB, what is the detection range for a 6 dBsm target? What is the detection range if the SNR threshold requirement is raised to 18 dB?

1.20. A radar has the following parameters: Peak power $P_t = 50KW$; total losses $L = 5dB$; operating frequency $f_0 = 5.6GHz$; noise figure $F = 10dB$ pulse width $\tau = 10\mu s$; PRF $f_r = 2KHz$; antenna beamwidth $\theta_{az} = 1°$ and $\theta_{el} = 5°$. (a) What is the antenna gain? (b) What is the effective aperture if the aperture efficiency is 60%? (c) Given a 14 dB threshold detection, what is the detection range for a target whose RCS is $\sigma = 1m^2$?

1.21. A certain radar has losses of 5 dB and a receiver noise figure of 10 dB. This radar has a detection coverage requirement that extends over 3/4 of a hemisphere and must complete it in 3 second. The base line target RCS is 6 dBsm and the minimum SNR is 15 dB. The radar detection range is less than 80 Km. What is the average power aperture product for this radar so that it can satisfy its mission?

1.22. Assume a bandwidth of $150KHz$. (a) Compute the noise figure for the three cascaded amplifiers. (b) Compute the effective temperature for the three cascaded amplifiers. (c) Compute the overall system noise figure.

1.23. An exponential expression for the index of refraction is given by $n = 1 + 315 \times 10^{-6}\exp(-0.136h)$ where the altitude h is in Km. Calculate the index of refraction for a well-mixed atmosphere at 10% and 50% of the troposphere.

1.24. A source with equivalent temperature $T_0 = 290K$ is followed by three amplifiers with specifications shown in the table below.

Amplifier	F, dB	G, dB	T_e
1	You must compute	12	350
2	10	22	
3	15	35	

1.25. Reproduce Figs. 1.30 and 1.31 by using $f = 8GHz$ and (a) $\varepsilon' = 2.8$ and $\varepsilon'' = 0.032$ (dry soil); (b) $\varepsilon' = 47$ and $\varepsilon'' = 19$ (sea water at $0°C$); (c) $\varepsilon' = 50.3$ and $\varepsilon'' = 18$ (lake water at $0°C$).

1.26. In reference to Fig. 8.16, assume a radar height of $h_r = 100m$ and a target height of $h_t = 500m$. The range is $R = 20Km$. (a) Calculate the lengths of the direct and indirect paths. (b) Calculate how long it will take a pulse to reach the target via the direct and indirect paths.

1.27. A radar at altitude $h_r = 10m$ and a target at altitude $h_t = 300m$, and assuming a spherical earth, calculate r_1, r_2, and ψ_g.

1.28. In the previous problem, assuming that you may be able to use the small grazing angle approximation: (a) Calculate the ratio of the direct to the indirect signal strengths at the target. (b) If the target is closing on the radar with velocity $v = 300m/s$, calculate the Doppler shift along the direct and indirect paths. Assume $\lambda = 3cm$.

1.29. Derive an asymptotic form for Γ_h and Γ_v when the grazing angle is very small.

1.30. In reference to Fig. 1.37, assume a radar height of $h_r = 100m$ and a target height of $h_t = 500m$. The range is $R = 20Km$. (a) Calculate the lengths of the direct and indirect paths. (b) Calculate how long it will take a pulse to reach the target via the direct and indirect paths.

1.31. Using the law of cosines, derive Eq. (1.138) from (1.137).

1.32. In the previous problem, assuming that you may be able to use the small grazing angle approximation. (a) Calculate the ratio of the direct to the indirect signal strengths at the target. (b) If the target is closing on the radar with velocity $v = 300m/s$, calculate the Doppler shift along the direct and indirect paths. Assume $\lambda = 3cm$.

1.33. In the previous problem, assuming that you may be able to use the small grazing angle approximation: (a) Calculate the ratio of the direct to the indirect signal strengths at the target. (b) If the target is closing on the radar with velocity $v = 300m/s$, calculate the Doppler shift along the direct and indirect paths. Assume $\lambda = 3cm$.

1.34. Calculate the range to the horizon corresponding to a radar at $5Km$ and $10Km$ of altitude. Assume 4/3 earth.

1.35. Develop a mathematical expression that can be used to reproduce Fig. 1.42.

1.36. Modify the MATLAB program *"multipath.m"* so that it uses the true spherical ground range between the radar and the target.

1.37. Modify the MATLAB program *"multipath.m"* so that it accounts for the radar antenna.

Chapter 2

Linear Systems and Complex Signal Representation

This chapter presents a top level overview of elements of signal theory that are relevant to radar detection and radar signal processing. It is assumed that the reader has sufficient and adequate background in signals and systems as well as in Fourier transform and its associated properties.

2.1. *Signal and System Classifications*

In general, electrical signals can represent either current or voltage and may be classified into two main categories: energy signals and power signals. Energy signals can be deterministic or random, while power signals can be periodic or random. A signal is said to be random if it is a function of a random parameter (such as random phase or random amplitude). Additionally, signals may be divided into lowpass or bandpass signals. Signals that contain very low frequencies (close to DC) are called lowpass signals; otherwise they are referred to as bandpass signals. Through modulation, lowpass signals can be mapped into bandpass signals.

The average power P for the current or voltage signal $x(t)$ over the interval (t_1, t_2) across a 1Ω resistor is

$$P = \frac{1}{t_2 - t_1} \int_{t_1}^{t_2} |x(t)|^2 \, dt \tag{2.1}$$

The signal $x(t)$ is said to be a power signal over a very large interval $T = t_2 - t_1$, if and only if it has finite power and satisfies the relation:

$$0 < \lim_{T \to \infty} \frac{1}{T} \int_{-T/2}^{T/2} |x(t)|^2 \, dt < \infty \tag{2.2}$$

Using Parseval's theorem, the energy E dissipated by the current or voltage signal $x(t)$ across a 1Ω resistor, over the interval (t_1, t_2), is

$$E = \int_{t_1}^{t_2} |x(t)|^2 \, dt \tag{2.3}$$

The signal $x(t)$ is said to be an energy signal if and only if it has finite energy,

$$E = \int_{-\infty}^{\infty} |x(t)|^2 \, dt \quad < \infty \tag{2.4}$$

A signal $x(t)$ is said to be periodic with period T if and only if

$$x(t) = x(t + nT) \qquad \text{for all } t \tag{2.5}$$

where n is an integer.

Example:

Classify each of the following signals as an energy signal, a power signal, or neither. All signals are defined over the interval $(-\infty < t < \infty)$: $x_1(t) = \cos t + \cos 2t$, $x_2(t) = \exp(-\alpha^2 t^2)$.

Solution:

$$P_{x_1} = \frac{1}{T} \int_{-T/2}^{T/2} (\cos t + \cos 2t)^2 \, dt = 1 \Rightarrow \quad power \ signal$$

Note that since the cosine function is periodic, the limit is not necessary.

$$E_{x_2} = \int_{-\infty}^{\infty} (e^{-\alpha^2 t^2})^2 \, dt = 2 \int_{0}^{\infty} e^{-2\alpha^2 t^2} \, dt = 2 \frac{\sqrt{\pi}}{2\sqrt{2}\alpha} = \frac{1}{\alpha}\sqrt{\frac{\pi}{2}} \Rightarrow \quad energy \ signal$$

2.2. The Fourier Transform

The Fourier Transform (FT) of the signal $x(t)$ is

$$F\{x(t)\} = X(\omega) = \int_{-\infty}^{\infty} x(t) e^{-j\omega t} \, dt \tag{2.6}$$

or

$$F\{x(t)\} = X(f) = \int_{-\infty}^{\infty} x(t)e^{-j2\pi ft} \, dt \qquad (2.7)$$

and the Inverse Fourier Transform (IFT) is

$$F^{-1}\{X(\omega)\} = x(t) = \frac{1}{2\pi} \int_{-\infty}^{\infty} X(\omega)e^{j\omega t} \, d\omega \qquad (2.8)$$

or

$$F^{-1}\{X(f)\} = x(t) = \int_{-\infty}^{\infty} X(f)e^{j2\pi ft} \, df \qquad (2.9)$$

where, in general, t represents time, while $\omega = 2\pi f$ and f represent frequency in radians per second and Hertz, respectively. In this book we will use both notations for the transform, as appropriate (i.e., $X(\omega)$ or $X(f)$).

2.3. Systems Classification

Any system can mathematically be represented as a transformation (mapping) of an input signal into an output signal. This transformation or mapping relationship between the input signal $x(t)$ and the corresponding output signal $y(t)$ can be written as

$$y(t) = f[x(t); \ (-\infty < t < \infty)] \qquad (2.10)$$

The relationship described in Eq. (2.10) can be linear or nonlinear, time invariant or time varying, causal or noncausal, and stable or nonstable systems. When the input signal is unit impulse (*Dirac delta function*) $\delta(t)$, the output signal is referred to as the system's impulse response $h(t)$.

2.3.1. Linear and Nonlinear Systems

A system is said to be linear if superposition holds true. More specifically, if

$$y_1(t) = f[x_1(t)]$$
$$y_2(t) = f[x_2(t)] \qquad (2.11)$$

then for a linear system

$$f[ax_1(t) + bx_2(t)] = ay_1(t) + by_2(t) \qquad (2.12)$$

for any constants (a, b). If the relationship in Eq. (2.12) is not true the system is said to be nonlinear.

2.3.2. Time Invariant and Time Varying Systems

A system is said to be time invariant (or shift invariant) if a time shift at its input produces the same time shift at its output. That is if

$$y(t) = f[x(t)] \tag{2.13}$$

then

$$y(t - t_0) = f[x(t - t_0)]; -\infty < t_0 < \infty \tag{2.14}$$

If the above relationship is not true the system is called time varying system.

Any Linear Time Invariant (LTI) system can be described using the convolution integral between the input signal and the system's impulse response, as

$$y(t) = \int_{-\infty}^{\infty} x(t - u)h(u) \ du = x \otimes h \tag{2.15}$$

where the operator $\otimes$ is used to symbolically describe the convolution integral. In the frequency domain convolution translates into multiplication. That is

$$Y(f) = X(f)H(f) \tag{2.16}$$

$H(f)$ is the FT for $h(t)$ and it is referred to as the system transfer function.

2.3.3. Stable and Nonstable Systems

A system is said to be stable if every bounded input signal produces a bounded output signal. From Eq. (2.15)

$$|y(t)| = \left| \int_{-\infty}^{\infty} x(t - u)h(u) \ du \right| \leq \int_{-\infty}^{\infty} |x(t - u)||h(u)| \ du \tag{2.17}$$

If the input signal is bounded, then there is some finite constant K such that

$$|x(t)| \leq K < \infty \tag{2.18}$$

Therefore,

$$y(t) \leq K \int_{-\infty}^{\infty} |h(u)| \ du \tag{2.19}$$

which can be finite if and only if

$$\int_{-\infty}^{\infty} |h(u)| \ du < \infty \tag{2.20}$$

Thus, the requirement for stability is that the impulse response must be absolutely integrable. Otherwise, the system is said to be unstable.

2.3.4. Causal and Noncausal Systems

A causal (or physically realizable) system is one whose output signal does not begin before the input signal is applied. Thus, the following relationship is true when the system is causal:

$$y(t_0) = f[x(t);t \le t_0];-\infty < t, t_0 < \infty \tag{2.21}$$

A system that does not satisfy Eq. (2.21) is said to be noncausal which means it cannot exist in real world.

2.4. Signal Representation Using the Fourier Series

A set of functions $S = \{\varphi_n(t) \ ; \ n = 1, ..., N\}$ is said to be orthogonal over the interval (t_1, t_2) if and only if

$$\int_{t_1}^{t_2} \varphi_i^*(t)\varphi_j(t)dt = \int_{t_1}^{t_2} \varphi_i(t)\varphi_j^*(t)dt = \left\{ \begin{matrix} 0 & i \ne j \\ \lambda_i & i = j \end{matrix} \right\} \tag{2.22}$$

where the asterisk indicates complex conjugation and λ_i are constants. If $\lambda_i = 1$ for all i, then the set S is said to be an orthonormal set. An electrical signal $x(t)$ can be expressed over the interval (t_1, t_2) as a weighted sum of a set of orthogonal functions as

$$x(t) \approx \sum_{n=1}^{N} X_n \varphi_n(t) \tag{2.23}$$

where X_n are, in general, complex constants and the orthogonal functions $\varphi_n(t)$ are called basis functions. If the integral-square error over the interval (t_1, t_2) is equal to zero as N approaches infinity, i.e.,

$$\lim_{N \to \infty} \int_{t_1}^{t_2} \left| x(t) - \sum_{n=1}^{N} X_n \varphi_n(t) \right|^2 dt = 0 \tag{2.24}$$

then the set $S = \{\varphi_n(t)\}$ is said to be complete, and Eq. (2.23) becomes an equality. The constants X_n are computed as

$$X_n = \frac{\displaystyle\int_{t_1}^{t_2} x(t)\varphi_n{}^*(t)dt}{\displaystyle\int_{t_1}^{t_2} |\varphi_n(t)|^2 dt} \tag{2.25}$$

Let the signal $x(t)$ be periodic with period T, and let the complete orthogonal set S be

$$S = \left\{ e^{\frac{j2\pi nt}{T}} \; ; \; n= -\infty, \infty \right\} \tag{2.26}$$

Then the complex exponential Fourier series of $x(t)$ is

$$x(t) = \sum_{n=-\infty}^{\infty} X_n e^{\frac{j2\pi nt}{T}} \tag{2.27}$$

Using Eq. (2.25) yields

$$X_n = \frac{1}{T} \int_{-T/2}^{T/2} x(t)e^{\frac{-j2\pi nt}{T}} dt \tag{2.28}$$

The FT of Eq. (2.27) is given by

$$X(\omega) = 2\pi \sum_{n=-\infty}^{\infty} X_n \delta\left(\omega - \frac{2\pi n}{T}\right) \tag{2.29}$$

where $\delta(\ )$ is delta function. When the signal $x(t)$ is real we can compute its trigonometric Fourier series from Eq. (2.27) as

$$x(t) = a_0 + \sum_{n=1}^{\infty} a_n \cos\left(\frac{2\pi nt}{T}\right) + \sum_{n=1}^{\infty} b_n \sin\left(\frac{2\pi nt}{T}\right) \tag{2.30}$$

$$a_0 = X_0 \tag{2.31a}$$

$$a_n = \frac{1}{T} \int\limits_{-T/2}^{T/2} x(t)\cos\left(\frac{2\pi n t}{T}\right) dt \qquad \text{(2.31b)}$$

$$b_n = \frac{1}{T} \int\limits_{-T/2}^{T/2} x(t)\sin\left(\frac{2\pi n t}{T}\right) dt \qquad \text{(2.31c)}$$

The coefficients a_n are all zeros when the signal $x(t)$ is an odd function of time. Alternatively, when the signal is an even function of time, then all b_n are equal to zero.

Consider the periodic energy signal defined in Eq. (2.30). The total energy associated with this signal is then given by

$$E = \frac{1}{T} \int\limits_{t_0}^{t_0+T} |x(t)|^2 dt = \frac{a_0^2}{4} + \sum_{n=1}^{\infty}\left(\frac{a_n^2}{2} + \frac{b_n^2}{2}\right) \qquad \text{(2.32)}$$

2.5. Convolution and Correlation Integrals

The convolution $\rho_{xh}(t)$ between the signals $x(t)$ and $h(t)$ is defined by

$$\rho_{xh}(t) = x(t) \otimes h(t) = \int\limits_{-\infty}^{\infty} x(\tau)h(t-\tau)d\tau \qquad \text{(2.33)}$$

where τ is a dummy variable. Convolution is commutative, associative, and distributive. More precisely,

$$x(t) \otimes h(t) = h(t) \otimes x(t)$$
$$x(t) \otimes (h(t) \otimes g(t)) = (x(t) \otimes h(t)) \otimes g(t) = x(t) \otimes (h(t) \otimes g(t)) \qquad \text{(2.34)}$$

For the convolution integral to be finite at least one of the two signals must be an energy signal. The convolution between two signals can be computed using the FT:

$$\rho_{xh}(t) = F^{-1}\{X(\omega)H(\omega)\} \qquad \text{(2.35)}$$

Consider an LTI system with impulse response $h(t)$ and input signal $x(t)$. It follows that the output signal $y(t)$ is equal to the convolution between the input signal and the system impulse response,

$$y(t) = \int_{-\infty}^{\infty} x(\tau)h(t-\tau)d\tau = \int_{-\infty}^{\infty} h(\tau)x(t-\tau)d\tau \qquad (2.36)$$

The cross-correlation function between the signals $x(t)$ and $g(t)$ is

$$R_{xg}(t) = \int_{-\infty}^{\infty} x^*(\tau)g(t+\tau)d\tau = R^*{}_{gx}(-t) = \int_{-\infty}^{\infty} g^*(\tau)x(t+\tau)d\tau \qquad (2.37)$$

Again, at least one of the two signals should be an energy signal for the correlation integral to be finite. The cross-correlation function measures the similarity between the two signals. The peak value of $R_{xg}(t)$ and its spread around this peak are an indication of how good this similarity is. This similarity is measured by a factor called *the correlation coefficient*, denoted by C_{xg}. For example, consider the signals $x(t)$ and $g(t)$, the correlation coefficient is

$$C_{xg} = \frac{\left|\int_{-\infty}^{\infty} x(t)\ g^*(t)dt\right|^2}{\int_{-\infty}^{\infty} |x(t)|^2 dt \int_{-\infty}^{\infty} |g(t)|^2 dt} = C_{gx} \qquad (2.38)$$

clearly the correlation coefficient is limited to $0 \le C_{xg} = C_{gx} \le 1$, with $C_{xg} = 0$ indicating no similarity while $C_{xg} = 1$ indicates 100% similarity between the signals $x(t)$ and $g(t)$.

The cross-correlation integral can be computed as

$$R_{xg}(t) = F^{-1}\{X^*(\omega)G(\omega)\} \qquad (2.39)$$

When $x(t) = g(t)$, we get the autocorrelation integral,

$$R_x(t) = \int_{-\infty}^{\infty} x^*(\tau)x(t+\tau)d\tau \qquad (2.40)$$

Note that the autocorrelation function is denoted by $R_x(t)$ rather than $R_{xx}(t)$. When the signals $x(t)$ and $g(t)$ are power signals, the correlation integral becomes infinite and, thus, time averaging must be included. More precisely,

$$\bar{R}_{xg}(t) = \lim_{T \to \infty} \frac{1}{T} \int_{-T/2}^{T/2} x^*(\tau)g(t+\tau)d\tau \tag{2.41}$$

2.5.1. Energy and Power Spectrum Densities

Consider an energy signal $x(t)$. From Parseval's theorem, the total energy associated with this signal is

$$E = \int_{-\infty}^{\infty} |x(t)|^2 dt = \frac{1}{2\pi} \int_{-\infty}^{\infty} |X(\omega)|^2 d\omega \tag{2.42}$$

When $x(t)$ is a voltage signal, the amount of energy dissipated by this signal when applied across a network of resistance R is

$$E = \frac{1}{R} \int_{-\infty}^{\infty} |x(t)|^2 dt = \frac{1}{2\pi R} \int_{-\infty}^{\infty} |X(\omega)|^2 d\omega \tag{2.43}$$

Alternatively, when $x(t)$ is a current signal, we get

$$E = R \int_{-\infty}^{\infty} |x(t)|^2 dt = \frac{R}{2\pi} \int_{-\infty}^{\infty} |X(\omega)|^2 d\omega \tag{2.44}$$

The quantity $\int |X(\omega)|^2 d\omega$ represents the amount of energy spread per unit frequency across a 1Ω resistor; therefore, the Energy Spectrum Density (ESD) function for the energy signal $x(t)$ is defined as

$$ESD = |X(\omega)|^2 \tag{2.45}$$

The ESD at the output of an LTI system when $x(t)$ is at its input is

$$|Y(\omega)|^2 = |X(\omega)|^2 |H(\omega)|^2 \tag{2.46}$$

where $H(\omega)$ is the FT of the system impulse response, $h(t)$. It follows that the energy present at the output of the system is

$$E_y = \frac{1}{2\pi} \int_{-\infty}^{\infty} |X(\omega)|^2 |H(\omega)|^2 d\omega \tag{2.47}$$

Example:

The voltage signal $x(t) = e^{-5t}$; $t \geq 0$ is applied to the input of a lowpass LTI system. The system bandwidth is $5\,Hz$, and its input resistance is $5\,\Omega$. If $H(\omega) = 1$ over the interval $(-10\pi < \omega < 10\pi)$ and zero elsewhere, compute the energy at the output.

Solution:

From Eqs. (2.43) and (2.47) we get

$$E_y = \frac{1}{2\pi R} \int\limits_{\omega = -10\pi}^{10\pi} |X(\omega)|^2 |H(\omega)|^2 d\omega$$

Using Fourier transform tables and substituting $R = 5$ yields

$$E_y = \frac{1}{5\pi} \int\limits_{0}^{10\pi} \frac{1}{\omega^2 + 25} d\omega$$

Completing the integration yields

$$E_y = \frac{1}{25\pi}[\operatorname{atanh}(2\pi) - \operatorname{atanh}(0)] = 0.01799 \ \ Joules$$

Note that an infinite bandwidth would give $E_y = 0.02$, only 11% larger.

The total power associated with a power signal $g(t)$ is

$$P = \lim_{T \to \infty} \frac{1}{T} \int\limits_{-T/2}^{T/2} |g(t)|^2 dt \tag{2.48}$$

The Power Spectrum Density (PSD) function for the signal $g(t)$ is $S_g(\omega)$, where

$$P = \lim_{T \to \infty} \frac{1}{T} \int\limits_{-T/2}^{T/2} |g(t)|^2 dt = \frac{1}{2\pi} \int\limits_{-\infty}^{\infty} S_g(\omega) d\omega \tag{2.49}$$

It can be shown that

$$S_g(\omega) = \lim_{T \to \infty} \frac{|G(\omega)|^2}{T} \tag{2.50}$$

Let the signals $x(t)$ and $g(t)$ be two periodic signals with period T. The complex exponential Fourier series expansions for those signals are, respectively, given by

$$x(t) = \sum_{n=-\infty}^{\infty} X_n e^{j\frac{2\pi nt}{T}} \tag{2.51}$$

$$g(t) = \sum_{m=-\infty}^{\infty} G_m e^{j\frac{2\pi mt}{T}} \tag{2.52}$$

The power cross-correlation function $\bar{R}_{gx}(t)$ was given in Eq. (2.41) and is repeated here as Eq. (2.53),

$$\bar{R}_{gx}(t) = \frac{1}{T} \int_{-T/2}^{T/2} g^*(\tau)x(t+\tau)d\tau \tag{2.53}$$

Note that because both signals are periodic the limit is no longer necessary. Substituting Eqs. (2.51) and (2.52) into Eq. (2.53), collecting terms, and using the definition of orthogonality, we get

$$\bar{R}_{gx}(t) = \sum_{n=-\infty}^{\infty} G_n^* X_n e^{j\frac{2\pi nt}{T}} \tag{2.54}$$

When $x(t) = g(t)$, Eq. (2.54) becomes the power autocorrelation function,

$$\bar{R}_x(t) = \sum_{n=-\infty}^{\infty} |X_n|^2 e^{j\frac{2\pi nt}{T}} = |X_0|^2 + 2\sum_{n=1}^{\infty} |X_n|^2 e^{j\frac{2\pi nt}{T}} \tag{2.55}$$

The power spectrum and cross-power spectrum density functions are then computed as the FT of Eqs. (2.55) and (2.54), respectively. More precisely,

$$\bar{S}_x(\omega) = 2\pi \sum_{n=-\infty}^{\infty} |X_n|^2 \delta\left(\omega - \frac{2n\pi}{T}\right)$$

$$\bar{S}_{gx}(\omega) = 2\pi \sum_{n=-\infty}^{\infty} G_n^* X_n \delta\left(\omega - \frac{2n\pi}{T}\right) \tag{2.56}$$

The line (or discrete) power spectrum is defined as the plot of $|X_n|^2$ versus n, where the lines are $\Delta f = 1/T$ apart. The DC power is $|X_0|^2$, and the total power is $\sum_{n = -\infty}^{\infty} |X_n|^2$.

Consider a signal $x(t)$ and its FT $X(f)$. The corresponding autocorrelation function and power spectrum density are, respectively $\bar{R}_x(t)$ and $\bar{S}_x(f)$. A few very useful relations that will be utilized often in this book include

$$x(0) = \int_{-\infty}^{\infty} X(f) df \qquad (2.57)$$

$$\int_{-\infty}^{\infty} x(t) dt = X(0) \qquad (2.58)$$

$$\bar{R}_x(0) = \int_{-\infty}^{\infty} |x(t)|^2 dt = \int_{-\infty}^{\infty} |X(f)|^2 df = \bar{S}_x(0) \qquad (2.59)$$

$$\int_{-\infty}^{\infty} |\bar{R}_x(t)|^2 dt = \int_{-\infty}^{\infty} |X(f)|^4 df \qquad (2.60)$$

Note that Eq. (2.57) or Eq. (2.58) represents the total DC power (in the case of a power signal) or voltage (in the case of an energy signal). Equation (2.59) represents the signal's total power (for power signals) or total energy (for energy signals).

2.6. Bandpass Signals

Signals that contain significant frequency composition at a low frequency band including DC are called lowpass (LP) signals. Signals that have significant frequency composition around some frequency away from the origin are called bandpass (BP) signals. A real BP signal $x(t)$ can be represented mathematically by

$$x(t) = r(t) \cos(2\pi f_0 t + \phi_x(t)) \qquad (2.61)$$

where $r(t)$ is the amplitude modulation or envelope, $\phi_x(t)$ is the phase modulation, f_0 is the carrier frequency, and both $r(t)$ and $\phi_x(t)$ have frequency components significantly smaller than f_0. The frequency modulation is

$$f_m(t) = \frac{1}{2\pi}\frac{d}{dt}\phi_x(t) \tag{2.62}$$

and the instantaneous frequency is

$$f_i(t) = \frac{1}{2\pi}\frac{d}{dt}(2\pi f_0 t + \phi_x(t)) = f_0 + f_m(t) \tag{2.63}$$

If the signal bandwidth is B and f_0 is very large compared to B, then the signal $x(t)$ is referred to as a narrow bandpass signal.

Bandpass signals can also be represented by two lowpass signals known as the quadrature components; in this case Eq. (2.61) can be rewritten as

$$x(t) = x_I(t)\cos 2\pi f_0 t - x_Q(t)\sin 2\pi f_0 t \tag{2.64}$$

where $x_I(t)$ and $x_Q(t)$ are real LP signals referred to as the quadrature components and are given, respectively, by

$$\begin{aligned} x_I(t) &= r(t)\cos\phi_x(t) \\ x_Q(t) &= r(t)\sin\phi_x(t) \end{aligned} \tag{2.65}$$

2.6.1. The Analytic Signal (Pre-Envelope)

Given a real valued signal $x(t)$ its Hilbert transform is

$$H\{x(t)\} = \hat{x}(t) = \frac{1}{\pi}\int_{-\infty}^{\infty}\frac{x(u)}{t-u}\,du \tag{2.66}$$

Observation of Eq. (2.66) indicates that the Hilbert transform is computed as the convolution between the signals $x(t)$ and $h(t) = 1/(\pi t)$. More precisely,

$$\hat{x}(t) = x(t) \otimes \frac{1}{\pi t} \tag{2.67}$$

The Fourier transform of $h(t)$ is

$$FT\{h(t)\} = FT\left\{\frac{1}{\pi t}\right\} = H(\omega) = e^{-j\frac{\pi}{2}}\text{sgn}(\omega) \tag{2.68}$$

where the function $\text{sgn}(\omega)$ is given by

$$\text{sgn}(\omega) = \frac{\omega}{|\omega|} = \begin{cases} 1 \ ; \ \omega > 0 \\ 0; \ \omega = 0 \\ -1 \ ; \ \omega < 0 \end{cases} \tag{2.69}$$

Thus, the effect of the Hilbert transform is to introduce a phase shift of $\pi/2$ on the spectra of $x(t)$. It follows that,

$$FT\{\hat{x}(t)\} = \hat{X}(\omega) = X(\omega) - j\,\text{sgn}(\omega)X(\omega) \tag{2.70}$$

The analytic signal $\psi(t)$ corresponding to the real signal $x(t)$ is obtained by cancelling the negative frequency contents of $X(\omega)$. Then, by definition

$$\Psi(\omega) = \begin{cases} 2X(\omega) & ;\omega > 0 \\ X(\omega) & ;\omega = 0 \\ 0 & ;\omega < 0 \end{cases} \tag{2.71}$$

or equivalently,

$$\Psi(\omega) = X(\omega)(1 + \text{sgn}(\omega)) \tag{2.72}$$

It follows that

$$\psi(t) = FT^{-1}\{\Psi(\omega)\} = x(t) + j\hat{x}(t) \tag{2.73}$$

The analytic signal is often referred to as the pre-envelope of $x(t)$ because the envelope of $x(t)$ can be obtained by simply taking the modulus of $\psi(t)$.

2.6.2. Pre-Envelope and Complex Envelope of Bandpass Signals

The Hilbert transform for the bandpass signal defined in Eq. (2.64) is

$$\hat{x}_{BP}(t) = x_I(t)\sin 2\pi f_0 t + x_Q(t)\cos 2\pi f_0 t \tag{2.74}$$

The subscript *BP* is used to indicate that $x(t)$ is a bandpass signal. The corresponding bandpass analytic signal (pre-envelope) is then given by

$$\psi_{BP}(t) = x_{BP}(t) + j\hat{x}_{BP}(t) \tag{2.75}$$

substituting Eq. (2.64) and Eq. (2.74) into Eq. (2.75) and collecting terms yield

$$\psi_{BP}(t) = [x_I(t) + jx_Q(t)]e^{j2\pi f_0 t} = \tilde{x}_{BP}(t)e^{j2\pi f_0 t} \tag{2.76}$$

The signal $\tilde{x}_{BP}(t) = x_I(t) + jx_Q(t)$ is the complex envelope of $x_{BP}(t)$. Thus, the envelope signal and associated phase deviation are given by

$$a(t) = |\tilde{x}_{BP}(t)| = |x_I(t) + jx_Q(t)| = |\psi_{BP}(t)| \tag{2.77}$$

$$\phi(t) = \arg(\tilde{x}_{BP}(t)) = \angle \tilde{x}_{BP}(t) \qquad (2.78)$$

In the remainder of this text, unless it is indicated to be otherwise, all signals will be considered to be bandpass signals and consequently the subscript *BP* will not be used. More specifically, a bandpass signal $x(t)$ and its corresponding pre-envelope (analytic signal) and complex envelope will shown as

$$x(t) = x_I(t)\cos 2\pi f_0 t - x_Q(t)\sin 2\pi f_0 t \qquad (2.79)$$

$$\psi(t) = x(t) + j\hat{x}(t) \equiv \tilde{x}(t)e^{j2\pi f_0 t} \qquad (2.80)$$
$$\tilde{x}(t) = x_I(t) + jx_Q(t) \qquad (2.81)$$

Obtaining the complex envelope for any bandpass signal requires extraction of the quadrature components. Figure 2.1 shows how the quadrature components can be extracted from a bandpass signal. First, the bandpass signal is split into two parts; one part is multiplied by $2\cos 2\pi f_0 t$ and the other is multiplied by $-2\sin 2\pi f_0 t$. From the figure the two signal $z_1(t)$ and $z_2(t)$ are,

$$z_1(t) = 2x_I(t)(\cos 2\pi f_0 t)^2 - 2x_Q(t)\cos(2\pi f_0 t)\sin(2\pi f_0 t) \qquad (2.82)$$

$$z_2(t) = -2x_I(t)\cos(2\pi f_0 t)\sin(2\pi f_0 t) + 2x_Q(t)(\sin 2\pi f_0 t)^2 \qquad (2.83)$$

Utilizing the appropriate trigonometry identities and after lowpass filtering the quadrature components are extracted.

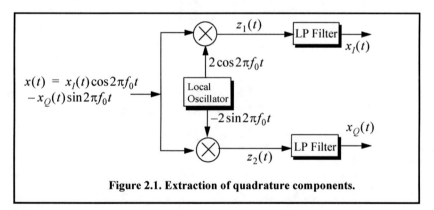

Figure 2.1. Extraction of quadrature components.

Example:

Extract the quadrature components, frequency modulation, instantaneous frequency, analytic signal, and complex envelope for the signals:

(a) $x(t) = Rect\left(\dfrac{t}{\tau}\right)\cos(2\pi f_0 t)$; *(b)* $x(t) = Rect\left(\dfrac{t}{\tau}\right)\cos\left(2\pi f_0 t + \dfrac{\pi B}{\tau}t^2\right)$

Solution:

(a) The quadrature components are extracted as described in Fig. 2.1. Define $z_1(t) = x(t) \times 2\cos(2\pi f_0 t)$, $z_2(t) = x(t) \times (-2)\sin(2\pi f_0 t)$, then

$$z_1(t) = Rect\left(\frac{t}{\tau}\right)\cos(2\pi f_0 t) \times 2\cos(2\pi f_0 t) =$$
$$Rect\left(\frac{t}{\tau}\right)\cos(0) + Rect\left(\frac{t}{\tau}\right)\cos(4\pi f_0 t)$$

$$z_2(t) = Rect\left(\frac{t}{\tau}\right)\cos(2\pi f_0 t) \times (-2)\sin(2\pi f_0 t) =$$
$$Rect\left(\frac{t}{\tau}\right)\sin(0) - Rect\left(\frac{t}{\tau}\right)\sin(4\pi f_0 t)$$

Thus, the output of the LPFs are

$$x_I(t) = Rect\left(\frac{t}{\tau}\right) \quad ; \quad x_Q(t) = 0$$

From Eq. (2.62) and Eq. (2.63) we get

$$f_m(t) = 0 \quad ; \quad f_i(t) = f_0$$

Finally the complex envelope and the analytic signal are given by

$$\tilde{x}(t) = x_I(t) + jx_Q(t) = x_I(t) = Rect\left(\frac{t}{\tau}\right)$$

$$\psi(t) = \tilde{x}(t)e^{j2\pi f_0 t} = Rect\left(\frac{t}{\tau}\right)e^{j2\pi f_0 t}$$

(b)

$$z_1(t) = Rect\left(\frac{t}{\tau}\right)\cos\left(2\pi f_0 t + \frac{\pi B}{\tau}t^2\right) \times 2\cos(2\pi f_0 t) =$$
$$Rect\left(\frac{t}{\tau}\right)\cos\left(\frac{\pi B}{\tau}t^2\right) + Rect\left(\frac{t}{\tau}\right)\cos\left(4\pi f_0 t + \frac{\pi B}{\tau}t^2\right)$$

$$z_2(t) = Rect\left(\frac{t}{\tau}\right)\cos\left(2\pi f_0 t + \frac{\pi B}{\tau}t^2\right) \times (-2)\sin(2\pi f_0 t) =$$
$$Rect\left(\frac{t}{\tau}\right)\sin\left(\frac{\pi B}{\tau}t^2\right) - Rect\left(\frac{t}{\tau}\right)\sin\left(4\pi f_0 t + \frac{\pi B}{\tau}t^2\right)$$

Thus, the outputs of the LPFs are

$$x_I(t) = Rect\left(\frac{t}{\tau}\right)\cos\left(\frac{\pi B}{\tau}t^2\right) \qquad ; \; x_Q(t) = Rect\left(\frac{t}{\tau}\right)\sin\left(\frac{\pi B}{\tau}t^2\right)$$

From Eq. (2.62) and Eq.(2.63) we get

$$f_m(t) = \frac{B}{\tau}t \qquad ; \; f_i(t) = f_0 + \frac{B}{\tau}t$$

The complex envelope is

$$\tilde{x}(t) = x_I(t) + jx_Q(t) = Rect\left(\frac{t}{\tau}\right)\cos\left(\frac{\pi B}{\tau}t^2\right) + jRect\left(\frac{t}{\tau}\right)\sin\left(\frac{\pi B}{\tau}t^2\right)$$

which can be written as

$$\tilde{x}(t) = Rect\left(\frac{t}{\tau}\right)e^{j\left(\frac{\pi B}{\tau}t^2\right)}$$

Finally, the analytic signal is

$$\psi(t) = \tilde{x}(t)e^{j2\pi f_0 t} = Rect\left(\frac{t}{\tau}\right)e^{j\left(\frac{\pi B}{\tau}t^2\right)}e^{j2\pi f_0 t} = Rect\left(\frac{t}{\tau}\right)e^{j\left(2\pi f_0 t + \frac{\pi B}{\tau}t^2\right)}$$

2.7. Spectra of a Few Common Radar Signals

The spectrum of a given signal describes the spread of its energy in the frequency domain. An energy signal (finite energy) can be characterized by its Energy Spectrum Density (ESD) function, while a power signal (finite power) is characterized by the Power Spectrum Density (PSD) function. The units of the ESD are Joules/Hertz and the PSD has units Watts/Hertz.

2.7.1. Frequency Modulation Signal

The discussion presented in this section will be restricted to sinusoidal modulating signals. In this case, the general formula for an FM waveform can be expressed by

$$x(t) = A\cos\left(2\pi f_0 t + k_f\int_0^t \cos 2\pi f_m u\, du\right) \tag{2.84}$$

f_0 is the radar operating frequency (carrier frequency), $\cos 2\pi f_m t$ is the modulating signal, A is a constant, and $k_f = 2\pi\Delta f_{peak}$, where Δf_{peak} is the peak frequency deviation. The phase is given by

$$\phi(t) = 2\pi f_0 t + 2\pi \Delta f_{peak} \int_0^t \cos 2\pi f_m u \, du = 2\pi f_0 t + \beta \sin 2\pi f_m t \qquad (2.85)$$

where β is the FM modulation index given by

$$\beta = (\Delta f_{peak})/f_m \qquad (2.86)$$

Let $x_r(t)$ be the received radar signal from a target at range R. It follows that

$$s_r(t) = A_r \cos(2\pi f_0(t - \Delta t) + \beta \sin 2\pi f_m(t - t_0)) \qquad (2.87)$$

where the delay t_0 is

$$t_0 = 2R/c \qquad (2.88)$$

c is the speed of light. Radar receivers utilize phase detectors in order to extract target range from the instantaneous frequency, as illustrated in Fig. 2.2. A good measurement of the phase detector output $x_o(t)$ implies a good measurement of t_0 and, hence, range. Consider the FM waveform $s(t)$ given by

$$x(t) = A \cos(2\pi f_0 t + \beta \sin 2\pi f_m t) \qquad (2.89)$$

which can be written as

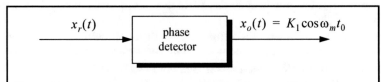

Figure 2.2. Extracting range from an FM signal return. K_1 is a constant.

$$x(t) = A Re\{e^{j2\pi f_0 t} e^{j\beta \sin 2\pi f_m t}\} \qquad (2.90)$$

where $Re\{ \ \}$ denotes the real part. Since the signal $\exp(j\beta \sin 2\pi f_m t)$ is periodic with period $T = 1/f_m$, it can be expressed using the complex exponential Fourier series as

$$e^{j\beta \sin 2\pi f_m t} = \sum_{n=-\infty}^{\infty} C_n e^{jn2\pi f_m t} \qquad (2.91)$$

where the Fourier series coefficients C_n are given by

$$C_n = \frac{1}{2\pi} \int_{-\pi}^{\pi} e^{j\beta \sin 2\pi f_m t} \, e^{-jn2\pi f_m t} \, dt \qquad (2.92)$$

Make the change of variable $u = 2\pi f_m t$, and recognize that the Bessel function of the first kind of order n is

$$J_n(\beta) = \frac{1}{2\pi} \int_{-\pi}^{\pi} e^{j(\beta \sin u - nu)} \, du \qquad (2.93)$$

Thus, the Fourier series coefficients are $C_n = J_n(\beta)$, and consequently Eq. (2.91) can now be written as

$$e^{j\beta \sin 2\pi f_m t} = \sum_{n = -\infty}^{\infty} J_n(\beta) e^{jn2\pi f_m t} \qquad (2.94)$$

which is known as the Bessel-Jacobi equation. Figure 2.3 shows a plot of Bessel functions of the first kind for $n = 0, 1, 2, 3$. The total power in the signal $x(t)$ is

$$P = \frac{1}{2}A^2 \sum_{n = -\infty}^{\infty} |J_n(\beta)|^2 = \frac{1}{2}A^2 \qquad (2.95)$$

Substituting Eq. (2.95) into Eq. (2.90) yields

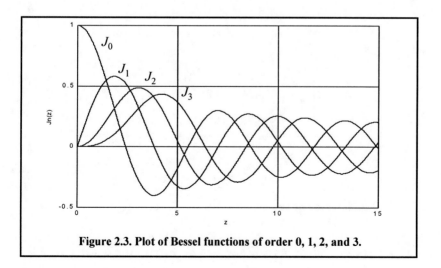

Figure 2.3. Plot of Bessel functions of order 0, 1, 2, and 3.

$$x(t) = A Re\left\{ e^{j2\pi f_0 t} \sum_{n=-\infty}^{\infty} J_n(\beta) e^{jn2\pi f_m t} \right\} \qquad (2.96)$$

Expanding Eq. (2.96) yields

$$x(t) = A \sum_{n=-\infty}^{\infty} J_n(\beta) \cos(2\pi f_0 + n2\pi f_m)t \qquad (2.97)$$

Finally, since $J_n(\beta) = J_{-n}(\beta)$ for n odd and $J_n(\beta) = -J_{-n}(\beta)$ for n even we can rewrite Eq. (2.97) as

$$x(t) = A\{J_0(\beta)\cos 2\pi f_0 t + \qquad (2.98)$$
$$J_1(\beta)[\cos(2\pi f_0 + 2\pi f_m)t - \cos(2\pi f_0 - 2\pi f_m)t]$$
$$+ J_2(\beta)[\cos(2\pi f_0 + 4\pi f_m)t + \cos(2\pi f_0 - 4\pi f_m)t]$$
$$+ J_3(\beta)[\cos(2\pi f_0 + 6\pi f_m)t - \cos(2\pi f_0 - 6\pi f_m)t]$$
$$+ J_4(\beta)[\cos((2\pi f_0 + 8\pi f_m)t + \cos(2\pi f_0 - 8\pi f_m)t)] + \ldots\}$$

The spectrum of $x(t)$ is composed of pairs of spectral lines centered at f_0, as sketched in Fig. 2.4. The spacing between adjacent spectral lines is f_m. The central spectral line has an amplitude equal to $AJ_0(\beta)$, while the amplitude of the *nth* spectral line is $AJ_n(\beta)$. As indicated by Eq. (2.98) the bandwidth of FM signals is infinite. However, the magnitudes of spectral lines of the higher orders are small, and thus the bandwidth can be approximated (i.e., effective bandlimited) using Carson's rule,

$$B \approx 2(\beta + 1)f_m \qquad (2.99)$$

When β is small, only $J_0(\beta)$ and $J_1(\beta)$ have significant values. Thus, we may approximate Eq. (2.99) by

$$x(t) \approx A\{J_0(\beta)\cos 2\pi f_0 t + J_1(\beta) \qquad (2.100)$$
$$[\cos(2\pi f_0 + 2\pi f_m)t - \cos(2\pi f_0 - 2\pi f_m)t]\}$$

Finally, for small β, the Bessel functions can be approximated by

$$J_0(\beta) \approx 1 \qquad (2.101)$$

$$J_1(\beta) \approx \frac{1}{2}\beta \qquad (2.102)$$

Thus, Eq. (2.100) may be approximated by

$$x(t) \approx A\left\{ \cos 2\pi f_0 t + \frac{1}{2}\beta[\cos(2\pi f_0 + 2\pi f_m)t - \cos(2\pi f_0 - 2\pi f_m)t] \right\} \qquad (2.103)$$

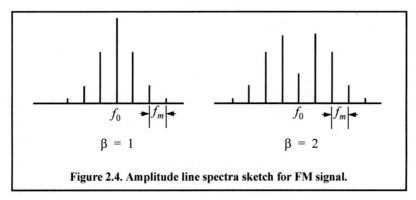

Figure 2.4. Amplitude line spectra sketch for FM signal.

Example:

If the modulation index is $\beta = 0.5$, *give an expression for the signal* $x(t)$.

Solution:

From Bessel function tables $J_0(0.5) = 0.9385$ *and* $J_1(0.5) = 0.2423$; *then using Eq. (2.100) yields*

$$x(t) \approx A\{(0.9385)\cos 2\pi f_0 t + (0.2423)$$
$$[\cos(2\pi f_0 + 2\pi f_m)t - \cos(2\pi f_0 - 2\pi f_m)t]\}$$

Example:

Consider an FM transmitter with output signal

$$x(t) = 100\cos(2000\pi t + \varphi(t)).$$

The frequency deviation is $4Hz$, *and the modulating waveform is* $x(t) = 10\cos 16\pi t$. *Determine the FM signal bandwidth. How many spectral lines will pass through a bandpass filter whose bandwidth is* $58Hz$ *centered at* $1000Hz$?

Solution:

The peak frequency deviation is $\Delta f_{peak} = 4 \times 10 = 40Hz$. *It follows that*

$$\beta = (\Delta f_{peak})/f_m = 40/8 = 5$$
$$B \approx 2(\beta + 1)f_m = 2 \times (5 + 1) \times 8 = 96Hz$$

However, only seven spectral lines pass through the bandpass filter as illustrated in the figure shown below

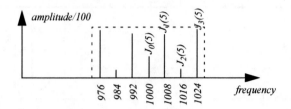

2.7.2. Continuous Wave Signal

Consider a Continuous Wave (CW) waveform given by

$$x_1(t) = \cos 2\pi f_0 t \tag{2.104}$$

The FT of $x_1(t)$ is

$$X_1(f) = \frac{1}{2}[\delta(f - f_0) + \delta(f + f_0)] \tag{2.105}$$

$\delta(\ )$ is the Dirac delta function. As indicated by the amplitude spectrum shown in Fig. 2.5, the signal $x_1(t)$ has infinitesimal bandwidth, located at $\pm f_0$.

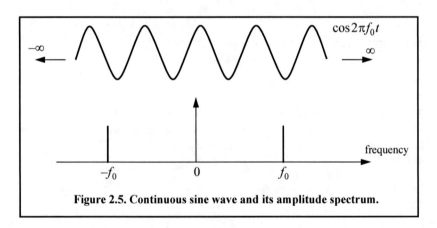

Figure 2.5. Continuous sine wave and its amplitude spectrum.

2.7.3. Finite Duration Pulse Signal

Consider the time-domain signal $x_2(t)$ given by

$$x_2(t) = x_1(t) Rect\left(\frac{t}{\tau_0}\right) = Rect\left(\frac{t}{\tau_0}\right) \cos 2\pi f_0 t \tag{2.106}$$

$$Rect\left(\frac{t}{\tau_0}\right) = \begin{cases} 1 & -\frac{\tau_0}{2} \le t \le \frac{\tau_0}{2} \\ 0 & otherwise \end{cases} \tag{2.107}$$

The Fourier transform of the *Rect* function is

$$FT\left\{Rect\left(\frac{t}{\tau_0}\right)\right\} = \tau_0 Sinc(f\tau_0) \tag{2.108}$$

where

$$Sinc(u) = \frac{\sin(\pi u)}{\pi u} \tag{2.109}$$

It follows that the FT is

$$X_2(f) = X_1(f) \otimes \tau_0 Sinc(f\tau_0) = \frac{1}{2}[\delta(f-f_0) + \delta(f+f_0)] \otimes \tau_0 Sinc(f\tau_0) \tag{2.110}$$

which can be written as

$$X_2(f) = \frac{\tau_0}{2}\{Sinc[(f-f_0)\tau_0] + Sinc[(f+f_0)\tau_0]\} \tag{2.111}$$

The amplitude spectrum of $x_2(t)$ is shown in Fig. 2.6. It is made up of two *Sinc* functions, as defined in Eq. (2.108), centered at $\pm f_0$.

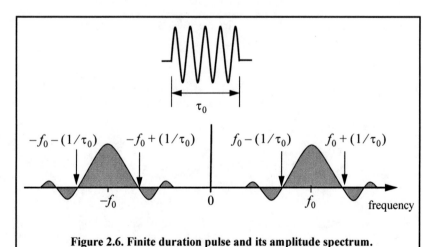

Figure 2.6. Finite duration pulse and its amplitude spectrum.

2.7.4. Periodic Pulse Signal

In this case, consider the coherent gated CW waveform $x_3(t)$ given by

$$x_3(t) = \sum_{n=-\infty}^{\infty} x_1(t) Rect\left(\frac{t-nT}{\tau_0}\right) = \cos 2\pi f_0 t \sum_{n=-\infty}^{\infty} Rect\left(\frac{t-nT}{\tau_0}\right) \quad (2.112)$$

The signal $x_3(t)$ is periodic, with period T (recall that $f_r = 1/T$ is the PRF), of course the condition $f_r \ll f_0$ is assumed. The FT of the signal $x_3(t)$ is

$$X_3(f) = X_1(f) \otimes FT\left\{ \sum_{n=-\infty}^{\infty} Rect\left(\frac{t-nT}{\tau_0}\right) \right\} = \quad (2.113)$$

$$\frac{1}{2}[\delta(f-f_0) + \delta(f+f_0)] \otimes FT\left\{ \sum_{n=-\infty}^{\infty} Rect\left(\frac{t-nT}{\tau_0}\right) \right\}$$

The complex exponential Fourier series of the summation inside Eq. (2.112) is

$$\sum_{n=-\infty}^{\infty} Rect\left(\frac{t-nT}{\tau_0}\right) = \sum_{n=-\infty}^{\infty} X_n e^{j\frac{nt}{T}} \quad (2.114)$$

where the Fourier series coefficients X_n are given by (see Eq. 2.28)

$$X_n = \frac{1}{T} FT\left\{ Rect\left(\frac{t}{\tau_0}\right) \right\}\Bigg|_{f=\frac{n}{T}} = \frac{\tau_0}{T} Sinc(f\tau_0)\Bigg|_{f=\frac{n}{T}} = \frac{\tau_0}{T} Sinc\left(\frac{n\tau_0}{T}\right) \quad (2.115)$$

It follows that

$$FT\left\{ \sum_{n=-\infty}^{\infty} X_n e^{j\frac{nt}{T}} \right\} = \left(\frac{\tau_0}{T}\right) \sum_{n=-\infty}^{\infty} Sinc(nf_r\tau_0)\delta(f-nf_r) \quad (2.116)$$

where the relation $f_r = 1/T$ was used in Eq. (2.116). Substituting Eq. (2.116) into Eq. (2.113) yields the FT of $x_3(t)$. That is

$$X_3(f) = \frac{\tau_0}{2T}[\delta(f-f_0) + \delta(f+f_0)] \otimes \sum_{n=-\infty}^{\infty} Sinc(nf_r\tau_0)\delta(f-nf_r) \qquad \textbf{(2.117)}$$

The amplitude spectrum of $x_3(t)$ has two parts centered at $\pm f_0$; each part corresponds to the spectrum of the second half of Eq. (2.117). The spectrum of the summation part is an infinite number of delta functions repeated every f_r, where the *nth* line is modulated in amplitude with the value corresponding to $Sinc(nf_r\tau_0)$. Therefore, the overall spectrum consists of an infinite number of lines separated by f_r and have $\sin u/u$ envelope that corresponds to X_n. This is illustrated in Fig. 2.7, for the positive portion of the spectrum only.

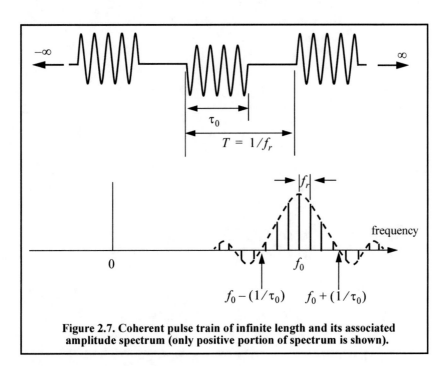

Figure 2.7. Coherent pulse train of infinite length and its associated amplitude spectrum (only positive portion of spectrum is shown).

2.7.5. Finite Duration Pulse Train Signal

Define the function $x_4(t)$ as

$$x_4(t) = \cos(2\pi f_0 t)\sum_{n=0}^{N-1} Rect\left(\frac{t-nT}{\tau_0}\right) = \cos 2\pi f_0 t \times g(t) \qquad \textbf{(2.118)}$$

where

$$g(t) = \sum_{n=0}^{N-1} Rect\left(\frac{t-nT}{\tau_0}\right) \tag{2.119}$$

The amplitude spectrum of the signal $x_4(t)$ is

$$X_4(f) = \frac{1}{2}G(f) \otimes [\delta(f-f_0) + \delta(f+f_0)] \tag{2.120}$$

where $G(f)$ is the FT of $g(t)$. This means that the amplitude spectrum of the signal $x_4(t)$ is equal to replicas of $G(f)$ centered at $\pm f_0$. Given this conclusion, we can then focus on computing $G(f)$.

The signal $g(t)$ can be written as (see top portion of Fig. 2.8)

$$g(t) = \sum_{n=-\infty}^{\infty} g_1(t)Rect\left(\frac{t-nT}{\tau_0}\right) \tag{2.121}$$

where

$$g_1(t) = Rect\left(\frac{t}{NT_t}\right) \tag{2.122}$$

It follows that the FT of Eq. (2.121) can be computed using similar analysis as that which led to Eq. (2.116). More precisely,

$$G(f) = \frac{\tau_0}{T}G_1(f) \otimes \sum_{n=-\infty}^{\infty} Sinc(nf_r\tau_0)\delta(f-nf_r) \tag{2.123}$$

and the FT of $g_1(t)$ is

$$G_1(f) = FT\left\{Rect\left(\frac{t}{T_t}\right)\right\} = T_tSinc(fT_t) \tag{2.124}$$

Using these results the FT of $x_4(t)$ can be written as

$$X_4(f) = \frac{T_t\tau_0}{2T}\left(Sinc(fT_t) \otimes \sum_{n=-\infty}^{\infty} Sinc(nf_r\tau_0)\delta(f-nf_r)\right) \tag{2.125}$$

$$\otimes [\delta(f-f_0) + \delta(f+f_0)]$$

Therefore, the overall spectrum of $x_4(t)$ consists of a two equal positive and negative portions, centered at $\pm f_0$. Each portion is made up of N $Sinc(fT_t)$ functions repeated every f_r with envelope corresponding to $Sinc(nf_r\tau_0)$. This is illustrated in Fig. 2.8, only positive portion of the spectrum is shown.

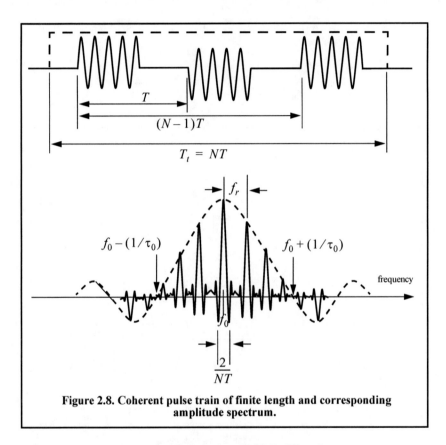

Figure 2.8. Coherent pulse train of finite length and corresponding amplitude spectrum.

2.7.6. Linear Frequency Modulation (LFM) Signal

Frequency or phase modulated signals can be used to achieve much wider operating bandwidths. Linear Frequency Modulation (LFM) is very commonly used in most modern radar systems. In this case, the frequency is swept linearly across the pulse width, either upward (up-chirp) or downward (down-chirp). Figure 2.9 shows a typical example of an LFM waveform. The pulse width is τ_0, and the bandwidth is B.

The LFM up-chirp instantaneous phase can be expressed by

$$\phi(t) = 2\pi\left(f_0 t + \frac{\mu}{2}t^2\right) \qquad -\frac{\tau_0}{2} \leq t \leq \frac{\tau_0}{2} \qquad \textbf{(2.126)}$$

where f_0 is the radar center frequency, and $\mu = B/\tau_0$ is the LFM coefficient. Thus, the instantaneous frequency is

$$f(t) = \frac{1}{2\pi}\frac{d}{dt}\phi(t) = f_0 + \mu t \qquad -\frac{\tau_0}{2} \le t \le \frac{\tau_0}{2} \qquad \text{(2.127)}$$

Similarly, the down-chirp instantaneous phase and frequency are given, respectively, by

$$\phi(t) = 2\pi\left(f_0 t - \frac{\mu}{2}t^2\right) \qquad -\frac{\tau_0}{2} \le t \le \frac{\tau_0}{2} \qquad \text{(2.128)}$$

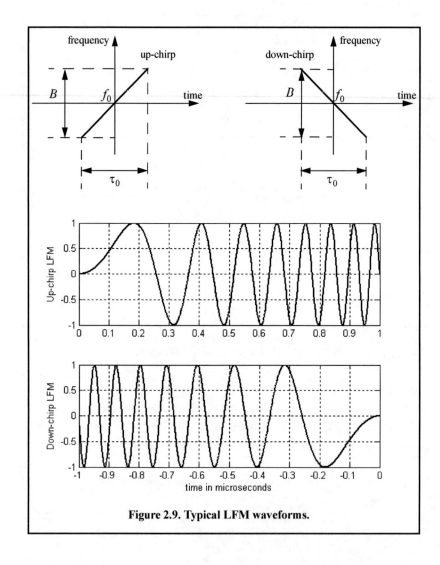

Figure 2.9. Typical LFM waveforms.

$$f(t) = \frac{1}{2\pi} \frac{d}{dt} \phi(t) = f_0 - \mu t \qquad -\frac{\tau_0}{2} \le t \le \frac{\tau_0}{2} \qquad (2.129)$$

A typical LFM waveform can be expressed by

$$x_1(t) = Rect\left(\frac{t}{\tau_0}\right) e^{j2\pi\left(f_0 t + \frac{\mu}{2} t^2\right)} \qquad (2.130)$$

where $Rect(t/\tau_0)$ denotes a rectangular pulse of width τ_0. Remember that the signal $x_1(t)$ is the analytic signal for the LMF waveform. It follows that

$$x_1(t) = \tilde{x}(t) e^{j2\pi f_0 t} \qquad (2.131)$$

$$\tilde{x}(t) = Rect\left(\frac{t}{\tau}\right) e^{j\pi\mu t^2} \qquad (2.132)$$

The spectrum of the signal $x_1(t)$ is determined from its complex envelope $\tilde{x}(t)$. The complex exponential term in Eq. (2.132) introduces a frequency shift about the center frequency f_o. Taking the FT of $\tilde{x}(t)$ yields

$$\tilde{X}(f) = \int_{-\infty}^{\infty} Rect\left(\frac{t}{\tau_0}\right) e^{j\pi\mu t^2} e^{-j2\pi f t} dt = \int_{-\frac{\tau_0}{2}}^{\frac{\tau_0}{2}} e^{j\pi\mu t^2} e^{-j2\pi f t} dt \qquad (2.133)$$

Let $\mu' = \pi\mu = \pi B/\tau_0$, and perform the change of variable

$$\left(z = \sqrt{\frac{2}{\pi}}\left(\sqrt{\mu'} t - \frac{\pi f}{\sqrt{\mu'}}\right)\right) \quad ; \quad \sqrt{\frac{\pi}{2\mu'}} \, dz = dt \qquad (2.134)$$

Thus, Eq. (2.133) can be written as

$$\tilde{X}(f) = \sqrt{\frac{\pi}{2\mu'}} \, e^{-j(\pi f)^2/\mu'} \int_{-z_1}^{z_2} e^{j\pi z^2/2} \, dz \qquad (2.135)$$

$$\tilde{X}(f) = \sqrt{\frac{\pi}{2\mu'}} \, e^{-j(\pi f)^2/\mu'} \left\{ \int_{0}^{z_2} e^{j\pi z^2/2} \, dz - \int_{0}^{-z_1} e^{j\pi z^2/2} \, dz \right\} \qquad (2.136)$$

$$z_1 = -\sqrt{\frac{2\mu'}{\pi}}\left(\frac{\tau_0}{2} + \frac{\pi f}{\mu'}\right) = \sqrt{\frac{B\tau_0}{2}}\left(1 + \frac{f}{B/2}\right) \qquad (2.137)$$

$$z_2 = \sqrt{\frac{\mu'}{\pi}\left(\frac{\tau_0}{2} - \frac{\omega}{\mu'}\right)} = \sqrt{\frac{B\tau_0}{2}\left(1 - \frac{f}{B/2}\right)} \qquad \textbf{(2.138)}$$

The Fresnel integrals, denoted by $C(z)$ and $S(z)$, are defined by

$$C(z) = \int_0^z \cos\left(\frac{\pi \upsilon^2}{2}\right) d\upsilon \text{ and } S(z) = \int_0^z \sin\left(\frac{\pi \upsilon^2}{2}\right) d\upsilon \qquad \textbf{(2.139)}$$

Fresnel integrals can be approximated by

$$C(z) \approx \frac{1}{2} + \frac{1}{\pi z}\sin\left(\frac{\pi}{2}z^2\right) \qquad ; z \gg 1 \qquad \textbf{(2.140)}$$

$$S(z) \approx \frac{1}{2} - \frac{1}{\pi z}\cos\left(\frac{\pi}{2}z^2\right) \qquad ; z \gg 1 \qquad \textbf{(2.141)}$$

Note that $C(-z) = -C(z)$ and $S(-z) = -S(z)$. Figure 2.10 shows a plot of both $C(z)$ and $S(z)$ for $0 \leq z \leq 4.0$. Using Eq. (2.139) into Eq. (2.136) and performing the integration yield

$$\tilde{X}(f) = \sqrt{\frac{\pi}{2\mu'}} \, e^{-j(\pi f)^2/(\mu')} \{[C(z_2) + C(z_1)] + j[S(z_2) + S(z_1)]\} \qquad \textbf{(2.142)}$$

Figure 2.11 shows typical plots for the LFM real part, imaginary part, and amplitude spectrum. The square-like spectrum shown in Fig. 2.11c is widely known as the Fresnel spectrum.

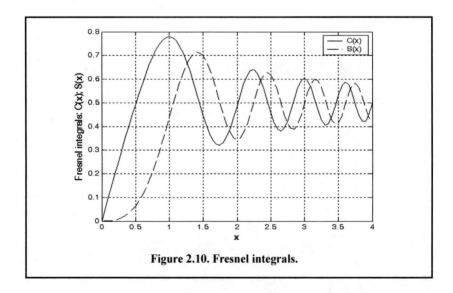

Figure 2.10. Fresnel integrals.

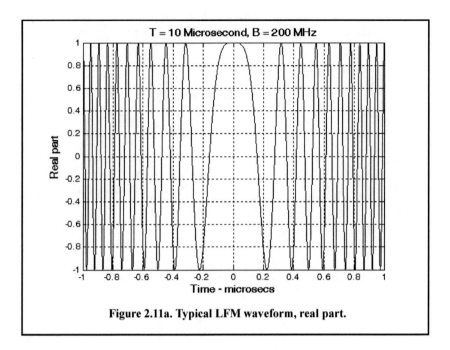

Figure 2.11a. Typical LFM waveform, real part.

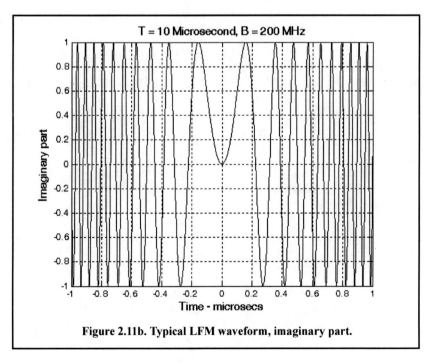

Figure 2.11b. Typical LFM waveform, imaginary part.

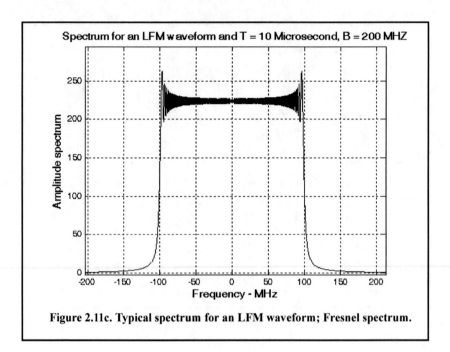

Figure 2.11c. Typical spectrum for an LFM waveform; Fresnel spectrum.

2.8. Signal Bandwidth and Duration

The signal bandwidth is the range of frequency over which the signal has a nonzero spectrum. In general, any signal can be defined using its duration (time domain) and bandwidth (frequency domain). A signal is said to be bandlimited if it has finite bandwidth. Signals that have finite durations (timelimited) will have infinite bandwidths, while bandlimited signals have infinite durations. The extreme case is a continuous sine-wave, whose bandwidth is infinitesimal.

Radar signal processing can be performed in either time domain or frequency domain. In either case, the radar signal processor assumes signals to be of finite duration (timelimited) and finite bandwidth (bandlimited). The trouble with this assumption is that timelimited and bandlimited signals cannot simultaneously exist. That is, a signal cannot have finite duration and have finite bandwidth. Because of this, it is customary to assume that radar signals are essentially limited in time and frequency.

Essentially timelimited signals are considered to be very small outside a certain finite time duration. If the FT of a signal is very small outside a certain finite frequency bandwidth, the signal is called essentially bandlimited signal.

A signal $x(t)$ over the time interval $\{T_1, T_2\}$ is said to be essentially timelimited relative to some very small signal level ε if and only if

$$\int_{T_1}^{T_2} |x(t)|^2 dt \geq (1 - \varepsilon) \int_{-\infty}^{\infty} |x(t)|^2 dt \qquad (2.143)$$

where the interval $\tau_e = T_2 - T_1$ is called the effective duration. The effective duration is defined as

$$\tau_e = \frac{\left(\int_{-\infty}^{\infty} |x(t)|^2 dt \right)^2}{\int_{-\infty}^{\infty} |x(t)|^4 dt} \qquad (2.144)$$

Similarly, a signal $x(t)$ over the frequency interval $\{B_1, B_2\}$ is said to be essentially bandlimited relative to some small signal level η if and only if

$$\int_{B_1}^{B_2} |X(f)|^2 df \geq (1 - \eta) \int_{-\infty}^{\infty} |X(f)|^2 df \qquad (2.145)$$

where $X(f)$ is the FT of $x(t)$ and the band $B_e = B_2 - B_1$ is called the effective bandwidth. The effective bandwidth is defined as

$$B_e = \frac{\left(\int_{-\infty}^{\infty} |X(f)|^2 df \right)^2}{\int_{-\infty}^{\infty} |X(f)|^4 df} \qquad (2.146)$$

Different, but equivalent, definitions for the effective bandwidth and effective duration can be found in the literature. In this book, the definitions cited in Burdic[1] are adopted. The quantity $B_e \tau_e$ is referred to as the time bandwidth product. In later chapters, it will be clear that large time bandwidth products are desirable in radar applications since they provide better pulse compression ratios (or compression gain).

---

1. Burdic, W. S., *Radar Signal Analysis*, Prentice-Hall, Englewood Cliffs, NJ, 1968.

Range resolution is defined as the reciprocal of the effective bandwidth. In Chapter 1, prior to introducing the concept of effective duration, the bandwidth was computed as the reciprocal of the pulsewidth, an approximation that is widely used and accepted, even though it is not quite 100% accurate. This is true since using one value or the other for the bandwidth does not make much difference in the overall calculation of the SNR when using the radar equation. Doppler resolution is computed as the reciprocal of the effective duration.

2.8.1. Effective Bandwidth and Duration Calculation

A few examples for computing the effective bandwidth and duration of most common radar signals are presented in this section.

Single Pulse

The single pulse was analyzed in the previous section. Consider the single pulse waveform given by

$$x(t) = Rect\left(\frac{t}{\tau_0}\right) \qquad ; \frac{-\tau_0}{2} < 0 < \frac{\tau_0}{2} \tag{2.147}$$

the effective bandwidth for this signal can be computed using Eq. (2.146). For this purpose, the denominator of Eq. (2.146) is

$$\int_{-\infty}^{\infty} |X(f)|^4 df = \int_{-\infty}^{\infty} |R_x(\tau)|^4 d\tau = \int_{-\infty}^{\infty} |\tau_0 Sinc(f\tau_0)|^4 df = \frac{2\tau_0^3}{3} \tag{2.148}$$

and its numerator is computed utilizing Eq. (2.59) as

$$\left(\int_{-\infty}^{\infty} |X(f)|^2 df\right)^2 = |R_x(0)|^2 = \tau_0^2 \tag{2.149}$$

Note that this value represents the square of the signal total energy. Therefore, the effective bandwidth is

$$B_e = \frac{\left(\int_{-\infty}^{\infty} |X(f)|^2 df\right)^2}{\int_{-\infty}^{\infty} |X(f)|^4 df} = \frac{(\tau_0^2)}{\left(\frac{2\tau_0^3}{3}\right)} = \frac{3}{2\tau_0} \tag{2.150}$$

The effective duration for the signal $x_2(t)$ is

$$\tau_e = \frac{\left(\displaystyle\int_{-\infty}^{\infty} |x(t)|^2 dt\right)^2}{\displaystyle\int_{-\infty}^{\infty} |x(t)|^4 dt} \qquad (2.151a)$$

$$\tau_e = \frac{\left(\displaystyle\int_{-\tau_0/2}^{\tau_0/2} (1)^2 dt\right)^2}{\displaystyle\int_{-\tau_0/2}^{\tau_0/2} (1)^4 dt} = \frac{\tau_0^2}{\tau_0} = \tau_0 \qquad (2.151b)$$

Finally, the time bandwidth product for this signal is

$$B_e \tau_e = \frac{3}{2\tau_0}\tau_0 = \frac{3}{2} \qquad (2.152)$$

Finite Duration Pulse Train Signal

The finite duration train signal was defined in the previous section; its complex envelope is given by

$$x(t) = Rect\left(\frac{t}{NT_r}\right) \sum_{n=-\infty}^{\infty} Rect\left(\frac{t-nT}{\tau_0}\right) \qquad (2.153)$$

The corresponding FT is

$$X(f) = \frac{T_t \tau_0}{T} Sinc(fT_t) \otimes \sum_{n=-\infty}^{\infty} Sinc(nf_r\tau_0)\delta(f-nf_r) \qquad (2.154)$$

The total energy for this signal is

$$\int_{-\infty}^{\infty} |X(f)|^2 df = \frac{T_t \tau_0}{T} \qquad (2.155)$$

It can be shown (see Problem 2.17) that

$$\int_{-\infty}^{\infty} |R_x(t)|^2 \, dt = \int_{-\infty}^{\infty} |X(f)|^4 \, df \approx \left(\frac{4}{3}\right)\left(\frac{T_i}{T}\right)^3 \left(\frac{2}{3}\right)(\tau_0)^3 \tag{2.156}$$

It follows that the effective bandwidth is

$$B_e \approx \frac{\left(\dfrac{T_i \tau_0}{T}\right)^2}{\left(\dfrac{4}{3}\right)\left(\dfrac{T_i}{T}\right)^3 \left(\dfrac{2}{3}\right)(\tau_0)^3} = \left(\frac{3T}{4T_i}\right)\left(\frac{3}{2\tau_0}\right) \tag{2.157}$$

The result of Eq. (2.157) clearly indicates that the effective bandwidth of the pulse train decreases as the length of the train is increased. This should intuitively make a lot of sense, since the bandwidth is inversely proportional to signal duration. Of course, when $T_i = T$ (i.e., single pulse case) Eq. (2.157) becomes identical to Eq. (2.150); note that in this case the factor $3/4$ will disappear from Eq. (2.156).

The effective duration of this signal can be computed using Eq. (2.144). Again the numerator of Eq. (2.144) represents the square of the total signal energy given in Eq. (2.155). The denominator of Eq. (2.144) is equal to unity (see Problem 2.18). Thus, the effective duration is

$$\tau_e = \frac{T_i \tau_0}{T} \tag{2.158}$$

and the time bandwidth product of this waveform is

$$B_e \tau_e \approx \left(\frac{3T}{4T_i}\right)\left(\frac{3}{2\tau_0}\right)\left(\frac{T_i \tau_0}{T}\right) = \frac{9}{8} \tag{2.159}$$

LFM Signal

In this case, the LFM complex envelope can be written as

$$x(t) = Rect\left(\frac{t}{\tau_0}\right) e^{j\mu\pi t^2} \tag{2.160}$$

where $\mu = B/\tau_0$ and B is the LFM bandwidth. Make a change of variables $\mu' = \pi\mu$, then the modulus of the FT of this signal can be approximated from Eq. (2.142) as

$$|X(f)| \approx \sqrt{\frac{\pi}{\mu'}} \; Rect\left(\frac{\pi f}{\mu' \tau_0}\right) \tag{2.161}$$

The FT of the autocorrelation function is equal to the square of the modulus of the signal FT, i.e.,

$$FT\{R_x(\tau)\} = |X(f)|^2 = \frac{\pi}{\mu'}Rect\left(\frac{\pi f}{\mu'\tau_0}\right) \tag{2.162}$$

Therefore,

$$\left(\int_{-\infty}^{\infty}|X(f)|^2 df\right)^2 \approx \tau_0^2 \tag{2.163}$$

also

$$\int_{-\infty}^{\infty}|X(f)|^4 df \approx \frac{\pi\tau_0}{\mu'} \tag{2.164}$$

Then the effective bandwidth is

$$B_e \approx \frac{\tau_0^2}{\dfrac{\pi\tau_0}{\mu'}} = \frac{\mu'\tau_0}{\pi} \tag{2.165}$$

The effective duration is

$$\tau_e = \frac{\left(\int\limits_{-\infty}^{\infty}|x(t)|^2 dt\right)^2}{\int\limits_{-\infty}^{\infty}|x(t)|^4 dt} = \frac{\left(\int\limits_{-\tau_0/2}^{\tau_0/2}(1)^2 dt\right)^2}{\int\limits_{-\tau_0/2}^{\tau_0/2}(1)^4 dt} = \frac{\tau_0^2}{\tau_0} = \tau_0 \tag{2.166}$$

And the time bandwidth product for LFM waveforms is computed as

$$B_e\tau_e \approx \frac{\mu'\tau_0}{\pi}\tau_0 = \frac{\mu'\tau_0^2}{\pi} = \frac{\pi\mu\tau_0^2}{\pi} = \frac{B\tau_0^2}{\tau_0} = B\tau_0 \tag{2.167}$$

2.9. Discrete Time Systems and Signals

Advances in computer hardware and in digital technologies completely revolutionized radar systems signal and data processing techniques. Virtually all modern radar systems use some form of a digital representation (signal samples) of their received signals for the purposes of signal and data processing. These samples of a timelimited signal are nothing more than a finite set of numbers (thought of as a vector) that represents discrete values of the continuous time domain signal. These samples are typically obtained by using Analog to Digital (A/D) conversion devices. Since in the digital world the radar receiver is now concerned with processing a set of finite numbers, its impulse response will also compose a set of finite numbers. Consequently, the radar receiver is now referred to as a discrete system. All input/output signal relationships are now carried out using discrete time samples. It must also be noted that just as in the case of continuous time domain systems, the discrete systems of interest to radar applications must also be causal, stable, and linear time invariant.

Consider a continuous lowpass signal that is essentially timelimited with duration τ and bandlimited with bandwidth B. This signal (as will be shown in the next section) can be completely represented by a set of $\{2\tau B\}$ samples. Since a finite set of discrete values (samples) is used to represent the signal, it is common to represent this signal by a finite dimensional vector of the same size. This vector is denoted by $\mathbf{x}$, or simply by the sequence $x[n]$,

$$\mathbf{x} \equiv x[n] = [x(0) \ x(1) \ ...x(N-2) \ x(N-1)]^t \qquad (2.168)$$

where the superscript t denotes transpose operation. The value N is at least $2\tau B$ for a real lowpass essentially limited signal $x(t)$ of duration τ and bandwidth B. If, however, the signal is complex, then N is at least τB and the components of the vector $\mathbf{x}$ are complex. The samples defined in Eq. (2.168) can be obtained from pulse to pulse samples at a fixed range (i.e., delay) of the radar echo signal. The PRF is denoted by f_r and the total observation interval is T_0; then N would be equal to $T_0 f_r$. Define the radar receiver transfer function as the discrete sequence $h[n]$ and the input signal sequence as $x[n]$; then the output sequence $y[n]$ is given by the convolution sum

$$y[n] = \sum_{m=0}^{M-1} h(m)x(n-m) \qquad (2.169)$$

where $\{h[n] = [h(0) \ h(1) \ ...h(M-2) \ h(M-1)]; \ M \le N\}$.

2.9.1. Sampling Theorem

Lowpass Sampling Theorem

In general, it is required to determine the necessary condition such that a signal can be fully reconstructed from its samples by filtering, or data processing in general. The answer to this question lies in the sampling theorem, which may be stated as follows: let the signal $x(t)$ be real-valued essentially bandlimited by the bandwidth B; this signal can be fully reconstructed from its samples if the time interval between samples is no greater than $1/(2B)$. Figure 2.12 illustrates the sampling process concept. The sampling signal $p(t)$ is periodic with period T_s, which is called the sampling interval.

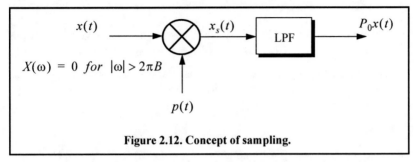

Figure 2.12. Concept of sampling.

The Fourier series expansion of $p(t)$ and the sampled signal $x_s(t)$ expressed using this Fourier series definition are, respectively, given by

$$p(t) = \sum_{n=-\infty}^{\infty} P_n e^{j\frac{2\pi nt}{T_s}} \qquad (2.170a)$$

$$p(t) = \sum_{n=-\infty}^{\infty} x(t) P_n e^{j\frac{2\pi nt}{T_s}} \qquad (2.170b)$$

$$p(t) = \sum_{n=-\infty}^{\infty} x(t) P_n e^{j\frac{2\pi nt}{T_s}} \qquad (2.170b)$$

Taking the FT of Eq. (2.170b) yields

$$X_s(\omega) = \sum_{n=-\infty}^{\infty} P_n X\left(\omega - \frac{2\pi n}{T_s}\right) = P_0 X(\omega) + \sum_{\substack{n=-\infty \\ n \neq 0}}^{\infty} P_n X\left(\omega - \frac{2\pi n}{T_s}\right) \quad (2.171)$$

where $X(\omega)$ is the FT of $x(t)$. Therefore, we conclude that the spectral density, $X_s(\omega)$, consists of replicas of $X(\omega)$ spaced $(2\pi/T_s)$ apart and scaled by the Fourier series coefficients P_n. A lowpass filter (LPF) of bandwidth B can then be used to recover the original signal $x(t)$.

When the sampling rate is increased (i.e., T_s decreases), the replicas of $X(\omega)$ move farther apart from each other. Alternatively, when the sampling rate is decreased (i.e., T_s increases), the replicas get closer to one another. The value of T_s such that the replicas are tangent to one another defines the minimum required sampling rate so that $x(t)$ can be recovered from its samples by using an LPF. It follows that

$$\frac{2\pi}{T_s} = 2\pi(2B) \Leftrightarrow T_s = \frac{1}{2B} \tag{2.172}$$

The sampling rate defined by Eq. (2.172) is known as the Nyquist sampling rate. When $T_s > (1/2B)$, the replicas of $X(\omega)$ overlap and, thus, $x(t)$ cannot be recovered cleanly from its samples. This is known as aliasing. In practice, ideal LPF cannot be implemented; hence, practical systems tend to over sample in order to avoid aliasing.

Example:

Assume that the sampling signal $p(t)$ is given by $p(t) = \sum\limits_{n=-\infty}^{\infty} \delta(t - nT_s)$.
Compute an expression for $X_s(\omega)$.

Solution:

The signal $p(t)$ is called the Comb function, with exponential Fourier series

$$p(t) = \sum_{n=-\infty}^{\infty} \frac{1}{T_s} e^{2\frac{\pi n t}{T_s}}$$

It follows that

$$x_s(t) = \sum_{n=-\infty}^{\infty} x(t) \frac{1}{T_s} e^{2\frac{\pi n t}{T_s}}$$

Taking the Fourier transform of this equation yields

$$X_s(\omega) = \frac{2\pi}{T_s} \sum_{n=-\infty}^{\infty} X\left(\omega - \frac{2\pi n}{T_s}\right)$$

It is desired to develop a general expression from which any lowpass signal can be recovered from its samples provided that Eq. (2.172) is satisfied. In order to do that, let $x(t)$ and $x_s(t)$ be the desired lowpass signal and its corresponding samples, respectively. Then an expression for $x(t)$ in terms of its samples can be derived as follows: First, obtain $X(\omega)$ by filtering the signal $X_s(\omega)$ using an ideal LPF whose transfer function is

$$H(\omega) = T_s Rect\left(\frac{\omega}{4\pi B}\right) \tag{2.173}$$

Thus,

$$X(\omega) = H(\omega)X_s(\omega) = T_s Rect\left(\frac{\omega}{4\pi B}\right)X_s(\omega) \qquad (2.174)$$

The signal $x(t)$ is now obtained from the inverse FT of Eq. (2.174) as

$$x(t) = FT^{-1}\{X(\omega)\} = FT^{-1}\left\{T_s Rect\left(\frac{\omega}{4\pi B}\right)X_s(\omega)\right\} = \qquad (2.175)$$

$$2BT_s Sinc(2\pi Bt) \otimes x_s(t)$$

The sampled signal $x_s(t)$ can be represented using an ideal sampling signal

$$p(t) = \sum_n \delta(t - nT_s) \qquad (2.176a)$$

thus,

$$x_s(t) = \sum_n x(nT_s)\delta(t - nT_s) \qquad (2.176b)$$

Substituting Eq. (2.176b) into Eq. (2.175) yields an expression for the signal $x(t)$ in terms of its samples

$$x(t) = 2BT_s \sum_n x(nT_s) \; Sinc(2\pi B(t - T_s)) \; ; T_s \le \frac{1}{2B} \qquad (2.177)$$

Bandpass Sampling Theorem

It was established in Section 2.6 that any bandpass signal can be expressed using the quadrature components as provided in Eq. (2.79) through Eq. (2.81). It follows that it is sufficient to construct the bandpass signal $x(t)$ from samples of the quadrature components $\{x_I(t), x_Q(t)\}$. Let the signal $x(t)$ be essentially bandlimited with bandwidth B, then each of the lowpass signals $x_I(t)$ and $x_Q(t)$ are also bandlimited each with bandwidth $B/2$. Hence, if either of these lowpass signal is sampled at a rate $f_s \le 1/B$ then the Nyquist criterion is not violated. Assume that both quadrature components are sampled synchronously, that is

$$x_I(t) = BT_s \sum_{n=-\infty}^{\infty} x_I(nT_s) \; Sinc(\pi B(t - nT_s)) \qquad (2.178)$$

$$x_Q(t) = BT_s \sum_{n=-\infty}^{\infty} x_Q(nT_s)\ Sinc(\pi B(t-nT_s)) \qquad (2.179)$$

where if the Nyquist rate is satisfied, then $BT_s = 1$ (unity time bandwidth product). Substituting Eq. (2.178) and Eq. (2.179) into Eq. (2.79) yields

$$x(t) = BT_s \left\{ \sum_{n=-\infty}^{\infty} [x_I(nT_s)\cos 2\pi f_0 t - x_Q(nT_s)\sin 2\pi f_0 t] \right. \qquad (2.180)$$

$$\left. Sinc(\pi B(t-nT_s)) \right\}$$

$$x(t) = Re\left\{ BT_s \sum_{n=-\infty}^{\infty} [x_I(nT_s) + jx_Q(nT_s)]e^{j2\pi f_0 t} Sinc(\pi B(t-nT_s)) \right\} \quad (2.181)$$

where, of course, $T_s \leq 1/B$ is assumed. This leads to the conclusion that if the total period over which the signal $x(t)$ is sampled is T_0, then $2BT_0$ samples are required, BT_0 samples for $x_I(t)$ and BT_0 samples for $x_Q(t)$.

2.9.2. The Z-Transform

The Z-transform is a transformation that maps samples of a discrete time-domain sequence into a new domain known as the z-domain. It is defined as

$$Z\{x(n)\} = X(z) = \sum_{n=-\infty}^{\infty} x(n)z^{-n} \qquad (2.182)$$

where $z = re^{j\omega}$, and for most cases, $r = 1$. It follows that Eq. (2.182) can be rewritten as

$$X(e^{j\omega}) = \sum_{n=-\infty}^{\infty} x(n)e^{-jn\omega} \qquad (2.183)$$

In the z-domain, the region over which $X(z)$ is finite is called the Region of Convergence (ROC).

Example:

Show that $Z\{nx(n)\} = -z\dfrac{d}{dz}X(z)$.

Solution:

Starting with the definition of the Z-transform,

$$X(z) = \sum_{n=-\infty}^{\infty} x(n)z^{-n}$$

Taking the derivative, with respect to z, of the above equation yields

$$\frac{d}{dz}X(z) = \sum_{n=-\infty}^{\infty} x(n)(-n)z^{-n-1}$$

$$= (-z^{-1}) \sum_{n=-\infty}^{\infty} nx(n)z^{-n}$$

It follows that

$$Z\{nx(n)\} = (-z)\frac{d}{dz}X(z)$$

A discrete LTI system has a transfer function $H(z)$ that describes how the system operates on its input sequence $x(n)$ in order to produce the output sequence $y(n)$. The output sequence $y(n)$ is computed from the discrete convolution between the sequences $x(n)$ and $h(n)$:

$$y(n) = \sum_{m=-\infty}^{\infty} x(m)h(n-m) \tag{2.184}$$

However, since practical systems require the sequence $x(n)$ and $h(n)$ to be of finite length, we can rewrite Eq. (2.184) as

$$y(n) = \sum_{m=0}^{N} x(m)h(n-m) \tag{2.185}$$

N denotes the input sequence length. The Z-transform of Eq. (2.185) is

$$Y(z) = X(z)H(z) \tag{2.186}$$

and the discrete system transfer function is

$$H(z) = \frac{Y(z)}{X(z)} \qquad (2.187)$$

Finally, the transfer function $H(z)$ can be written as

$$H(z)\Big|_{z = e^{j\omega}} = \left|H(e^{j\omega})\right|e^{\angle H(e^{j\omega})} \qquad (2.188)$$

where $\left|H(e^{j\omega})\right|$ is the amplitude response, and $\angle H(e^{j\omega})$ is the phase response.

2.9.3. The Discrete Fourier Transform

The Discrete Fourier Transform (DFT) is a mathematical operation that transforms a discrete sequence, usually from the time domain into the frequency domain, in order to explicitly determine the spectral information for the sequence. The time-domain sequence can be real or complex. The DFT has finite length N and is periodic with period equal to N. The discrete Fourier transform pairs for the finite sequence $x(n)$ are defined by

$$X(k) = \sum_{n=0}^{N-1} x(n)e^{-j\frac{2\pi nk}{N}} \qquad ; \; k = 0, ..., N-1 \qquad (2.189)$$

$$x(n) = \frac{1}{N}\sum_{k=0}^{N-1} X(k)e^{j\frac{2\pi nk}{N}} \qquad ; \; n = 0, ..., N-1 \qquad (2.190)$$

The Fast Fourier Transform (FFT) is not a new kind of transform different from the DFT. Instead, it is an algorithm used to compute the DFT more efficiently. There are numerous FFT algorithms that can be found in the literature. In this book we will interchangeably use the DFT and the FFT to mean the same thing. Furthermore, we will assume radix-2 FFT algorithm, where the FFT size is equal to $N = 2^m$ for some integer m.

2.9.4. Discrete Power Spectrum

Practical discrete systems utilize DFTs of finite length as a means of numerical approximation for the Fourier transform. The input signals must be truncated to a finite duration (denoted by T) before they are sampled. This is necessary so that a finite length sequence is generated prior to signal processing. Unfortunately, this truncation process may cause some serious problems.

To demonstrate this difficulty, consider the time-domain signal $x(t) = \sin 2\pi f_0 t$. The spectrum of $x(t)$ consists of two spectral lines at $\pm f_0$. Now, when $x(t)$ is truncated to length T seconds and sampled at a rate $T_s = T/N$, where N is the number of desired samples, we produce the sequence $\{x(n); \; n = 0, 1, \ldots, N-1\}$.

The spectrum of $x(n)$ would still be composed of the same spectral lines if T is an integer multiple of T_s and if the DFT frequency resolution Δf is an integer multiple of f_0. Unfortunately, those two conditions are rarely met, and as a consequence, the spectrum of $x(n)$ spreads over several lines (normally the spread may extend up to three lines). This is known as spectral leakage. Since f_0 is normally unknown, this discontinuity caused by an arbitrary choice of T cannot be avoided. Windowing techniques can be used to mitigate the effect of this discontinuity by applying smaller weights to samples close to the edges.

A truncated sequence $x(n)$ can be viewed as one period of some periodic sequence with period N. The discrete Fourier series expansion of $x(n)$ is

$$x(n) = \sum_{k=0}^{N-1} X_k e^{j\frac{2\pi nk}{N}} \tag{2.191}$$

It can be shown that the coefficients X_k are given by

$$X_k = \frac{1}{N}\sum_{n=0}^{N-1} x(n) e^{-j\frac{2\pi nk}{N}} = \frac{1}{N}X(k) \tag{2.192}$$

where $X(k)$ is the DFT of $x(n)$. Therefore, the Discrete Power Spectrum (DPS) for the bandlimited sequence $x(n)$ is the plot of $|X_k|^2$ versus k, where the lines are Δf apart,

$$P_0 = \frac{1}{N^2}|X(0)|^2 \tag{2.193a}$$

$$P_k = \frac{1}{N^2}\{|X(k)|^2 + |X(N-k)|^2\} \qquad ; \; k = 1, 2, \ldots, \frac{N}{2}-1 \tag{2.193b}$$

$$P_{N/2} = \frac{1}{N^2}|X(N/2)|^2 \tag{2.193c}$$

Before proceeding to the next section, we will show how to select the FFT parameters. For this purpose, consider a bandlimited signal $x(t)$ with bandwidth B. If the signal is not bandlimited, an LPF can be used to eliminate frequencies greater than B. In order to satisfy the sampling theorem, one must choose a sampling frequency $f_s = 1/T_s$, such that

$$f_s \geq 2B \qquad (2.194)$$

The truncated sequence duration T and the total number of samples N are related by

$$T = NT_s \qquad (2.195)$$

or equivalently,

$$f_s = N/T \qquad (2.196)$$

It follows that

$$f_s = \frac{N}{T} \geq 2B \qquad (2.197)$$

and the frequency resolution is

$$\Delta f = \frac{1}{NT_s} = \frac{f_s}{N} = \frac{1}{T} \geq \frac{2B}{N} \qquad (2.198)$$

2.9.5. Windowing Techniques

Truncation of the sequence $x(n)$ can be accomplished by computing the product

$$x_w(n) = x(n)w(n) \qquad (2.199)$$

where

$$w(n) = \left\{ \begin{array}{ll} f(n) & ; n = 0, 1, ..., \quad N-1 \\ 0 & otherwise \end{array} \right\} \qquad (2.200)$$

where $f(n) \leq 1$. The finite sequence $w(n)$ is called a windowing sequence, or simply a window. The windowing process should not impact the phase response of the truncated sequence. Consequently, the sequence $w(n)$ must retain linear phase. This can be accomplished by making the window symmetrical with respect to its central point.

If $f(n) = 1$ for all n, we have what is known as the rectangular window. It leads to the Gibbs phenomenon, which manifests itself as an overshoot and a ripple before and after a discontinuity. Figure 2.13 shows the amplitude spectrum of a rectangular window. Note that the first side-lobe is at $-13.46 dB$ below the main lobe. Windows that place smaller weights on the samples near the edges will have less overshoot at the discontinuity points (lower side-lobes); hence, they are more desirable than a rectangular window. However, reduction of the sidelobes is offset by a widening of the main lobe. Therefore, the proper choice of a windowing sequence is continuous trade-off between side-lobe reduction and main-lobe widening. Table 2.1 gives a summary of some commonly used windows with the corresponding impact on main beam widening and peak reduction.

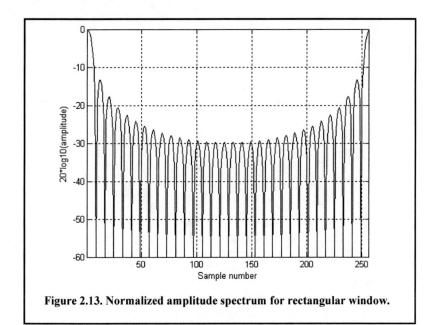

Figure 2.13. Normalized amplitude spectrum for rectangular window.

TABLE 2.1. Common windows

Window	Null-to-Null Beamwidth Rectangular Window is the Reference	Peak Reduction
Rectangular	*1*	*1*
Hamming	*2*	*0.73*
Hanning	*2*	*0.664*
Blackman	*6*	*0.577*
Kaiser (β = 6)	*2.76*	*0.683*
Kaiser (β = 3)	*1.75*	*0.882*

The multiplication process defined in Eq. (2.199) is equivalent to cyclic convolution in the frequency domain. It follows that $X_w(k)$ is a smeared (distorted) version of $X(k)$. To minimize this distortion, we would seek windows that have a narrow main lobe and small side-lobes. Additionally, using a window other than a rectangular window reduces the power by a factor P_w, where

$$P_w = \frac{1}{N}\sum_{n=0}^{N-1} w^2(n) = \sum_{k=0}^{N-1}|W(k)|^2 \qquad (2.201)$$

It follows that the DPS for the sequence $x_w(n)$ is now given by

$$P_0^w = \frac{1}{P_w N^2}|X(0)|^2 \qquad (2.202)$$

$$P_k^w = \frac{1}{P_w N^2}\{|X(k)|^2 + |X(N-k)|^2\} \qquad ; \ k = 1, 2, ..., \frac{N}{2}-1 \qquad (2.202b)$$

$$P_{N/2}^w = \frac{1}{P_w N^2}|X(N/2)|^2 \qquad (2.202c)$$

where P_w is defined in Eq. (2.193). Table 2.2 lists some common windows. Figures 2.14 through 2.16 show the frequency domain characteristics for these windows. These plots can be reproduced using the following MATLAB code.

TABLE 2.2. Some common windows. $n = 0, N-1$

Window	Expression	First Side-lobe	Main Lobe Width
Rectangular	$w(n) = 1$	$-13.46dB$	1
Hamming	$w(n) = 0.54 - 0.46\cos\left(\dfrac{2\pi n}{N-1}\right)$	$-41dB$	2
Hanning	$w(n) = 0.5\left[1 - \cos\left(\dfrac{2\pi n}{N-1}\right)\right]$	$-32dB$	2
Kaiser	$w(n) = \dfrac{I_0[\beta\sqrt{1-(2n/N)^2}]}{I_0(\beta)}$ I_0 is the zero-order modified Bessel function of the first kind	$-46dB$ for $\beta = 2\pi$	$\sqrt{5}$ for $\beta = 2\pi$

```
%Use this program to reproduce figures 2.14 through 2.16.
clear all;
close all;
eps = 0.001;
N = 32;
win_rect (1:N) = 1;
win_ham = hamming(N);
win_han = hanning(N);
win_kaiser = kaiser(N, pi);
win_kaiser2 = kaiser(N, 5);
Yrect = abs(fft(win_rect, 256));
Yrectn = Yrect ./ max(Yrect);
Yham = abs(fft(win_ham, 256));
Yhamn = Yham ./ max(Yham);
Yhan = abs(fft(win_han, 256));
Yhann = Yhan ./ max(Yhan);
YK = abs(fft(win_kaiser, 256));
YKn = YK ./ max(YK);
YK2 = abs(fft(win_kaiser2, 256));
YKn2 = YK2 ./ max(YK2);
figure (1)
plot(20*log10(Yrectn+eps), 'k')
xlabel('Sample number');
ylabel('20*log10(amplitude)')
axis tight;
grid
figure(2)
plot(20*log10(Yhamn + eps), 'k')
xlabel('Sample number');
 ylabel('20*log10(amplitude)')
grid;
axis tight
figure (3)
plot(20*log10(Yhann+eps), 'k')
xlabel('Sample number');
ylabel('20*log10(amplitude)'); grid
axis tight
figure(4)
plot(20*log10(YKn+eps), 'k')
grid; hold on
plot(20*log10(YKn2+eps), 'k--')
xlabel('Sample number');
ylabel('20*log10(amplitude)')
legend('Kaiser par. = \pi', 'Kaiser par. =5')
axis tight;
hold off
```

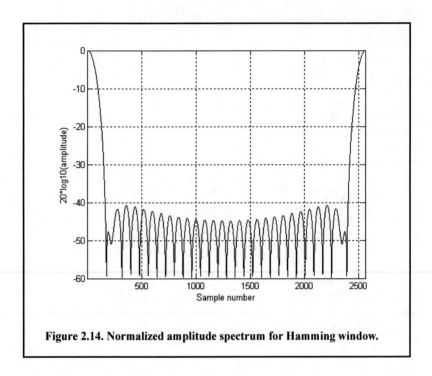

Figure 2.14. Normalized amplitude spectrum for Hamming window.

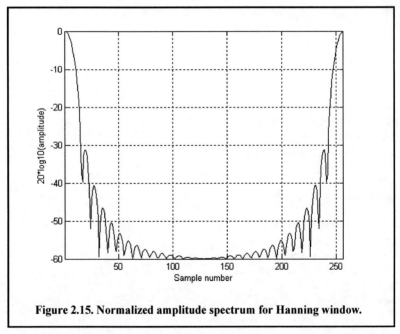

Figure 2.15. Normalized amplitude spectrum for Hanning window.

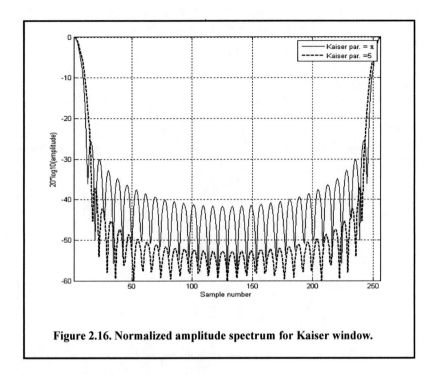

Figure 2.16. Normalized amplitude spectrum for Kaiser window.

2.9.6. Decimation and Interpolation

Decimation

Typically, radar systems use many signals for different functions, such as search, track, and discrimination to name a few. All signals are assumed to be essentially limited; however, since these signals have different functions, they do not have the same time and bandwidth durations (τ, B). Earlier in this chapter, it was established that the number of samples required to sufficiently recover any signal from its samples is $N \geq 2\tau B$. Therefore, it is important to use an A/D with high enough sampling rate to account for the largest possible number of samples required. As a result, it is often the case that some radar signals are sampled at a much higher rate than actually needed.

The process for decreasing the number of samples for a given sequence is called decimation. This is because the original data set has been reduced (decimated) in number. The process that increases the number of data samples is referred to as interpolation. The typical implementation for either operation is to alter the sampling rate, without violating the Nyquist sampling rate, of the input sequence. In decimation, the sampling rate is decreased by increasing the time steps between successive samples. More precisely, if the t_1 is the original sampling interval and t_2 is the decimated sampling interval, then

$$t_2 = Dt_1 \qquad (2.203)$$

D is the decimation ratio and it is greater than unity. If D is an integer, then decimation effectively decreases the original sequence by discarding $(D-1)$ samples of D samples. This is illustrated in Fig. 2.17 for $D = 3$.

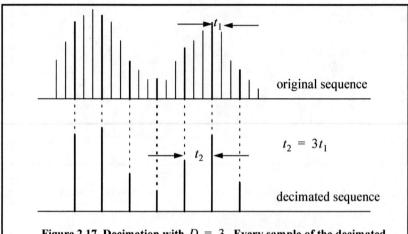

Figure 2.17. Decimation with $D = 3$. Every sample of the decimated sequence coincides with every third sample of the original sequence.

When D is not an integer, it is then necessary to first perform interpolation to determine new values for the new sequence. For example, if $D = 2.2$, then four out of every five samples in the decimated sequence are between samples in the original sequence and must be found by interpolation. This is illustrated in Fig. 2.18 for $D = 2.2$. In this example,

$$\left(t_2 = 2.2t_1 = \frac{11}{5}t_1\right) \Rightarrow 5t_2 = 11t_1 \qquad (2.204)$$

which indicates that there are five samples in the decimated sequence for every eleven samples of the original sequence. Additionally, every fifth sample in the decimated sequence is equal to every eleventh sample of the original sequence.

Interpolation

Suppose that a signal $x(t)$ whose duration is T seconds has been sampled at a sampling rate t_1 to obtain a sequence

$$\mathbf{x} = x[n] = \{x(nt_1), n = 0, 1, ..., N_1 - 1\} \qquad (2.205)$$

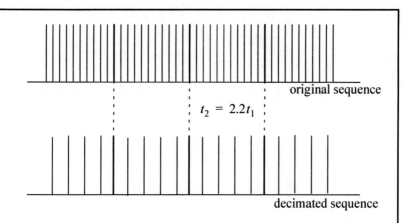

Figure 2.18. Decimation with $D = 2.2$. **Every fifth sample of the decimated sequence coincides with a sample in the original sequence.**

in this case, $N_1 = T/t_1$. Suppose you want to interpolate between the samples of $x[n]$ to generate a new sequence of size N_2 and sampling interval t_2, where $t_2 = t_1/k$. This effectively corresponds to a new sampling frequency $f_{s2} = kf_{s1}$ where $f_{s1} = 1/t_1$. This can be accomplished using Eq. (2.177) (see Problem 2.33); however, a more efficient interpolation can be performed using the FFT as will be described in the rest of this section.

Denote the FFT of the sequences $x_1[n]$ and $x_2[n]$ by $X_1[l]$ and $X_2[l]$. Assume that the signal $x(t)$ is essentially bandlimited with bandwidth $B = M\Delta f$ where M is an integer and $\Delta f = 1/T$. It follows that in order not to violate the sampling theorem

$$M\Delta f < f_{s1}/2 < f_{s2}/2 \qquad (2.206)$$

It is clear that the coefficients of $X_1[l]$ and $X_2[l]$ are zero for all $|l| > M$. More precisely,

$$\begin{aligned} X_1[l] &= 0; \quad l = M+1, M+2, \ldots, N_1 - 3 \\ X_2[l] &= 0; \quad l = M+1, M+2, \ldots, N_2 - 3 \end{aligned} \qquad (2.207)$$

Therefore, one can easily obtain the new sequence $X_2[l]$ from $X_1[l]$ by adding zeros in between the negative and positive frequencies from

$$N_1 - (2M+1) \quad to \quad N_2 - (2M+1) \qquad (2.208)$$

and the sequence $x_2[n]$ is simply generated by computing the inverse DFT of the sequence $X_2[l]$. Interpolation can also be applied to the frequency domain sequence. For this purpose, one can simply zero pad the time domain sequence to the desired size then take the DFT of the newly interpolated sequence.

Problems

2.1. Classify each of the following signals as an energy signal, a power signal, or neither.

(a) $\exp(0.5t)$ $(t \geq 0)$,

(b) $\exp(-0.5t)$ $(t \geq 0)$,

(c) $\cos t + \cos 2t$ $(-\infty < t < \infty)$,

(d) $e^{-a|t|}$ $(a > 0)$.

2.2. A definition for the instantaneous frequency was given in Eq. (2.58). A more general definition is

$$f_i(t) = \frac{1}{2\pi} Im\left\{\frac{d}{dt}\ln\psi(t)\right\}$$

where $Im\{.\}$, indicates imaginary part. Using this definition, calculate the instantaneous frequency for

(a) $x(t) = Rect\left(\frac{t}{\tau}\right)\cos(2\pi f_0 t)$

(b) $x(t) = Rect\left(\frac{t}{\tau}\right)\cos\left(2\pi f_0 t + \frac{B}{2\tau}t^2\right)$

2.3. Consider the two bandpass signals $x(t) = r_x(t)\cos(2\pi f_0 t + \phi_x(t))$ and $h(t) = r_h(t)\cos(2\pi f_0 t + \phi_h(t))$. Derive an expression for the complex envelope for the signal $s(t) = x(t) + h(t)$.

2.4. Consider the bandpass signal $x(t)$ whose complex envelope is equal to $\tilde{x}(t) = x_I(t) + jx_Q(t)$. Derive an expression for the autocorrelation function and the power spectrum density for $x(t)$ and $\tilde{x}(t)$.

2.5. Find the autocorrelation integral of the pulse train

$$y(t) = Rect(t/T) - Rect\left(\frac{t-T}{T}\right) + Rect\left(\frac{t-2T}{T}\right).$$

2.6. Compute the discrete convolution $y(n) = x(m) \bullet h(m)$ where

$$\{x(k), k = -1, 0, 1, 2\} = [-1.9, 0.5, 1.2, 1.5]$$
$$\{h(k), k = 0, 1, 2\} = [-2.1, 1.2, 0.8].$$

2.7. Define $\{x_I(n) = 1, -1, 1\}$ and $\{x_Q(n) = 1, 1, -1\}$. (a) Compute the discrete correlations: R_{x_I}, R_{x_Q}, $R_{x_I x_Q}$, and $R_{x_Q x_I}$. (b) A certain radar transmits the signal $s(t) = x_I(t)\cos 2\pi f_0 t - x_Q(t)\sin 2\pi f_0 t$. Assume that the autocorrelation $s(t)$ is equal to $y(t) = y_I(t)\cos 2\pi f_0 t - y_Q(t)\sin 2\pi f_0 t$. Compute and sketch $y_I(t)$ and $y_Q(t)$.

2.8. Compute the energy associated with the signal $x(t) = A Rect(t/\tau)$.

2.9. (a) Prove that $\varphi_1(t)$ and $\varphi_2(t)$, shown in figure below, are orthogonal over the interval $(-2 \le t \le 2)$. (b) Express the signal $x(t) = t$ as a weighted sum of $\varphi_1(t)$ and $\varphi_2(t)$ over the same time interval

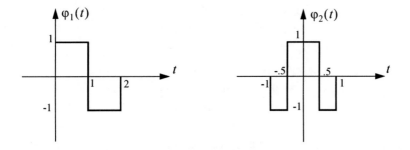

2.10. A periodic signal $x_p(t)$ is formed by repeating the pulse $x(t) = 2\Delta((t-3)/5)$ every 10 seconds. (a) What is the Fourier transform of $x(t)$? (b) Compute the complex Fourier series of $x_p(t)$. (c) Give an expression for the autocorrelation function $\bar{R}_{x_p}(t)$ and the power spectrum density $\bar{S}_{x_p}(\omega)$

2.11. If the Fourier series is

$$x(t) = \sum_{n=-\infty}^{\infty} X_n e^{j2\pi n t/T}$$

define $y(t) = x(t - t_0)$. Compute an expression for the complex Fourier series expansion of $y(t)$.

2.12. Show that (a) $\bar{R}_x(-t) = \bar{R}_x^*(t)$, (b) If $x(t) = f(t) + m_1$ and $y(t) = g(t) + m_2$, show that $\bar{R}_{xy}(t) = m_1 m_2$, where the average values for $f(t)$ and $g(t)$ are zeroes.

2.13. What is the power spectral density for the signal
$$x(t) = A\cos(2\pi f_0 t + \theta_0)?$$

2.14. Consider the signal
$$x(t) = Rect(t/\tau)\cos(\omega_0 t - Bt^2/2\tau)$$
and let $\tau = 15\mu s$ and $B = 10 MHz$. What are the quadrature components?

2.15. Determine the quadrature components for the signal
$$h(t) = \delta(t) - \left(\frac{\omega_0}{\omega_d}\right)e^{-2t}\sin(\omega_0 t)u(t).$$

2.16. If $x(t) = x_1(t) - 2x_1(t-5) + x_1(t-10)$, determine the autocorrelation functions $R_{x_1}(t)$ and $R_x(t)$ when $x_1(t) = \exp(-t^2/2)$.

2.17. Derive Eq. (2.156).

2.18. Prove that the effective duration of a finite pulse train is equal to $(T_t \tau_0)/T$, where τ_0 is the pulsewidth, T is the PRI, and T_t is as defined in Fig. 2.8.

2.19. A certain bandlimited signal has bandwidth $B = 20KHz$. Find the FFT size required so that the frequency resolution is $\Delta f = 50Hz$. Assume radix 2 FFT and a record length of 1 second.

2.20. Write an expression for the autocorrelation function $R_y(t)$, where
$$y(t) = \sum_{n=1}^{5} Y_n Rect\left(\frac{t-n5}{2}\right) \text{ and } \{Y_n\} = \{0.8, 1, 1, 1, 0.8\}.$$
Give an expression for the density function $S_y(\omega)$.

2.21. An LTI system has impulse response
$$h(t) = \begin{cases} \exp(-2t) & t \geq 0 \\ 0 & t < 0 \end{cases}$$
(a) Find the autocorrelation function $R_h(\tau)$. (b) Assume the input of this system is $x(t) = 3\cos(100t)$. What is the output?

2.22. Let $\bar{S}_X(\omega)$ be the PSD function for the stationary random process $X(t)$. Compute an expression for the PSD function of
$$Y(t) = X(t) - 2X(t - T).$$

2.23.   Assume that a certain sequence is determined by its FFT. If the record length is $2ms$ and the sampling frequency is $f_s = 10KHz$, find N.

2.24.   Prove that

$$\sum_{n=-\infty}^{\infty} J_n(z) = 1.$$

2.25.   Show that $J_{-n}(z) = (-1)^n J_n(z)$. Hint: You may utilize the relation

$$J_n(z) = \frac{1}{\pi} \int_0^\pi \cos(z\sin y - ny) dy.$$

2.26.   Compute the Z-transform for

(a) $x_1(n) = \dfrac{1}{n!}u(n)$,

(b) $x_2(n) = \dfrac{1}{(-n)!}u(-n)$.

2.27.   (a) Write an expression for the FT of $x(t) = Rect(t/3)$. (b) Assume that you want to compute the modulus of the FT using a DFT of size *512* with a sampling interval of *1* second. Evaluate the modulus at frequency $(80/512)Hz$. Compare your answer to the theoretical value and compute the error.

2.28.   Generate *512* samples of the signal $x(t) = 2.0e^{-5t}\sin(4\pi t)$, using sampling interval equal to 0.002. Compute the resultant spectrum and then truncate the spectrum at *15* Hz. Generate the time-domain sequence for the truncated spectrum. Determine the sampling rate of the new sequence.

2.29.   Assume that a time-domain sequence generated by using a sampling interval equal to *0.01* is given by $v(k) = \{0, 2, 5, 12, 5, 3, 3, -1, 1, 0\}$. Decimate this sequence so that the sampling interval is *0.02*.

2.30.   Write a MATLAB program to decimate any sequence of finite length and demonstrate it using the previous problem.

2.31.   You are given a sequence of samples $\{x(kT), k = -\infty, ..., \infty\}$ where the sampling interval T corresponds to twice the Nyquist rate. Give an expression to compute the samples of $x(t)$ at a new sampling rate corresponding to $T' = 0.7T$.

2.32.   Write a short argument to explain why the matched filter used in radar application ought to be an LTI filter.

2.33. A certain bandlimited signal has bandwidth $B = 20KHz$. Find the FFT size required so that the frequency resolution is $\Delta f = 50Hz$. Assume radix 2 FFT and a record length of 1 second.

2.34. Assume that a certain sequence is determined by its FFT. If the record length is $2ms$ and the sampling frequency is $f_s = 10KHz$, find N.

Chapter 3   *Random Variables and Processes*

3.1. Random Variables

Consider an experiment with outcomes defined by a certain sample space. The rule or functional relationship that maps each point in this sample space into a real number is called a random variable. Random variables are designated by capital letters (e.g., X, Y, ...), and a particular value of a random variable is denoted by a lowercase letter (e.g., x, y, ...).

The Cumulative Distribution Function (*cdf*) associated with the random variable X is denoted as $F_X(x)$ and is interpreted as the total probability that the random variable X is less than or equal to the value x. More precisely,

$$F_X(x) = Pr\{X \le x\} \tag{3.1}$$

The probability that the random variable X is in the interval (x_1, x_2) is then given by

$$F_X(x_2) - F_X(x_1) = Pr\{x_1 \le X \le x_2\} \tag{3.2}$$

The probability that a random variable X has values in the interval (x_1, x_2) is

$$F_X(x_2) - F_X(x_1) = Pr\{x_1 \le X \le x_2\} = \int_{x_1}^{x_2} f_X(x)dx \tag{3.3}$$

It is often practical to describe a random variable by the derivative of its *cdf*, which is called the Probability Density Function *(pdf)*. The *pdf* of the random variable X is

$$f_X(x) = \frac{d}{dx}F_X(x)$$

$$\tag{3.4}$$

or, equivalently,

$$F_X(x) = Pr\{X \le x\} = \int_{-\infty}^{x} f_X(\lambda)\,d\lambda \qquad (3.5)$$

The *cdf* has the following properties:

$$0 \le F_X(x) \le 1$$
$$F_X(-\infty) = 0$$
$$F_X(\infty) = 1 \qquad (3.6)$$
$$F_X(x_1) \le F_X(x_2) \Leftrightarrow x_1 \le x_2$$

Define the *nth* moment for the random variable X as

$$E[X^n] = \overline{X^n} = \int_{-\infty}^{\infty} x^n f_X(x)\,dx \qquad (3.7)$$

The first moment, $E[X]$, is called the mean value, while the second moment, $E[X^2]$, is called the mean squared value. When the random variable X represents an electrical signal across a 1Ω resistor, then $E[X]$ is the DC component, and $E[X^2]$ is the total average power.

The *nth* central moment is defined as

$$E[(X - \overline{X})^n] = \overline{(X - \overline{X})^n} = \int_{-\infty}^{\infty} (x - \overline{x})^n f_X(x)\,dx \qquad (3.8)$$

and, thus, the first central moment is zero. The second central moment is called the variance and is denoted by the symbol σ_X^2,

$$\sigma_X^2 = \overline{(X - \overline{X})^2} \qquad (3.9)$$

In practice, the random nature of an electrical signal may need to be described by more than one random variable. In this case, the joint *cdf* and *pdf* functions need to be considered. The joint *cdf* and *pdf* for the two random variables X and Y are, respectively, defined by

$$F_{XY}(x, y) = Pr\{X \le x; Y \le y\} \qquad (3.10)$$

$$f_{XY}(x, y) = \frac{\partial^2}{\partial x \partial y} F_{XY}(x, y) \qquad (3.11)$$

The marginal *cdf*s are obtained as follows:

$$F_X(x) = \int\limits_{-\infty}^{\infty} \int\limits_{-\infty}^{x} f_{UV}(u, v) du dv = F_{XY}(x, \infty)$$

$$F_Y(y) = \int\limits_{-\infty}^{\infty} \int\limits_{-\infty}^{y} f_{UV}(u, v) dv du = F_{XY}(\infty, y)$$

(3.12)

If the two random variables are statistically independent, then the joint *cdf*s and *pdf*s are, respectively, given by

$$F_{XY}(x, y) = F_X(x)F_Y(y)$$

(3.13)

$$f_{XY}(x, y) = f_X(x)f_Y(y)$$

(3.14)

Let us now consider a case when the two random variables X and Y are mapped into two new variables U and V through some transformations T_1 and T_2 defined by

$$U = T_1(X, Y) \qquad ; \quad V = T_2(X, Y)$$

(3.15)

The joint *pdf*, $f_{UV}(u, v)$, may be computed based on the invariance of probability under the transformation. One must first compute the matrix of derivatives; then the new joint *pdf* is computed as

$$f_{UV}(u, v) = f_{XY}(x, y)|J|$$

(3.16)

$$|J| = \begin{vmatrix} \dfrac{\partial x}{\partial u} & \dfrac{\partial x}{\partial v} \\ \dfrac{\partial y}{\partial u} & \dfrac{\partial y}{\partial v} \end{vmatrix}$$

(3.17)

where the determinant of the matrix of derivatives $|J|$ is called the Jacobian. The characteristic function for the random variable X is defined as

$$C_X(\omega) = E[e^{j\omega X}] = \int\limits_{-\infty}^{\infty} f_X(x)e^{j\omega x} dx$$

(3.18)

The characteristic function can be used to compute the *pdf* for a sum of independent random variables. More precisely, let the random variable Y be

$$Y = X_1 + X_2 + \dots + X_N$$

(3.19)

where $\{X_i \; ; \; i = 1, \dots, N\}$ is a set of independent random variables. It can be shown that

$$C_Y(\omega) = C_{X_1}(\omega)C_{X_2}(\omega)...C_{X_N}(\omega) \tag{3.20}$$

and the *pdf* $f_Y(y)$ is computed as the inverse Fourier transform of $C_Y(\omega)$ (with the sign of y reversed):

$$f_Y(y) = \frac{1}{2\pi} \int_{-\infty}^{\infty} C_Y(\omega)e^{-j\omega y}d\omega \tag{3.21}$$

The characteristic function may also be used to compute the *nth* moment for the random variable X as

$$E[X^n] = (-j)^n \frac{d^n}{d\omega^n}C_X(\omega)\Bigg|_{\omega=0} \tag{3.22}$$

3.2. Multivariate Gaussian Random Vector

Consider a joint probability for m random variables, $X_1, X_2, ..., X_m$. These variables can be represented as components of an $m \times 1$ random column vector, **X**. More precisely,

$$\mathbf{X} = \begin{bmatrix} X_1 & X_2 & ... & X_m \end{bmatrix}^t \tag{3.23}$$

where the superscript t indicates the transpose operation. The joint *pdf* for the vector **X** is

$$f_{\underline{X}}(\underline{x}) = f_{X_1, X_2, ..., X_m}(x_1, x_2, ..., x_m) \tag{3.24}$$

The mean vector is defined as

$$\mu_X = \begin{bmatrix} E[X_1] & E[X_2] & ... & E[X_m] \end{bmatrix}^t \tag{3.25}$$

and the covariance is an $m \times m$ matrix given by

$$\mathbf{C}_X = E[\mathbf{X}\,\mathbf{X}^t] - \mu_X\,\mu_X^t \tag{3.26}$$

Note that if the elements of the vector **X** are independent, then the covariance matrix is a diagonal matrix.

A random vector **X** is multivariate Gaussian if its *pdf* is of the form

$$f_X(\underline{x}) = \frac{1}{\sqrt{(2\pi)^m|\mathbf{C}_X|}}\exp\left(-\frac{1}{2}(\mathbf{x}-\mu_X)^t\mathbf{C}_X^{-1}(\mathbf{x}-\mu_X)\right) \tag{3.27}$$

where μ_x is the mean vector, $\boldsymbol{C}_x$ is the covariance matrix, $\boldsymbol{C}_x^{-1}$ is inverse of the covariance matrix and $|\boldsymbol{C}_x|$ is its determinant, and $\boldsymbol{X}$ is of dimension m. If $\boldsymbol{A}^1$ is a $k \times m$ matrix of rank k, then the random vector $\boldsymbol{Y} = \boldsymbol{AX}$ is a k-variate Gaussian vector with

$$\mu_Y = \boldsymbol{A}\mu_X \tag{3.28}$$

$$\boldsymbol{C}_Y = \boldsymbol{A}\Lambda_X \boldsymbol{A}^t \tag{3.29}$$

The characteristic function for a multivariate Gaussian *pdf* is defined by

$$C_X = E[\exp\{j(\omega_1 X_1 + \omega_2 X_2 + \dots + \omega_m X_m)\}] = \tag{3.30}$$

$$\exp\left\{j\mu_X^t\omega - \frac{1}{2}\omega^t \boldsymbol{C}_X\omega\right\}$$

Then the moments for the joint distribution can be obtained by partial differentiation. For example,

$$E[X_1 X_2 X_3] = \frac{\partial^3}{\partial\omega_1 \partial\omega_2 \partial\omega_3} C_X(\omega_1, \omega_2, \omega_3) \qquad at \quad \omega = \boldsymbol{0} \tag{3.31}$$

Example:

The vector $\boldsymbol{X}$ *is a 4-variate Gaussian with*

$$\mu_X = \begin{bmatrix} 2 & 1 & 1 & 0 \end{bmatrix}^t \ and \ \boldsymbol{C}_X = \begin{bmatrix} 6 & 3 & 2 & 1 \\ 3 & 4 & 3 & 2 \\ 2 & 3 & 4 & 3 \\ 1 & 2 & 3 & 3 \end{bmatrix}$$

Define

$$\boldsymbol{X}_1 = \begin{bmatrix} X_1 \\ X_2 \end{bmatrix} \qquad \boldsymbol{X}_2 = \begin{bmatrix} X_3 \\ X_4 \end{bmatrix}$$

Find the distribution of $\boldsymbol{X}_1$ *and the distribution of*

---

1. Note that matrices are denoted by italicized upper case bold face letters, while vectors are denoted by lower and upper regular (not italicized) letters.

$$Y = \begin{bmatrix} 2\mathbf{X}_1 \\ \mathbf{X}_1 + 2\mathbf{X}_2 \\ \mathbf{X}_3 + \mathbf{X}_4 \end{bmatrix}$$

Solution:

$\mathbf{X}_1$ has a bivariate Gaussian distribution with

$$\mu_{X_1} = \begin{bmatrix} 2 \\ 1 \end{bmatrix} \qquad \mathbf{C}_{X_1} = \begin{bmatrix} 6 & 3 \\ 3 & 4 \end{bmatrix}$$

The vector $\mathbf{Y}$ can be expressed as

$$Y = \begin{bmatrix} 2 & 0 & 0 & 0 \\ 1 & 2 & 0 & 0 \\ 0 & 0 & 1 & 1 \end{bmatrix} \begin{bmatrix} \mathbf{X}_1 \\ \mathbf{X}_2 \\ \mathbf{X}_3 \\ \mathbf{X}_4 \end{bmatrix} = \mathbf{AX}$$

It follows that

$$\mu_Y = \mathbf{A}\mu_X = \begin{bmatrix} 4 & 4 & 1 \end{bmatrix}^t$$

$$\mathbf{C}_Y = \mathbf{A}\mathbf{C}_X \mathbf{A}^t = \begin{bmatrix} 24 & 24 & 6 \\ 24 & 34 & 13 \\ 6 & 13 & 13 \end{bmatrix}$$

A special case of Eq. (3.29) is when the matrix $\mathbf{A}$ is given by

$$\mathbf{A} = \begin{bmatrix} a_1 a_2 & \cdots & a_m \end{bmatrix} \tag{3.32}$$

It follows that $\mathbf{Y} = \mathbf{AX}$ is a sum of random variables X_i, that is

$$Y = \sum_{i=1}^{m} a_i X_i \tag{3.33}$$

The finding in Eq. (3.33) leads to the conclusion that the linear sum of Gaussian variables is also a Gaussian variable with mean and variance given by

$$\bar{Y} = a_1 \bar{X}_1 + a_2 \bar{X}_2 + \ldots + a_m \bar{X}_m \tag{3.34}$$

$$\sigma_Y^2 = E[(X - \bar{X})^2] = \tag{3.35}$$
$$E[a_1(X_1 - \bar{X}_1) + a_2(X_2 - \bar{X}_2) + \dots + a_m(X_m - \bar{X}_m)]$$

and if the variables X_i are independent then Eq.(3.35) reduces to

$$\sigma_Y^2 = a_1^2 \sigma_{X_1}^2 + a_2^2 \sigma_{X_2}^2 + \dots + a_m^2 \sigma_{X_m}^2 \tag{3.36}$$

finally, in this case, the probability density function $f_Y(y)$ is given by (which can also be derived from Eq. (3.20))

$$f_Y(y) = f_{X_1}(x_1) \otimes f_{X_2}(x_2) \otimes \dots \otimes f_{X_m}(x_m) \tag{3.37}$$

where $\otimes$ indicates convolution.

3.2.1. Complex Multivariate Gaussian Random Vector

Consider the complex vector random variable

$$\tilde{\mathbf{X}} = \mathbf{X}_I + j\mathbf{X}_Q \tag{3.38}$$

where $\mathbf{X}_I$ and $\mathbf{X}_Q$ are real random multivariate Gaussian random vectors. The joint *pdf* for the complex random vector $\tilde{\mathbf{X}}$ is computed from the joint *pdf* of the two real vectors. The mean for the vector $\tilde{\mathbf{X}}$ is

$$E[\tilde{\mathbf{X}}] = E[\mathbf{X}_I] + jE[\mathbf{X}_Q] \tag{3.39}$$

The covariance matrix is also defined by

$$\tilde{\mathbf{C}} = E[(\tilde{\mathbf{X}} - E[\tilde{\mathbf{X}}])(\tilde{\mathbf{X}} - E[\tilde{\mathbf{X}}])^\dagger] \tag{3.40}$$

where the operator $\dagger$ indicates complex conjugate transpose.

The *pdf* for the vector $\tilde{\mathbf{X}}$ is

$$f_{\tilde{X}}(\tilde{\mathbf{x}}) = \frac{\exp[-(\tilde{\mathbf{x}} - E[\mathbf{x}])^\dagger \tilde{\mathbf{C}}^{-1} (\tilde{\mathbf{x}} - E[\mathbf{x}])]}{\pi^N |\tilde{\mathbf{C}}|} \tag{3.41}$$

with the following three conditions holding true

$$E[(\mathbf{X}_{I_i} - E[\mathbf{X}_{I_i}])(\mathbf{X}_{Q_i} - E[\mathbf{X}_{Q_i}])^\dagger] = \mathbf{0} \tag{3.42}$$

$$E[(\mathbf{X}_{I_i} - E[\mathbf{X}_{I_i}])(\mathbf{X}_{I_k} - E[\mathbf{X}_{I_k}])^\dagger] = \tag{3.43}$$
$$E[(\mathbf{X}_{Q_i} - E[\mathbf{X}_{Q_i}])(\mathbf{X}_{Q_k} - E[\mathbf{X}_{Q_k}])^\dagger] \quad ;all \ i, k$$

$$E[(\mathbf{X}_{I_i} - E[\mathbf{X}_{I_i}])(\mathbf{X}_{Q_k} - E[\mathbf{X}_{Q_k}])^\dagger] =$$
$$-E[(\mathbf{X}_{Q_i} - E[\mathbf{X}_{Q_i}])(\mathbf{X}_{I_k} - E[\mathbf{X}_{I_k}])^\dagger] \quad ;all \ \ i \neq k \tag{3.44}$$

3.3. Rayleigh Random Variables

Let X_I and X_Q be zero mean independent Gaussian random variables with zero mean and variance σ^2. Define two new random variables R and Φ as

$$X_I = R\cos\Phi$$
$$X_Q = R\sin\Phi \tag{3.45}$$

The joint *pdf* of the two random variables $X_I;X_Q$ is

$$f_{X_I X_Q}(x_I, x_Q) = \frac{1}{2\pi\sigma^2}\exp\left(-\frac{x_I^2 + x_Q^2}{2\sigma^2}\right) = \frac{1}{2\pi\sigma^2}\exp\left(-\frac{(r\cos\varphi)^2 + (r\sin\varphi)^2}{2\sigma^2}\right) \tag{3.46}$$

The joint *pdf* for the two random variables $R;\Phi$ is given by

$$f_{R\Phi}(r, \varphi) = f_{X_I X_Q}(x_I, x_Q)|\mathcal{J} \tag{3.47}$$

where $[\mathcal{J}]$ is a matrix of derivatives defined by

$$[\mathcal{J}] = \begin{bmatrix} \frac{\partial x_I}{\partial r} & \frac{\partial x_I}{\partial \varphi} \\ \frac{\partial x_Q}{\partial r} & \frac{\partial x_Q}{\partial \varphi} \end{bmatrix} = \begin{bmatrix} \cos\varphi & -r\sin\varphi \\ \sin\varphi & r\cos\varphi \end{bmatrix} \tag{3.48}$$

The determinant of the matrix of derivatives is called the Jacobian, and in this case it is equal to

$$|\mathcal{J}| = r \tag{3.49}$$

Substituting Eqs. (3.46) and (3.49) into Eq. (3.47) and collecting terms yield

$$f_{R\Phi}(r, \varphi) = \frac{r}{2\pi\sigma^2}\exp\left(-\frac{(r\cos\varphi)^2 + (r\sin\varphi)^2}{2\sigma^2}\right) = \frac{r}{2\pi\sigma^2}\exp\left(-\frac{r^2}{2\sigma^2}\right) \tag{3.50}$$

The *pdf* for R alone is obtained by integrating Eq. (3.50) over φ

$$f_R(r) = \int_0^{2\pi} f_{R\Phi}(r, \varphi)d\varphi = \frac{r}{\sigma^2}\exp\left(-\frac{r^2}{2\sigma^2}\right) \frac{1}{2\pi}\int_0^{2\pi} d\varphi \qquad (3.51)$$

where the integral inside Eq. (3.51) is equal to 2π; thus,

$$f_R(r) = \frac{r}{\sigma^2}\exp\left(-\frac{r^2}{2\sigma^2}\right) \quad ; r \geq 0 \qquad (3.52)$$

The *pdf* described in Eq. (3.52) is referred to as a Rayleigh probability density function.

The density function for the random variable Φ is obtained from

$$f_\Phi(\varphi) = \int_0^r f(r, \varphi)\ dr \qquad (3.53)$$

substituting Eq. (3.50) into Eq. (3.53) and performing integration by parts yields

$$f_\Phi(\varphi) = \frac{1}{2\pi} \quad ; \ 0 < \varphi < 2\pi \qquad (3.54)$$

which is a uniform probability density function.

3.4. The Chi-Square Random Variables

3.4.1. Central Chi-Square Random Variable with N Degrees of Freedom

Let the random variables $\{X_1, X_2, ..., X_N\}$ be zero mean, statistically independent Gaussian random variable with unity variance. The variable

$$\chi_N^2 = \sum_{i=1}^N X_i^2 \qquad (3.55)$$

is called a central chi-square random variable with N degrees of freedom. The chi-square *pdf* is

$$f_{\chi_N^2}(x) = \begin{cases} \dfrac{x^{(N-2)/2}\ e^{(-x/2)}}{2^{N/2}\ \Gamma(N/2)} & x \geq 0 \\[2mm] 0 & x < 0 \end{cases} \qquad (3.56)$$

where the Gamma function is define as

$$\Gamma(n) = \int_0^\infty \lambda^{n-1} e^{-\lambda} \, d\lambda \; ; \; n > 0 \tag{3.57}$$

with the following recursion

$$\Gamma(n+1) = n\Gamma(n) \tag{3.58}$$

and

$$\Gamma(n+1) = n! \quad ; \; n = 0, 1, 2, ..., \text{ and } 0! = 1 \tag{3.59}$$

The mean and variance for the central chi-square are, respectively given by

$$E[\chi_N^2] = N \tag{3.60}$$

$$\sigma_{\chi_N^2} = 2N \tag{3.61}$$

Hence, the degrees of freedom N is the ratio of twice the squared mean to the variance

$$N = (2E^2[\chi_N^2])/\sigma_{\chi_N^2} \tag{3.62}$$

3.4.2. Noncentral Chi-Square Random Variable with N Degrees of Freedom

In the general case, the chi-square random variable requires that the Gaussian random variables $\{X_1, X_2, ..., X_N\}$ do not have zero means. Define a multivariate random variable **Y** such that

$$Y_i = X_i + \mu_{X_i} \; ; i = 1, 2, ..., N \tag{3.63}$$

Consider the random variable

$$\chi_N'^2 = \sum_{i=1}^N Y_i^2 = \sum_{i=1}^N (X_i + \mu_{X_i})^2 \tag{3.64}$$

the variable $\chi_N'^2$ is called the noncentral chi-square random variable with N degrees of freedom and with a noncentral parameter λ, where

$$\lambda = \sum_{i=1}^{N} \mu_{X_i}^2 = \sum_{i=1}^{N} E^2[Y_i] \qquad (3.65)$$

The noncentral chi-square *pdf* is

$$f_{\chi_N'^2}(x) = \begin{cases} \left(\dfrac{1}{2}\right)\left(\dfrac{x}{\lambda}\right)^{(N-2)/4} e^{[-(x+\lambda)/2]}I_{(N-2)/2}(\sqrt{\lambda x}) & x \geq 0 \\[4mm] 0 & x < 0 \end{cases} \qquad (3.66)$$

where I is the modified Bessel function (or occasionally called the hyperbolic Bessel function) of the first kind; and the subscripts is referred to as its order.

3.5. Random Processes

A random variable X is by definition a mapping of all possible outcomes of a random experiment to numbers. When the random variable becomes a function of both the outcomes of the experiment time, it is called a random process and is denoted by $X(t)$. Thus, one can view a random process as an ensemble of time-domain functions that are the outcome of a certain random experiment, as compared with single real numbers in the case of a random variable.

Since the *cdf* and *pdf* of a random process are time dependent, we will denote them as $F_X(x;t)$ and $f_X(x;t)$, respectively. The *nth* moment for the random process $X(t)$ is

$$E[X^n(t)] = \int_{-\infty}^{\infty} x^n f_X(x;t)dx \qquad (3.67)$$

A random process $X(t)$ is referred to as stationary to order one if all its statistical properties do not change with time. Consequently, $E[X(t)] = \bar{X}$, where $\bar{X}$ is a constant. A random process $X(t)$ is called stationary to order two (or wide-sense stationary) if

$$f_X(x_1, x_2;t_1, t_2) = f_X(x_1, x_2;t_1 + \Delta t, t_2 + \Delta t) \qquad (3.68)$$

for all t_1, t_2 and Δt.

Define the statistical autocorrelation function for the random process $X(t)$ as

$$\Re_X(t_1, t_2) = E[X(t_1)X(t_2)] \qquad (3.69)$$

The correlation $E[X(t_1)X(t_2)]$ is, in general, a function of (t_1, t_2). As a consequence of the wide-sense stationary definition, the autocorrelation function depends on the time difference $\tau = t_2 - t_1$, rather than on absolute time; and thus, for a wide-sense stationary process we have

$$E[X(t)] = \bar{X}$$
$$\Re_X(\tau) = E[X(t)X(t+\tau)]$$

(3.70)

If the time average and time correlation functions are equal to the statistical average and statistical correlation functions, the random process is referred to as an ergodic random process. The following is true for an ergodic process:

$$\lim_{T \to \infty} \frac{1}{T} \int_{-T/2}^{T/2} x(t)dt = E[X(t)] = \bar{X}$$

(3.71)

$$\lim_{T \to \infty} \frac{1}{T} \int_{-T/2}^{T/2} x^*(t)x(t+\tau)dt = \Re_X(\tau)$$

(3.72)

The covariance of two random processes $X(t)$ and $Y(t)$ is defined by

$$C_{XY}(t, t+\tau) = E[\{X(t) - E[X(t)]\}\{Y(t+\tau) - E[Y(t+\tau)]\}]$$

(3.73)

which can also be written as

$$C_{XY}(t, t+\tau) = \Re_{XY}(\tau) - \bar{X}\bar{Y}$$

(3.74)

3.6. Bandpass Gaussian Random Process

It is customary to define the bandpass Gaussian random process through its complex envelope as

$$\tilde{X}(t) = X_I(t) + jX_Q(t)$$

(3.75)

where both $X_I(t)$ and $X_Q(t)$ are lowpass Gaussian random processes with zero mean and variance σ^2. The *pdf* for a sample $\tilde{X}(t_0)$ of the complex envelope is the joint *pdf* for $X_I(t)$ and $X_Q(t)$. That is,

$$f_{\tilde{X}}(\tilde{x}(t_0)) = \frac{1}{2\pi\sigma^2}\exp\left[-\frac{x_I^2(t_0) + x_Q^2(t_0)}{2\sigma^2}\right] = \frac{1}{2\pi\sigma^2}\exp\left[-\frac{|\tilde{x}(t_0)|^2}{2\sigma^2}\right]$$

(3.76)

Now, if both lowpass processes do not have zero mean and instead have a mean defined by

$$\mu(t) = \mu_I(t)\cos(2\pi f_0 t) + j\mu_Q(t)\sin(2\pi f_0 t) \qquad (3.77)$$

the mean complex envelope is

$$\tilde{\mu}(t) = \mu_I(t) + j\mu_Q(t) \qquad (3.78)$$

It follows that Eq. (3.76) can be rewritten as

$$f_{\tilde{X}}(\tilde{x}(t_0)) = \frac{1}{2\pi\sigma^2}\exp\left[-\frac{[x_I(t_0) - \mu_I(t_0)]^2 + [x_Q(t_0) - \mu_Q(t_0)]^2}{2\sigma^2}\right] = \qquad (3.79)$$

$$\frac{1}{2\pi\sigma^2}\exp\left[\frac{|\tilde{x}(t_0) - \tilde{\mu}(t_0)|^2}{2\sigma^2}\right]$$

Consider a duration of the process than spans the interval $\{0, T_0\}$. Then this segment of the complex envelope of the random process can be represented using a complex random variable vector of at least $N = BT_0$ elements where B is the bandwidth of the process. Define

$$\tilde{X}_i = \tilde{X}\left(\frac{i}{B}\right) \quad ; i = 1, 2, \ldots, BT_0 \qquad (3.80)$$

$$\tilde{\mathbf{X}}^\dagger = \left[\tilde{X}_1\ \tilde{X}_2\ \ldots \tilde{X}_{BT_0}\right] \qquad (3.81)$$

By definition the covariance matrix $\tilde{\boldsymbol{C}}$ is

$$\tilde{\boldsymbol{C}} = E[(\tilde{\mathbf{X}} - \tilde{\mu})(\tilde{\mathbf{X}} - \tilde{\mu})^\dagger] = 2(\tilde{\boldsymbol{C}}_I + j\tilde{\boldsymbol{C}}_{IQ}) \qquad (3.82)$$

where

$$\tilde{\boldsymbol{C}}_I = E[(\tilde{\mathbf{X}}_I - \tilde{\mu}_I)(\tilde{\mathbf{X}}_I - \tilde{\mu}_I)^\dagger] \qquad (3.83)$$

$$\tilde{\boldsymbol{C}}_{IQ} = E[(\tilde{\mathbf{X}}_I - \tilde{\mu}_I)(\tilde{\mathbf{X}}_Q - \tilde{\mu}_Q)^\dagger] \qquad (3.84)$$

Therefore, the *pdf* for the segment $\{\tilde{X}(t)\ ; 0 < t < T_0\}$ is

$$f_{\tilde{X}}(\tilde{\mathbf{x}}) = \frac{\exp[-(\tilde{\mathbf{x}} - \tilde{\mu})^\dagger \tilde{\boldsymbol{C}}^{-1}(\tilde{\mathbf{x}} - \tilde{\mu})]}{\pi^N |\tilde{\boldsymbol{C}}|} \qquad (3.85)$$

3.6.1. The Envelope of a Bandpass Gaussian Process

Consider the *pdf* of a segment of the envelope of a bandpass Gaussian random process. This process can expressed as

$$X(t) = X_I(t)\cos(2\pi f_0 t) - X_Q(t)\sin(2\pi f_0 t) \qquad (3.86)$$

where $X_I(t)$ and $X_Q(t)$ are zero mean independent lowpass Gaussian processes. The envelope and phase are respectively denoted by $R(t)$ and $\Phi(t)$, where

$$R(t) = \sqrt{X_I(t)^2 + X_Q(t)^2} \qquad (3.87)$$

and

$$\Phi(t) = \left[\tan\left(\frac{X_Q(t)}{X_I(t)}\right)\right]^{-1} \qquad (3.88)$$

where

$$\begin{aligned} X_I(t) &= R(t)\cos(\Phi(t)) \\ X_Q(t) &= R(t)\sin(\Phi(t)) \end{aligned} \qquad (3.89)$$

The two processes $R(t)$ and $\Phi(t)$ are also independent, and their respective *pdfs* were derived in Section 3.3 and were given in Eqs. (3.52) and (3.54), respectively.

Problems

3.1. Suppose you want to determine an unknown DC voltage v_{dc} in the presence of additive white Gaussian noise $n(t)$ of zero mean and variance σ_n^2. The measured signal is $x(t) = v_{dc} + n(t)$. An estimate of v_{dc} is computed by making three independent measurements of $x(t)$ and computing the arithmetic mean, $v_{dc} \approx (x_1 + x_2 + x_3)/3$. (a) Find the mean and variance of the random variable v_{dc}. (b) Does the estimate of v_{dc} get better by using ten measurements instead of three? Why?

3.2. Assume the X and Y miss distances of darts thrown at a bulls-eye dart board are Gaussian with zero mean and variance σ^2. (a) Determine the probability that a dart will fall between 0.8σ and 1.2σ. (b) Determine the radius of a circle about the bull's-eye that contains 80% of the darts thrown. (c) Consider a square with side s in the first quadrant of the board. Determine s so that the probability that a dart will fall within the square is 0.07.

3.3. (a) A random voltage $v(t)$ has an exponential distribution function $f_V(v) = a\exp(-av)$, where $(a>0);(0 \le v < \infty)$. The expected value $E[V] = 0.5$. Determine $Pr\{V > 0.5\}$. Consider the network shown in figure below, where $x(t)$ is a random voltage with zero mean and autocorrelation function $\mathfrak{R}_x(\tau) = 1 + \exp(-a|t|)$. Find the power spectrum $S_x(\omega)$. What is the transfer function? Find the power spectrum $S_v(\omega)$.

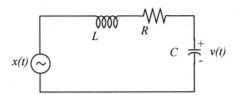

3.4. Let $\bar{S}_X(\omega)$ be the PSD function for the stationary random process $X(t)$. Compute an expression for the PSD function of

$$Y(t) = X(t) - 2X(t-T).$$

3.5. Let X be a random variable with

$$f_X(x) = \begin{cases} \dfrac{1}{\sigma} t^3 e^{-t} & t \geq 0 \\ 0 & elsewhere \end{cases}$$

(a) Determine the characteristic function $C_X(\omega)$. (b) Using $C_X(\omega)$, validate that $f_X(x)$ is a proper *pdf*. (c) Use $C_X(\omega)$ to determine the first two moments of X. (d) Calculate the variance of X.

3.6. Let the random variable Z be written in terms of two other random variables X and Y as follows: $Z = X + 3Y$. Find the mean and variance for the new random variable in terms of the other two.

3.7. Suppose you have the following sequences of statistically independent Gaussian random variables with zero means and variances σ^2. if

$$X_1, X_2, ..., X_N \; ; \; X_i = A_i \cos\Theta_i \text{ and } Y_1, Y_2, ..., Y_N \; ; \; Y_i = A_i \sin\Theta_i.$$

Define $Z = \displaystyle\sum_{i=1}^{N} A_i^2$. Find an expression that Z exceeds a threshold value v_T.

3.8. Repeat the previous problem when two single delay line cancellers are cascaded to produce a double delay line canceller. Let $X(t)$ be a stationary random process, $E[X(t)] = 1$ and the autocorrelation $\Re_x(\tau) = 3 + \exp(-|\tau|)$. Define a new random variola Y as

$$Y = \int_0^z x(t)dt$$

Compute $E[Y(t)]$ and σ_Y^2.

3.9. Consider the single delay line canceller in the figure below. The input $x(t)$ is a wide sense stationary random process with variance σ_x^2 and mean μ_x and a covariance matrix Λ. Find the mean and variance and the autocorrelation function of the output $y(t)$.

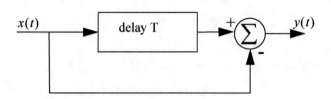

Chapter 4   *The Matched Filter*

4.1. The Matched Filter SNR

The topics of matched filtering and pulse compression (see Chapter 8) are central to almost all radar systems. In this chapter the focus is the matched filter. The unique characteristic of the matched filter is that it produces the maximum achievable instantaneous SNR at its output when a signal plus additive white noise is present at the input. Maximizing the SNR is key in all radar applications, as was described in Chapter 1 in the context of the radar equation and as will be discussed in Chapter 7 in the context of target detection.

Therefore, it is important to use a radar receiver which can be modeled as an LTI system that maximizes the signal's SNR at its output. For this purpose, the basic radar receiver of interest is often referred to as the matched filter receiver. The matched filter is an optimum filter in the sense of SNR because the SNR at its output is maximized at some delay t_0 that corresponds to the true target range R_0 (i.e., $t_0 = (2R_0)/c$). Figure 4.1 shows a simplified block diagram for the radar receiver of interest.

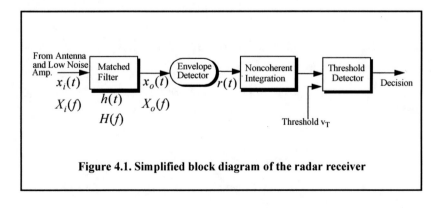

Figure 4.1. Simplified block diagram of the radar receiver

157

In order to derive the general expression for the transfer function and the impulse response of this optimum filter, adopt the following notation:

$h(t)$ is the optimum filter impulse response

$H(f)$ is the optimum filter transfer function

$x_i(t)$ is the input signal

$X_i(f)$ is the FT of the input signal

$x_o(t)$ is the output signal

$X_o(f)$ is the FT of the output signal

$n_i(t)$ is the input noise signal

$N_i(f)$ is the input noise PSD

$n_o(t)$ is the out noise signal

$N_o(f)$ is the output noise PSD

The optimum filter input signal can then be represented by

$$s_i(t) = x_i(t - t_0) + n_i(t) \tag{4.1}$$

where t_0 is an unknown time delay proportional to the target range. The optimum filter output signal is

$$s_o(t) = x_o(t - t_0) + n_o(t) \tag{4.2}$$

where

$$n_o(t) = n_i(t) \otimes h(t) \tag{4.3}$$

$$x_o(t) = x_i(t) \otimes h(t) \tag{4.4}$$

The operator ($\otimes$) indicates convolution. The FT of Eq. (4.4) is

$$X_o(f) = X_i(f)H(f) \tag{4.5}$$

Consequently the signal output at time t_0 can be calculated using the inverse FT, evaluated at t_0, as

$$x_o(t_0) = \int_{-\infty}^{\infty} X_i(f)H(f)e^{j2\pi f t_0} \, df \tag{4.6}$$

Additionally, the total noise power at the output of the filter is calculated using Parseval's theorem as

$$N_o = \int_{-\infty}^{\infty} N_i(f)|H(f)|^2 \, df \tag{4.7}$$

Since the output signal power at time t_0 is equal to the modulus square of Eq. (4.6), then the instantaneous SNR at time t_0 is

$$SNR(t_0) = \frac{\left| \int_{-\infty}^{\infty} X_i(f) H(f) e^{j2\pi f t_0} \, df \right|^2}{\int_{-\infty}^{\infty} N_i(f) |H(f)|^2 \, df} \qquad (4.8)$$

Remember Schawrz's inequality which has the form

$$\frac{\left| \int_{-\infty}^{\infty} X_1(f) X_2(f) \, df \right|^2}{\int_{-\infty}^{\infty} |X_1(f)|^2 \, df} \leq \int_{-\infty}^{\infty} |X_2(f)|^2 \, df \qquad (4.9)$$

The equal sign in Eq. (4.9) applies when $X_1(f) = K X_2^*(f)$ for some arbitrary constant K. Apply Schawrz's inequality to Eq. (4.8) with the following assumptions

$$X_1(f) = H(f) \sqrt{N_i(f)} \qquad (4.10)$$

$$X_2(f) = \frac{X_i(f) e^{j2\pi f t_0}}{\sqrt{N_i(f)}} \qquad (4.11)$$

It follows that the SNR is maximized when

$$H(f) = K \frac{X_i^*(f) e^{-j2\pi f t_0}}{N_i(f)} \qquad (4.12)$$

An alternative way of writing Eq. (4.12) is

$$X_i(f) H(f) e^{j2\pi f t_0} = K N_i(f) |X_i(f)|^2 \qquad (4.13)$$

The optimum filter impulse response is computed using inverse FT integral

$$h(t) = \int_{-\infty}^{\infty} K \frac{X_i^*(f) e^{-j2\pi f t_0}}{N_i(f)} e^{j2\pi f t} \, df \qquad (4.14)$$

A special case of great interest to radar systems is when the input noise is bandlimited white noise with PSD given by

$$N_i(f) = \eta_0/2 \qquad \text{(4.15)}$$

η_0 is a constant. The transfer function for this optimum filter is then given by

$$H(f) = X_i^*(f)e^{-j2\pi f t_0} \qquad \text{(4.16)}$$

where the constant K was set equal to $\eta_0/2$. It follows that

$$h(t) = \int_{-\infty}^{\infty} [X_i^*(f)e^{-j2\pi f t_0}] \, e^{j2\pi f t} \, df \qquad \text{(4.17)}$$

which can be written as

$$h(t) = x_i^*(t_0 - t) \qquad \text{(4.18)}$$

Observation of Eq. (4.18) indicates that the impulse response of the optimum filter is matched to the input signal, and thus, the term *matched filter* is used for this special case. Under these conditions, the maximum instantaneous SNR at the output of the matched filter is

$$SNR(t_0) = \frac{\left| \int_{-\infty}^{\infty} X_i(f)H(f)e^{j2\pi f t_0} \, df \right|^2}{\eta_0/2} \qquad \text{(4.19)}$$

and using Parseval's theorem the numerator in Eq. (4.19) is equal to the input signal energy, E_x; consequently one can write the output peak instantaneous SNR as

$$SNR(t_0) = \frac{2E_x}{\eta_0} \qquad \text{(4.20)}$$

Note that Eq. (4.20) is unitless since the unit for η_0 are in Watts per Hertz (or Joules). Finally, one can draw the conclusion that the peak instantaneous SNR depends only on the signal energy and input noise power, and is independent of the waveform utilized by the radar.

As indicated by Eq. (4.18) the impulse response $h(t)$ may not be causal if the value for t_0 is less than the signal duration. Thus, an additional time delay term $\tau_0 \geq T$ is added to ensure causality, where T is the signal duration. Thus, a realizable matched filter response is given by

$$h(t) = \begin{cases} x_i^*(\tau_0 + t_0 - t) & ;t > 0, \tau_0 \geq T \\ 0 & ;t < 0 \end{cases} \tag{4.21}$$

The transfer function for this casual filter is

$$H(f) = \int_{-\infty}^{\infty} x_i^*(\tau_0 + t_0 - t)e^{-j2\pi ft}dt = \int_{\infty}^{-\infty} x_i^*(t + \tau_0 + t_0)e^{j2\pi ft}dt \tag{4.22}$$

$$= X_i^*(f)e^{-j2\pi f(\tau_0 + t_0)}$$

Substituting the right-hand side of Eq. (4.22) into Eq. (4.6) yields

$$x_o(\tau_0) = \int_{-\infty}^{\infty} X_i(f)X_i^*(f)e^{-j2\pi f(\tau_0 + t_0)}e^{j2\pi ft_0}\,df = \int_{-\infty}^{\infty} |X_i(f)|^2 e^{-j2\pi f\tau_0}\,df \tag{4.23}$$

which has a maximum value when τ_0. This result leads to the following conclusion: The peak value of the matched filter output is obtained by sampling its output at times equal to the filter delay after the start of the input signal, and the minimum value for τ_0 is equal to the signal duration T.

Example:

Compute the maximum instantaneous SNR at the output of a linear filter whose impulse response is matched to the signal $x(t) = \exp(-t^2/2T)$.

Solution:

The signal energy is

$$E_x = \int_{-\infty}^{\infty} |x(t)|^2 dt = \int_{-\infty}^{\infty} e^{(-t^2)/T} dt = \sqrt{\pi T} \ Joules$$

It follows that the maximum instantaneous SNR is

$$SNR = \frac{\sqrt{\pi T}}{\eta_0/2}$$

where $\eta_0/2$ is the input noise power spectrum density.

4.1.1. The Replica

Again, consider a radar system that uses a finite duration energy signal $x(t)$, and assume that a matched filter receiver is utilized. From Eq. (4.1) the input signal can be written as,

$$s(t) = x(t - t_0) + n(t) \qquad (4.24)$$

The matched filter output $s_o(t)$ can be expressed by the convolution integral between the filter's impulse response and $s(t)$:

$$s_0(t) = \int_{-\infty}^{\infty} s(u)h(t-u)du \qquad (4.25)$$

Substituting Eq. (4.21) into Eq. (4.25) yields

$$s_o(t) = \int_{-\infty}^{\infty} s(u)x^*(t - \tau_0 - t_0 + u)du = \bar{R}_{sx}(t - T_0) \qquad (4.26)$$

where $T_0 = \tau_0 + t_0$ and $\bar{R}_{sx}(t - T_0)$ is a cross-correlation between $s(t)$ and $x(T_0 - t)$. Therefore, the matched filter output can be computed from the cross-correlation between the radar received signal and a delayed replica of the transmitted waveform. If the input signal is the same as the transmitted signal, the output of the matched filter would be the autocorrelation function of the received (or transmitted) signal. In practice, replicas of the transmitted waveforms are normally computed and stored in memory for use by the radar signal processor when needed.

4.2. Mean and Variance of the Matched Filter Output

Since the matched filter is an LTI filter, then when its input's statistics is Gaussian, its output statistics is also Gaussian, as discussed in Chapter 3. For this purpose, consider the following two hypotheses. Hypothesis H_0 is when the input to the matched filter consists of noise only. That is,

$$H_0 \Leftrightarrow s(t) = n_i(t) \qquad (4.27)$$

where $n_i(t)$ is zero mean Gaussian bandlimited white noise with PSD $\eta_0/2$. Hypothesis H_1 is when the input consists of signal plus noise. That is,

$$H_1 \Leftrightarrow s(t) = x_i(t) + n_i(t) \qquad (4.28)$$

Denote the conditional means and variances for both hypotheses by $E[s_o/H_0]$, the mean value of $s_0(\tau_0)$, when the signal is absent; $E[s_o/H_1]$ is the mean value of $s_0(\tau_0)$ when the signal is present; $Var[s_o/H_0]$ is the variance of $s_0(\tau_0)$ when the signal is absent; and $Var[s_o/H_1]$ is the variance of $s_0(\tau_0)$ when the signal is present. It follows that

$$E[s_o/H_0] = 0 \tag{4.29}$$

$$E[s_o/H_1] = \int_{-\infty}^{\infty} |x_i(t)|^2 dt = E_x \tag{4.30}$$

where E_x is the signal energy. Finally,

$$Var[s_o/H_0] = Var[s_o/H_1] = E_x \eta_0/2 \tag{4.31}$$

4.3. General Formula for the Output of the Matched Filter

Two cases are analyzed; the first is when a stationary target is present. The second case is concerned with a moving target whose velocity is constant. Assume the range to the target is

$$R(t) = R_0 - v(t - t_0) \tag{4.32}$$

where v is the target radial velocity (i.e. the target velocity component on the radar line of sight.) The initial detection range R_0 is given by

$$t_0 = \frac{2R_0}{c} \tag{4.33}$$

where c is the speed of light and t_0 is the round trip delay it takes a certain radar pulse to travel from the radar to the target at range R_0 and back.

The general expression for the radar bandpass signal is

$$s(t) = s_I(t)\cos 2\pi f_0 t - s_Q(t)\sin 2\pi f_0 t \tag{4.34}$$

which can be written using its pre-envelope (analytic signal) as

$$s(t) = Re\{\psi(t)\} = Re\{\tilde{s}(t)e^{j2\pi f_0 t}\} \tag{4.35}$$

where $Re\{\ \}$ indicates "the real part of." Again $\tilde{s}(t)$ is the complex envelope.

4.3.1. Stationary Target Case

In this case, the received radar return is given by

$$s_r(t) = s\left(t - \frac{2R_0}{c}\right) = s(t - t_0) = Re\{\tilde{s}(t - t_0)e^{j2\pi f_0(t - t_0)}\} \quad \text{(4.36)}$$

It follows that the received analytic and complex envelope signals are, respectively, given by

$$\psi_r(t) = \tilde{s}(t - t_0)e^{-j2\pi f_0 t_0}e^{j2\pi f_0 t} \quad \text{(4.37)}$$

$$\tilde{s}_r(t) = \tilde{s}(t - t_0)e^{-j2\pi f_0 t_0} \quad \text{(4.38)}$$

Observation of Eq. (4.38) clearly indicates that the received complex envelope is more than just a delayed version of the transmitted complex envelope. It actually contains an additional phase shift φ_0 which represents the phase corresponding to the two-way optical length for the target range. That is,

$$\varphi_0 = -2\pi f_0 t_0 = -2\pi f_0 2\frac{R_0}{c} = -\frac{2\pi}{\lambda}2R_0 \quad \text{(4.39)}$$

where λ is the radar wavelength and is equal to c/f_0. Since a very small change in range can produce significant change in this phase term, this phase is often treated as a random variable with uniform probability density function over the interval $\{0, 2\pi\}$. Furthermore, the radar signal processor will first attempt to remove (correct for) this phase term through a process known as phase unwrapping.

Substituting Eq. (4.38) into Eq. (4.25) provides the output of the matched filter. It is given by

$$s_o(t) = \int_{-\infty}^{\infty} \tilde{s}_r(u)h(t - u)du \quad \text{(4.40)}$$

where the impulse response $h(t)$ is in Eq. (4.18). It follows that

$$s_o(t) = \int_{-\infty}^{\infty} \tilde{s}(u - t_0)e^{-j2\pi f_0 t_0}\tilde{s}^*(t - t_0 + u)du \quad \text{(4.41)}$$

Make the following change of variables:

$$z = u - t_0 \Rightarrow dz = du \quad \text{(4.42)}$$

Therefore, the output of the matched filter when a stationary target is present is computed from Eq (4.41) as

$$s_o(t) = e^{-j2\pi f_0 t_0} \int_{-\infty}^{\infty} \tilde{s}(z)\tilde{s}^*(t-z)dz = e^{-j2\pi f_0 t_0}\bar{R}_s(t) \tag{4.43}$$

where $\bar{R}_s(t)$ is the autocorrelation function for the signal $\tilde{s}(t)$.

4.3.2. Moving Target Case

In this case, the received signal only not is delayed in time by t_0 but also has a Doppler frequency shift f_d corresponding to the target velocity, where

$$f_d = 2vf_0/c = 2v/\lambda \tag{4.44}$$

The pre-envelope of the received signal can be written as

$$\psi_r(t) = \psi\left(t - \frac{2R(t)}{c}\right) = \tilde{s}\left(t - \frac{2R(t)}{c}\right)e^{j2\pi f_0\left(t - \frac{2R(t)}{c}\right)} \tag{4.45}$$

Substituting Eq. (4.32) into Eq. (4.45) yields

$$\psi_r(t) = \tilde{s}\left(t - \frac{2R_0}{c} + \frac{2vt}{c} - \frac{2vt_0}{c}\right)e^{j2\pi f_0\left(t - \frac{2R_0}{c} + \frac{2vt}{c} - \frac{2vt_0}{c}\right)} \tag{4.46}$$

Collecting terms yields

$$\psi_r(t) = \tilde{s}\left(t\left(1 + \frac{2v}{c}\right) - t_0\left(1 + \frac{2v}{c}\right)\right)e^{j2\pi f_0\left(t - \frac{2R_0}{c} + \frac{2vt}{c} - \frac{2vt_0}{c}\right)} \tag{4.47}$$

Define the scaling factor γ as

$$\gamma = 1 + \frac{2v}{c} \tag{4.48}$$

then Eq. (4.47) can be written as

$$\psi_r(t) = \tilde{s}(\gamma(t - t_0))e^{j2\pi f_0\left(t - \frac{2R_0}{c} + \frac{2vt}{c} - \frac{2vt_0}{c}\right)} \tag{4.49}$$

Since $c \gg v$, the following approximation can be used

$$\tilde{s}(\gamma(t - t_0)) \approx \tilde{s}(t - t_0) \tag{4.50}$$

It follows that Eq. (4.49) can now be rewritten as

$$\psi_r(t) = \tilde{s}(t - t_0)e^{j2\pi f_0 t}e^{-j2\pi f_0 \frac{2R_0}{c}}e^{j2\pi f_0 \frac{2vt}{c}}e^{-j2\pi f_0 \frac{2vt_0}{c}} \tag{4.51}$$

Recognizing that $f_d = (2vf_0)/c$ and $t_0 = (2R_0)/c$, the received pre-envelope signal is

$$\psi_r(t) = \tilde{s}(t-t_0)e^{j2\pi f_0 t}e^{-j2\pi f_0 t_0}e^{j2\pi f_d t}e^{-j2\pi f_d t_0} = \tilde{s}(t-t_0)e^{j2\pi(f_0+f_d)(t-t_0)} \tag{4.52}$$

or

$$\psi_r(t) = \{\tilde{s}(t-t_0)e^{j2\pi f_d t}e^{-j2\pi(f_0+f_d)t_0}\}e^{j2\pi f_0 t} \tag{4.53}$$

Then by inspection the complex envelope of the received signal is

$$\tilde{s}_r(t) = \tilde{s}(t-t_0)e^{j2\pi f_d t}e^{-j2\pi(f_0+f_d)t_0} \tag{4.54}$$

Finally, it is concluded that the complex envelope of the received signal when the target is moving at a constant velocity v is a delayed (by t_0) version of the complex envelope signal of the stationary target case except that:

1. An additional phase shift term corresponding to the target's Doppler frequency is present, and

2. The phase shift term $(-2\pi f_d t_0)$ is present.

The output of the matched filter was derived in Eq. (4.25). Substituting Eq. (4.54) into Eq. (4.25) yields

$$s_o(t) = \int_{-\infty}^{\infty} \tilde{s}(u-t_0)e^{j2\pi f_d u}e^{-j2\pi(f_0+f_d)t_0}\tilde{s}^*(t-t_0+u) \, du \tag{4.55}$$

Applying the change of variables given in Eq. (4.42) and collecting terms provide

$$s_o(t) = e^{-j2\pi f_0 t_0}\int_{-\infty}^{\infty} \tilde{s}(z)\tilde{s}^*(t-z)e^{j2\pi f_d z}e^{j2\pi f_d t_0}e^{-j2\pi f_d t_0} \, dz \tag{4.56}$$

Observation of Eq. (4.56) shows that the output is a function of both t and f_d. Thus, it is more appropriate to rewrite the output of the matched filter as a two-dimensional function of both variables. That is,

$$s_o(t;f_d) = e^{-j2\pi f_0 t_0}\int_{-\infty}^{\infty} \tilde{s}(z)\tilde{s}^*(t-z)e^{j2\pi f_d z} \, dz \tag{4.57}$$

It is customary but not necessary to set $t_0 = 0$. Note that if the causal impulse response is used (i.e., Eq. (4.21)), the same analysis will hold true. However, in this case, the phase term is equal to $\exp(-j2\pi f_0 T_0)$, instead of $\exp(-j2\pi f_0 t_0)$ where $T_0 = \tau_0 + t_0$.

4.4. Waveform Resolution and Ambiguity

As indicated by Eq. (4.20), the radar sensitivity (in the case of white additive noise) depends only on the total energy of the received signal and is independent of the shape of the specific waveform. This leads to the following question: If the radar sensitivity is independent of the waveform, what is the best choice for the transmitted waveform? The answer depends on many factors; however, the most important consideration lies in the waveform's range and Doppler resolution characteristics, which can be determined from the output of the matched fitter.

As discussed in Chapter 1, range resolution implies separation between distinct targets in range. Alternatively, Doppler resolution implies separation between distinct targets in frequency. Thus, ambiguity and accuracy of this separation are closely associated terms.

4.4.1. Range Resolution

Consider radar returns from two stationary targets (zero Doppler) separated in range by distance ΔR. What is the smallest value of ΔR so that the returned signal is interpreted by the radar as two distinct targets? In order to answer this question, assume that the radar transmitted bandpass pulse is denoted by $x(t)$,

$$x(t) = r(t)\cos(2\pi f_0 t + \phi(t)) \tag{4.58}$$

where f_0 is the carrier frequency, $r(t)$ is the amplitude modulation, and $\phi(t)$ is the phase modulation. The signal $x(t)$ can then be expressed as the real part of the pre-envelope signal $\psi(t)$, where

$$\psi(t) = r(t)e^{j(2\pi f_0 t - \phi(t))} = \tilde{x}(t)e^{2\pi f_0 t} \tag{4.59}$$

and the complex envelope is

$$\tilde{x}(t) = r(t)e^{-j\phi(t)} \tag{4.60}$$

It follows that

$$x(t) = Re\{\psi(t)\} \tag{4.61}$$

The returns from two close targets are, respectively, given by

$$x_1(t) = \psi(t - \tau_0) \tag{4.62}$$

$$x_2(t) = \psi(t - \tau_0 - \tau) \tag{4.63}$$

where τ is the difference in delay between the two target returns. One can assume that the reference time is τ_0, and thus without any loss of generality, one may set $\tau_0 = 0$. It follows that the two targets are distinguishable by how large or small the delay τ can be.

In order to measure the difference in range between the two targets, consider the integral square error between $\psi(t)$ and $\psi(t-\tau)$. Denoting this error as ε_R^2, it follows that

$$\varepsilon_R^2 = \int_{-\infty}^{\infty} |\psi(t) - \psi(t-\tau)|^2 \ dt \tag{4.64}$$

which can be written as

$$\varepsilon_R^2 = \int_{-\infty}^{\infty} |\psi(t)|^2 \ dt + \int_{-\infty}^{\infty} |\psi(t-\tau)|^2 \ dt - \tag{4.65}$$

$$\int_{-\infty}^{\infty} \{(\psi(t)\psi^*(t-\tau) + \psi^*(t)\psi(t-\tau)) \ dt\}$$

Using Eq. (4.59) into Eq. (4.65) yields

$$\varepsilon_R^2 = 2\int_{-\infty}^{\infty} |\tilde{x}(t)|^2 \ dt - 2Re\left\{\int_{-\infty}^{\infty} \psi^*(t)\psi(t-\tau) \ dt\right\} = \tag{4.66}$$

$$2\int_{-\infty}^{\infty} |\tilde{x}(t)|^2 \ dt - 2Re\left\{e^{-j\omega_0\tau}\int_{-\infty}^{\infty} \tilde{x}^*(t)\tilde{x}(t-\tau) \ dt\right\}$$

This squared error is minimum when the second portion of Eq. (4.66) is positive and maximum. Note that the first term in the right-hand side of Eq. (4.66) represents the total signal energy, and is assumed to be constant. The second term is a varying function of τ with its fluctuation tied to the carrier frequency. The integral inside the right most side of this equation is defined as the range ambiguity function,

$$\chi_R(\tau) = \int_{-\infty}^{\infty} \tilde{x}^*(t)\tilde{x}(t-\tau) \ dt \tag{4.67}$$

This range ambiguity function is equivalent to the integral given in Eq. (4.43) with $t_0 = 0$. Comparison between Eq. (4.67) and Eq. (4.43) indicates that the output of the matched filter and the range ambiguity function have the same envelope (in this case the Doppler shift f_d is set to zero). This indicates that the matched filter, in addition to providing the maximum instantaneous SNR at its output, also preserves the signal range resolution properties. The value of $\chi_R(\tau)$ that minimizes the squared error in Eq. (4.66) occurs when $\tau = 0$.

Target resolvability in range is measured by the squared magnitude $|\chi_R(\tau)|^2$. It follows that if $|\chi_R(\tau)| = \chi_R(0)$ for some nonzero value of τ, then the two targets are indistinguishable. Alternatively, if $|\chi_R(\tau)| \neq \chi_R(0)$ for some non-zero value of τ, then the two targets may be distinguishable (resolvable). As a consequence, the most desirable shape for $\chi_R(\tau)$ is a very sharp peak (thumb tack shape) centered at $\tau = 0$ and falling very quickly away from the peak. The minimum range resolution corresponding to a time duration τ_e or effective bandwidth B_e is

$$\Delta R = \frac{c\tau_e}{2} = \frac{c}{2B_e} \tag{4.68}$$

The effective time duration and the effective bandwidth for any waveform were defined in Chapter 2 and are repeated here as Eq. (4.69) and Eq. (4.70), respectively

$$\tau_e = \left[\int_{-\infty}^{\infty} |\tilde{x}(t)|^2 dt\right]^2 \Big/ \int_{-\infty}^{\infty} |\tilde{x}(t)|^4 dt \tag{4.69}$$

$$B_e = \left[\int_{-\infty}^{\infty} |\tilde{X}(f)|^2 \, df\right]^2 \Big/ \left(\int_{-\infty}^{\infty} |\tilde{X}(f)|^4 \, df\right) \tag{4.70}$$

4.4.2. Doppler Resolution

The Doppler shift corresponding to the target radial velocity is

$$f_d = \frac{2v}{\lambda} = \frac{2vf_0}{c} \tag{4.71}$$

where v is the target radial velocity, λ is the wavelength, f_0 is the frequency, and c is the speed of light.

The FT of the pre-envelope is

$$\Psi(f) = \int_{-\infty}^{\infty} \psi(t) e^{-j2\pi ft} \, dt \qquad (4.72)$$

Due to the Doppler shift associated with the target, the received signal spectrum will be shifted by f_d. In other words, the received spectrum can be represented by $\Psi(f-f_d)$. In order to distinguish between the two targets located at the same range but having different velocities, one may use the integral square error. More precisely,

$$\varepsilon_f^2 = \int_{-\infty}^{\infty} |\Psi(f) - \Psi(f-f_d)|^2 \, df \qquad (4.73)$$

Using similar analysis as that which led to Eq. (4.66), one should maximize

$$Re\left\{ \int_{-\infty}^{\infty} \Psi^*(f)\Psi(f-f_d) \, df \right\} \qquad (4.74)$$

Taking the FT of the pre-envelope (analytic signal) defined in Eq. (4.59) yields

$$\Psi(f) = \tilde{X}(2\pi f - 2\pi f_0) \qquad (4.75)$$

Thus,

$$\int_{-\infty}^{\infty} \tilde{X}^*(2\pi f)\tilde{X}(2\pi f - 2\pi f_d) \, df = \qquad (4.76)$$

$$\int_{-\infty}^{\infty} \tilde{X}^*(2\pi f - 2\pi f_0)\tilde{X}(2\pi f - 2\pi f_0 - 2\pi f_d) \, df$$

The complex frequency correlation function is then defined as

$$\chi_f(f_d) = \int_{-\infty}^{\infty} \tilde{X}^*(2\pi f)\tilde{X}(2\pi f - 2\pi f_d) \, df = \int_{-\infty}^{\infty} |\tilde{x}(t)|^2 e^{j2\pi f_d t} \, dt \qquad (4.77)$$

The velocity resolution (Doppler resolution) is by definition

$$\Delta v = (c\Delta f_d)/(2f_0) \qquad (4.78)$$

where Δf_d is the minimum resolvable Doppler difference between the Doppler frequencies corresponding to two moving targets, i.e., $\Delta f_d = f_{d1} - f_{d2}$, where

f_{d1} and f_{d2} are the two individual Doppler frequencies for targets 1 and 2, respectively. The Doppler resolution Δf_d is equal to the inverse of the total effective duration of the waveform. Thus,

$$\Delta f_d = \left(\int_{-\infty}^{\infty} |\chi_f(f_d)|^2 df_d \right) / (\chi_f^2(0)) = \left(\int_{-\infty}^{\infty} |\tilde{x}(t)|^4 dt \right) / \left[\int_{-\infty}^{\infty} |\tilde{x}(t)|^2 dt \right]^2 = \frac{1}{\tau_e} \quad \text{(4.79)}$$

4.4.3. Combined Range and Doppler Resolution

In this general case, one needs to use a two-dimensional function in the pair of variables (τ, f_d). For this purpose, assume that the pre-envelope of the transmitted waveform is

$$\psi(t) = \tilde{x}(t) e^{j2\pi f_0 t} \quad \text{(4.80)}$$

Then the delayed and Doppler-shifted signal is (see Eq. (4.53))

$$\psi(t-\tau) = \tilde{x}(t-\tau) e^{j2\pi(f_0 - f_d)(t-\tau)} \quad \text{(4.81)}$$

Computing the integral square error between Eq. (4.80) and Eq. (4.81) yields

$$\varepsilon^2 = \int_{-\infty}^{\infty} |\psi(t) - \psi(t-\tau)|^2 dt \quad \text{(4.82a)}$$

$$\varepsilon^2 = 2 \int_{-\infty}^{\infty} |\psi(t)|^2 dt - 2 Re \left\{ \int_{-\infty}^{\infty} \psi^*(t) - \psi(t-\tau) dt \right\} \quad \text{(4.82b)}$$

which can be written as

$$\varepsilon^2 = 2 \int_{-\infty}^{\infty} |\tilde{x}(t)|^2 dt - 2 Re \left\{ e^{j2\pi(f_0 - f_d)\tau} \int_{-\infty}^{\infty} \tilde{x}(t)\tilde{x}^*(t-\tau) e^{j2\pi f_d t} dt \right\} \quad \text{(4.83)}$$

Again, in order to maximize this squared error for $\tau \neq 0$, one must minimize the last term of Eq. (4.83). Define the combined range and Doppler correlation function as

$$\chi(\tau, f_d) = \int_{-\infty}^{\infty} \tilde{x}(t)\tilde{x}^*(t-\tau) e^{j2\pi f_d t} dt \quad \text{(4.84)}$$

In order to achieve the most range and Doppler resolution, the modulus square of this function must be minimized at $\tau \neq 0$ and $f_d \neq 0$. Note that the output of the matched filter, except for a phase term, is identical to that given in Eq. (4.84). This means that the output of the filter exhibits maximum instantaneous SNR as well as the most achievable range and Doppler resolutions. The modulus square of Eq. (4.84) is often referred to as the ambiguity function:

$$|\chi(\tau, f_d)|^2 = \left| \int_{-\infty}^{\infty} \tilde{x}(t)\tilde{x}^*(t-\tau) e^{j2\pi f_d t} dt \right|^2 \tag{4.85}$$

The ambiguity function is often used by radar designers and analysts to determine the *goodness* of a given radar waveform, where this *goodness* is measured by its range and Doppler resolutions. Remember that since the matched filter is used, maximum SNR is guaranteed.

4.5. Range and Doppler Uncertainty

The formula derived in Eq. (4.84) represents the output of the matched filter when the signal at its input comprises target returns only and has no noise components, an assumption that cannot be true in practical situations. In general, the input at the matched filter contains both target and noise returns. The noise signal is assumed to be an additive random process that is uncorrelated with the target and has bandlimited white spectrum. Referring to Eq. (4.84), a peak at the output of the matched filter at (τ_1, f_{d1}) represents a target whose delay (range) corresponds to τ_1 and Doppler frequency equal to f_{d1}. Therefore, measuring targets' exact range and Doppler frequency is determined from measuring peak locations occurring in the two-dimensional space (τ, f_d). This last statement, however, is correct only if noise is not present at the input of the matched filter. When noise is present and because noise is random, it will generate ambiguity (uncertainty) about the exact location of the ambiguity function peaks in the (τ, f_d) space.

4.5.1. Range Uncertainty

Consider the case when the return signal complex envelope is (assuming stationary target)

$$\tilde{s}_r(t) = \tilde{x}_r(t) + \tilde{n}(t) \tag{4.86}$$

where $\tilde{x}_r(t)$ is the target return signal complex envelope and $\tilde{n}(t)$ is the noise signal complex envelope. The integral squared error between the total received signal (target plus noise) and the shifted (delayed) transmitted waveform is

$$\varepsilon^2 = \int_0^{T_{max}} \left| \tilde{x}(t-\tau) - \tilde{s}_r(t) \right|^2 dt \qquad (4.87)$$

where T_{max} corresponds to maximum range under consideration. Expanding this squared error yields

$$\varepsilon^2 = 2 \int_0^{T_{max}} \left| \tilde{x}(t) \right|^2 dt + 2 \int_0^{T_{max}} \left| \tilde{n}(t) \right|^2 dt - 2Re\left\{ \int_0^{T_{max}} \tilde{x}^*(t-\tau)\tilde{s}_r(t)dt \right\} \qquad (4.88)$$

which can be written as

$$\varepsilon^2 = E_x + E_n - 2Re\left\{ \int_0^{T_{max}} \tilde{x}^*(t-\tau)\tilde{x}_r(t)dt + \int_0^{T_{max}} \tilde{x}^*(t-\tau)\tilde{n}(t)dt \right\} \qquad (4.89)$$

This expression is minimum at some value τ that makes the integral term inside Eq. (4.88) maximum and positive. More precisely, the following correlation functions must be maximized

$$R_{x_r x}(\tau) = \int_0^{T_{max}} \tilde{x}^*(t-\tau)\tilde{x}_r(t)dt \qquad (4.90)$$

$$R_{nx}(\tau) = \int_0^{T_{max}} \tilde{x}^*(t-\tau)\tilde{n}(t)dt \qquad (4.91)$$

Therefore, Eq. (4.89) can be written as

$$\varepsilon^2 = E - 2Re\{R_{x_r x}(\tau) + R_{nx}(\tau)\} \qquad (4.92)$$

Expanding the quantity $\{R_{x_r x}(\tau)\}$ using Taylor series expansion about the point $\tau = t_0$, where $t_0 = 2R/c$, and R is the exact target range leads to

$$R_{x_r x}(\tau) = R_{x_r x}(t_0) + R'_{x_r x}(t_0)(\tau - t_0) + \frac{R''_{x_r x}(t_0)(\tau - t_0)^2}{2!} + \dots \qquad (4.93)$$

where R' and R'', respectively, indicate the first and second derivatives with respect to delay. Remember that since the real part of the correlation function is an even function, all its odd number derivatives are equal to zero. Now, approximate Eq. (4.93) by using the first three terms (terms 1 and 3 are, of course, equal to zero) to get

$$Re\{R_{x,x}(\tau)\} \approx R_{x,x}(t_0) + \frac{R''_{x,x}(t_0)(\tau - t_0)^2}{2} \tag{4.94}$$

There is some value τ_1 close to the exact target range, t_0, that will minimize the expression in Eq. (4.92). In order to find this minimum value, differentiate the quantity $Re\{R_{x,x}(\tau) + R_{nx}(\tau)\}$ with respect to τ and set the result equal to zero to find τ_1. More specifically,

$$Re\left\{\frac{d}{d\tau}R_{x,x}(\tau) + \frac{d}{d\tau}R_{nx}(\tau)\right\} = Re\{R'_{x,x}(\tau) + R'_{nx}(\tau)\} = 0 \tag{4.95}$$

The derivative of the $Re\{R_{x,x}(\tau)\}$ can be found from Eq. (4.94) as

$$Re\left\{\frac{d}{d\tau}R_{x,x}(\tau)\right\} = \frac{d}{d\tau}\left(R_{x,x}(t_0) + \frac{R''_{x,x}(t_0)(\tau-t_0)^2}{2!}\right) = R''_{x,x}(t_0)(\tau-t_0) \tag{4.96}$$

Substituting the result of Eq. (4.96) into Eq. (4.95) and collecting terms yield

$$(\tau_1 - t_0) = -\frac{Re\{R'_{nx}(\tau_1)\}}{R''_{x,x}(t_0)} \tag{4.97}$$

The value $(\tau_1 - t_0)$ represent the amount of target range error measurement. It is more meaningful, since noise is random, to compute this error in terms of the standard deviation of its rms value. Hence, the standard deviation for range measurement error is

$$\sigma_\tau = (\tau_1 - t_0)_{rms} = -\frac{Re\{R'_{nx}(\tau_1)\}_{rms}}{R''_{x,x}(t_0)} \tag{4.98}$$

By using the differentiation property of the Fourier transform and Parseval's theorem the denominator of Eq. (4.89) can be determined by

$$R''_{x,x}(t_0) = (2\pi)^2 \int_{-\infty}^{\infty} f^2 |X(f)|^2 df \tag{4.99}$$

Next, from relations developed in Chapter 2, one can write the FT of $R_{nx}(\tau)$ as

$$FT\{R_{nx}(\tau)\} = X^*(f)\frac{\eta_0}{2} \tag{4.100}$$

where $\eta_0/2$ is the noise power spectrum density value (white noise). From the Fourier transform properties, the FT of the derivative of $R_{nx}(\tau)$ is

$$FT\{R'_{nx}(\tau)\} = (j2\pi f)\left(X^*(f)\frac{\eta_0}{2}\right) = (j2\pi f)S_{nx}(f) \qquad \textbf{(4.101)}$$

The rms value for $R'_{nx}(\tau)$ is by definition

$$\{R'_{nx}(\tau)\}_{rms} = \sqrt{\lim_{T_{max}} \frac{1}{T_{max}} \int_0^{T_{max}} R'_{nx}(\tau)\ d\tau} \qquad \textbf{(4.102)}$$

which can be rewritten using Parseval's theorem as

$$\{R'_{nx}(\tau)\}_{rms} = \sqrt{\int_0^{T_{max}} |FT\{R'_{nx}(\tau)\}|^2\ df} \qquad \textbf{(4.103)}$$

substituting Eq. (4.101) into Eq. (4.103) yields

$$\{R'_{nx}(\tau)\}_{rms} = \sqrt{\frac{\eta_0}{2}(2\pi)^2 \int_0^{T_{max}} f^2\ |X(f)|^2\ df} \qquad \textbf{(4.104)}$$

Finally, the standard deviation for range measurement error can be written as

$$\sigma_\tau = \frac{\sqrt{\eta_0/2}}{\sqrt{(2\pi)^2 \int_{-\infty}^{\infty} f^2\ |X(f)|^2 df}} \qquad \textbf{(4.105)}$$

Define the bandwidth rms value, B_{rms}^2, as

$$B_{rms}^2 = \frac{(2\pi)^2 \int_{-\infty}^{\infty} f^2\ |X(f)|^2 df}{\int_{-\infty}^{\infty} |X(f)|^2 df} \qquad \textbf{(4.106)}$$

It follows that Eq. (4.105) can now be written as

$$\sigma_\tau = \frac{\sqrt{\eta_0/2}}{B_{rms}\sqrt{\displaystyle\int_{-\infty}^{\infty}|X(f)|^2\,df}} = \frac{\sqrt{\eta_0/2}}{B_{rms}\sqrt{E_x}} = \frac{1}{B_{rms}\sqrt{2E_x/\eta_0}} \qquad \textbf{(4.107)}$$

which leads to the conclusion that the uncertainty in range measurement is inversely proportional to the rms bandwidth and the square root of the ratio of signal energy to the noise power density (square root of the SNR).

4.5.2. Doppler (Velocity) Uncertainty

For this purpose, assume that the target range is completely known. In the next section the case where both target range and target Doppler are not known will be analyzed. Denote the signal transmitted by the radar as $x(t)$ and the received signal (target plus noise) as $x_r(t)$. The integral square difference between the two returns can be written as

$$\varepsilon^2 = \int_0^{f_{max}} |X(f-f_c) - X_r(f)|^2\,df \qquad \textbf{(4.108)}$$

where $X(f)$ is the FT of $x(t)$, $X_r(f)$ is the FT of $x_r(t)$, and f_{max} is the maximum anticipated target Doppler. Again expand Eq. (4.108) to get

$$\varepsilon^2 = \int_0^{f_{max}} |X(f)|^2\,df + \int_0^{f_{max}} |X_r(f)|^2\,df - 2Re\left\{\int_0^{f_{max}} |X^*(f-f_c)X_r(f)|^2\,df\right\} \quad \textbf{(4.109)}$$

Minimizing the error squared in Eq. (4.109) requires maximizing the value

$$Re\left\{\int_0^{f_{max}} |X^*(f-f_c)X_r(f)|^2\,df\right\}$$

Conducting similar analysis as that performed in the previous section, the duration rms, τ_{rms}^2, value can be defined as

$$\tau_{rms}^2 = \left((2\pi)^2\int_{-\infty}^{\infty} t^2\,|x(t)|^2\,dt\right)\Bigg/\left(\int_{-\infty}^{\infty}|x(t)|^2\,dt\right) \qquad \textbf{(4.110)}$$

The standard deviation in the Doppler measurement can be derived as

$$\sigma_{f_d} = \frac{1}{\tau_{rms}\sqrt{2E_x/\eta_0}} \qquad (4.111)$$

Comparison of Eq. (4.111) and Eq. (4.107) indicates that the error in estimating Doppler is inversely proportional to the signal duration, while the error in estimating range is inversely proportional to the signal bandwidth. Therefore, and as expected, larger bandwidths minimize the range measurement errors and longer integration periods minimize the Doppler measurement errors.

4.5.3. Range-Doppler Coupling

In the previous two sections, range estimate error and Doppler estimate error were derived by assuming that they are uncoupled estimates. In other words, range error was derived assuming stationary target, while Doppler error was derived assuming completely known target range. In this section a more general formula for the combined range and Doppler errors is derived.

The analytic signal for this case was derived in Section 4.3 and was given in Eq. (4.52) which is repeated here as Eq. (4.112) for easy reference:

$$\psi_r(t) = \tilde{s}(t - t_0)e^{j2\pi f_0 t}e^{-j2\pi f_0 t_0}e^{j2\pi f_d t}e^{-j2\pi f_d t_0} = \tilde{s}(t - t_0)e^{j2\pi (f_0 + f_d)(t - t_0)} \qquad (4.112)$$

One can assume with any loss of generality that $t_0 = 0$, thus, Eq. (4.112) can be expressed as

$$\psi_r(t) = \tilde{s}(t)e^{j2\pi(f_0 + f_d)t} = r(t)e^{j\varphi(t)}e^{j2\pi(f_0 + f_d)t} \qquad (4.113)$$

where the complex envelope signal, $\tilde{s}(t)$, can be expressed as

$$\tilde{s}(t) = r(t)e^{j\varphi(t)} \qquad (4.114)$$

Range Error Estimate

From the analysis performed in Section 4.5.1, the estimate for the range error is determined by maximizing the function

$$Re\{R_{ss}(\tau, f_d) + R_{ns}(\tau)\} \qquad (4.115)$$

It follows that for some fixed value f_{d1} there is a value τ_1 close to $t_0 = 0$ that will maximize Eq. (4.115); that is,

$$Re\{R'_{ss}(\tau_1, f_{d1}) + R'_{ns}(\tau_1)\} = 0 \qquad (4.116)$$

Again the Taylor series expansion of R_{ss} about $\tau = 0$ is

$$R_{ss}(\tau, f_d) = Re\left\{R_{ss}(0, f_{d1}) + R'_{ss}(0, f_{d1})(\tau) + \frac{R''_{ss}(0, f_{d1})\tau^2}{2!} + \dots\right\} \quad \textbf{(4.117)}$$

Thus,

$$Re\left\{\frac{d}{d\tau}R_{ss}(\tau, f_d)\right\} \approx Re\{R'_{ss}(0, f_{d1}) + R''_{ss}(0, f_{d1})\tau\} \quad \textbf{(4.118)}$$

Substituting Eq. (4.118) into Eq. (4.116) and solving for τ_1 yields

$$\tau_1 = -\frac{Re\{R'_{ns}(\tau_1) + R'_{ss}(0, f_{d1})\}}{Re\{R''_{ss}(0, f_{d1})\}} \quad \textbf{(4.119)}$$

The value of $R''_{ss}(0, f_{d1})$ is not much different from $R''_{ss}(0, 0)$; thus,

$$\tau_1 \approx -\frac{Re\{R'_{ns}(\tau_1) + R'_{ss}(0, f_{d1})\}}{R''_{ss}(0, 0)} \quad \textbf{(4.120)}$$

To evaluate the term $R'_{ss}(0, f_{d1})$, start with the definition of $R_{ss}(\tau, f_d)$,

$$R_{ss}(\tau, f_d) = \int_{-\infty}^{\infty} r(t - \tau)e^{-j\varphi(t - \tau)}r(t)e^{j(\varphi(t) + 2\pi f_d t)}\,dt \quad \textbf{(4.121)}$$

Compute the derivative of Eq. (4.121) with respect to τ

$$R'_{ss}(\tau, f_d) = -\int_{-\infty}^{\infty}\{r'(t - \tau)r(t) - j\varphi'(t - \tau)r(t - \tau)r(t)\} \times \quad \textbf{(4.122)}$$

$$e^{j[\varphi(t) - \varphi(t - \tau) + 2\pi f_d t]}\,dt$$

Evaluating Eq. (4.122) at $\tau = 0$ and $f_d = f_{d1}$ gives

$$R'_{ss}(0, f_{d1}) = -\int_{-\infty}^{\infty}\{r'(t)r(t) - j\varphi'(t)r^2(t)\} \times e^{j[2\pi f_{d1} t]}\,dt \quad \textbf{(4.123)}$$

The complex exponential term in Eq. (4.123) can be approximated using small angle approximation as

$$e^{j[2\pi f_{d1} t]} = \cos(2\pi f_{d1} t) + j\sin(2\pi f_{d1} t) \approx 1 + 2\pi f_{d1} t \quad \textbf{(4.124)}$$

Next substitute Eq. (4.124) into Eq. (4.123), collect terms, and compute its real part to get

$$Re\{R'_{ss}(0,f_{d1})\} = -\int_{-\infty}^{\infty} r'(t)r(t)dt - 2\pi f_{d1}\int_{-\infty}^{\infty} t\varphi'(t)r^2(t)dt \qquad (4.125)$$

The first integral is evaluated (using FT properties and Parseval's theorem) as

$$\int_{-\infty}^{\infty} r'(t)r(t)dt = (j2\pi)\int_{-\infty}^{\infty} f_d|R(f)|^2 df \qquad (4.126)$$

Remember that since the envelope function $r(t)$ is a real lowpass signal, its Fourier transform is an even function; thus, Eq. (4.126) is equal to zero. Using this result, Eq. (4.125) becomes

$$Re\{R'_{ss}(0,f_{d1})\} = -2\pi f_{d1}\int_{-\infty}^{\infty} t\varphi'(t)r^2(t)dt \qquad (4.127)$$

Substitute Eq. (4.127) into Eq. (4.120) to get

$$\tau_1 = -\frac{Re\{R'_{ns}(\tau_1)\} - 2\pi f_{d1}\int_{-\infty}^{\infty} t\varphi'(t)r^2(t)dt}{R''_{ss}(0,0)} \qquad (4.128)$$

Equation (4.128) provides a measure for the degree of coupling between range and Doppler estimates. Clearly, if $\varphi(t) = 0 \Rightarrow \varphi'(t) = 0$, then there is zero coupling between the two estimates. Define the range-Doppler coupling constant as

$$\rho_{\tau RDC} = \left(2\pi\int_{-\infty}^{\infty} t\varphi'(t)|\tilde{s}(t)|^2 dt\right) \Big/ \left(\int_{-\infty}^{\infty} |\tilde{s}(t)|^2 dt\right) \qquad (4.129)$$

Doppler Error Estimate

Applying similar analysis as that performed in the preceding section to the spectral cross correlation function yields an expression for the range-Doppler coupling term. It is given by

$$\rho_{f_d RDC} = \frac{2\pi \int\limits_{-\infty}^{\infty} f \ \Phi'(f) |\tilde{S}(f)|^2 df}{\int\limits_{-\infty}^{\infty} |\tilde{S}(f)|^2 df} \tag{4.130}$$

where $\Phi(f)$ is the FT of $\varphi(t)$.

It can be shown that Eq. (4.129) and Eq. (4.130) are equal (see Problem 4.15). Given this result, the subscripts τ and f_d in Eq. (4.129) and Eq. (4.130) are dropped and the range-Doppler term is simply referred to as ρ_{RDC}.

4.5.4. Range-Doppler Coupling in LFM Signals

Referring to Eq. (4.113) and Eq. (4.114), the phase for an LFM signal can be expressed as

$$\varphi(t) = \mu' t^2 \tag{4.131}$$

where $\mu' = (\pi B)/\tau_0$, B is the LFM bandwidth, and τ_0 is the pulsewidth. Substituting Eq. (4.131) into Eq. (4.129) yields

$$\rho_{RDC} = \frac{4\pi\mu' \int\limits_{-\infty}^{\infty} t^2 |\tilde{s}(t)|^2 dt}{\int\limits_{-\infty}^{\infty} |\tilde{s}(t)|^2 dt} = \frac{\mu'}{\pi} \tau_e^2 \tag{4.132}$$

where τ_e is the effective duration. Thus,

$$\sigma_\tau^2 = \frac{(\eta_0/2)}{B_e^2 2E_x} + \frac{f_{d1}^2 \rho_{RDC}^2}{B_e^4} \tag{4.133}$$

Similarly,

$$\sigma_{f_d}^2 = \frac{(\eta_0/2)}{\tau_e^2 2E_x} + \frac{t_1^2 \rho_{RDC}^2}{\tau_e^4} \tag{4.134}$$

where f_{d1} and t_1 are constants. Since estimates of range or Doppler when noise is present cannot be 100% exact, it is better to replace these constants with their equivalent mean-squared errors. That is, let

$$f_{d1}^2 = \sigma_{fd}^2 \qquad , \qquad t_1^2 = \sigma_\tau^2 \tag{4.135}$$

where σ_τ is as in Eq. (4.133) and σ_{fd} is in Eq. (4.134). Thus, Eq. (4.133) can be written as

$$\sigma_{\tau_{RDC}}^2 = \frac{(\eta_0/2)}{B_e^2 2E_x} + \frac{\rho_{RDC}^2}{B_e^4}\left(\frac{(\eta_0/2)}{\tau_e^2 2E_x} + \frac{\rho_{RDC}^2 \sigma_\tau^2}{\tau_e^4}\right) \tag{4.136}$$

which can be algebraically manipulated to get

$$\sigma_{\tau_{RDC}}^2 = \frac{(\eta_0/2)}{B_e^2 2E_x} \frac{1}{(1 - (\rho_{RDC}^2/B_e^2 \tau_e^2))} \tag{4.137}$$

Using similar analysis,

$$\sigma_{f_{dRDC}}^2 = \frac{(\eta_0/2)}{\tau_e^2 2E_x} \frac{1}{(1 - (\rho_{RDC}^2/B_e^2 \tau_e^2))} \tag{4.138}$$

These results lead to the conclusion that one can estimate target range and Doppler simultaneously only when the product of the rms bandwidth and rms duration is very large (i.e., very large time bandwidth products). This is the reason radars using LFM waveforms cannot estimate target Doppler accurately unless very large time bandwidth products are utilized. Often, the LFM waveforms are referred to as "Doppler insensitive" waveforms.

4.6. *Target Parameter Estimation*

Target parameters of interest to radar applications include, but are not limited to, target range (delay), amplitude, phase, Doppler, and angular location (azimuth and elevation). Target information (parameters) is typically embedded in the return signals amplitude and phase. Different classes waveforms are used by the radar signal and data processors to extract different target parameters more efficiently than others. Since radar echoes typically comprise signal plus additive noise, most if not all the target information is governed by the statistics of the input noise, whose statistical parameters most likely are not known but can be estimated. Thus, statistical estimates of the target parameters (amplitude, phase, delay, Doppler, etc.) are utilized instead of the actual corresponding measurements. The general form of the radar signal can be expressed in the following form

$$x(t) = Ar(t - t_0)\cos[2\pi(f_0 + f_d)(t - t_0) + \phi(t - t_0) + \phi_0] \tag{4.139}$$

where A is the signal amplitude, $r(t)$ is the envelope lowpass signal, ϕ_0 is some constant phase, f_0 is the carrier frequency, t_0 and f_d are the target delay and Doppler, respectively. The analysis in this section closely follows Melsa and Cohen[1].

4.6.1. What Is an Estimator?

In the case of radar systems it always safe to assume, due to the central limit theorem, that the input noise is always Gaussian with mainly unknown parameters. Furthermore, one can assume that this noise is bandlimited white noise. Consequently, the primary question that needs to be answered is as follows: Given that the probability density function of the observation is known (Gaussian in this case) and given a finite number of independent measurements, can one determine an estimate of a given parameter (such as range, Doppler, amplitude, or phase)?

Let $f_X(x;\theta)$ be the *pdf* of a random variable X with an unknown parameter θ. Define the values $\{x_1, x_2, ..., x_N\}$ as N observed independent values of the variable X. Define the function or estimator $\hat{\theta}(x_1, x_2, ..., x_N)$ as an estimate of the unknown parameter θ. The bias of estimation is defined as

$$E[\hat{\theta} - \theta] = b \tag{4.140}$$

where $E[\ ]$ represents the "expected value of." The estimator $\hat{\theta}$ is referred to as an unbiased estimator if and only if

$$E[\hat{\theta}] = \theta \tag{4.141}$$

One of the most popular and common measures of the quality or effectiveness of an estimator is the Mean Square Deviation (MSD) referred to symbolically as $\Delta^2(\theta)$. For an unbiased estimator

$$\Delta^2(\hat{\theta}) = \sigma_{\hat{\theta}}^2 \tag{4.142}$$

where $\sigma_{\hat{\theta}}^2$ is the estimator variance. It can be shown that the Cramer-Rao bound for this MSD is given by

$$\sigma^2(\hat{\theta}) \geq \sigma_{min}^2(\theta) = \frac{1}{N \int\limits_{-\infty}^{\infty} \left(\frac{\partial}{\partial\theta}\log\{f_X(x;\theta)\}\right)^2 f_X(x;\theta)\ dx} \tag{4.143}$$

The efficiency of this unbiased estimator is defined by

---

1. Melsa, J. L. Cohen, D. L., *Decision and Estimation Theory*, McGraw-Hill, New York, 1978.

$$\varepsilon(\hat{\theta}) = \frac{\sigma^2_{min}(\theta)}{\sigma^2(\hat{\theta})} \qquad (4.144)$$

when $\varepsilon(\hat{\theta}) = 1$ the unbiased estimator is called an efficient estimate.

Consider an essentially timelimited signal $x(t)$ with effective duration τ_e and assume a bandlimited white noise with PSD $\eta_0/2$. In this case, Eq. (4.144) is equivalent to

$$\sigma^2(\hat{\theta}_i) \geq 1 \bigg/ \left(\frac{2}{\eta_0} \int_0^{NT_r} \left(\frac{\partial}{\partial \theta_i} x(t) \right)^2 dt \right) \qquad (4.145)$$

where $\hat{\theta}_i$ is the estimate for the ith parameter of interest and T_r is the pulse repetition interval for the pulsed sequence. In the next two sections, estimates of the target amplitude and phase are derived. It must be noted that since these estimates represent independent random variables, they are referred to as uncoupled estimates; that is, the computation of one estimate does not depend on apriori knowledge of the other estimates.

4.6.2. Amplitude Estimation

The signal amplitude A in Eq. (4.139) is the parameter of interest, in this case. Taking the partial derivative of Eq. (4.139) with respect to A and squaring the result yields

$$\left(\frac{\partial}{\partial t_0} x(t) \right)^2 = (r(t-t_0) \cos[2\pi(f_0 + f_d)(t-t_0) + \phi(t-t_0) + \phi_0])^2 \qquad (4.146)$$

Thus,

$$\int_0^{NT_r} \left(\frac{\partial}{\partial A} x(t) \right)^2 dt = \int_0^{NT_r} (x(t))^2 \ dt = NE_x \qquad (4.147)$$

where E_x is the signal energy (from Parseval's theorem). Substituting Eq. (4.147) into Eq. (4.145) and collecting terms yield the variance for the amplitude estimate as

$$\sigma_A^2 \geq \frac{1}{\frac{2}{\eta_0} NE_x} = \frac{1}{N \ SNR} \qquad (4.148)$$

In this case Eq. (4.20) used in Eq. (4.148) and *SNR* is the signal to noise ratio of the signal at the output of the matched filter. This clearly indicates that the signal amplitude estimate is improved as the SNR is increased.

4.6.3. Phase Estimation

In this case, it is desired to compute the best estimate for the signal phase ϕ_0. Again taking the partial derivative of the signal in Eq. (4.139) with respect to ϕ_0 and squaring the result yield

$$\left(\frac{\partial}{\partial\phi_0}x(t)\right)^2 = (-r(t-t_0)\sin[2\pi(f_0+f_d)(t-t_0)+\phi(t-t_0)+\phi_0])^2 \quad \text{(4.149)}$$

It follows that

$$\int_0^{NT_r}\left(\frac{\partial}{\partial\phi_0}x(t)\right)^2 dt = \int_0^{NT_r}(x(t))^2 dt = NE_x \quad \text{(4.150)}$$

Thus, the variance of the phase estimate is

$$\sigma_{\phi_0}^2 \geq \frac{1}{\frac{2}{\eta_0}NE_x} = \frac{1}{N\ SNR} \quad \text{(4.151)}$$

Problems

4.1. Show that the SNR at the output of the matched filter can be written as

$$SNR = \frac{2}{\alpha\pi}(S_i(\alpha))^2$$

where $\alpha = (\pi BT)/2$, B is the bandwidth, T is the pulsewidth. Assume that the radar is using unmodulated rectangular pulse of width T and that there is a target detected at range R. The value S_i is the signal power at the input of the matched filter.

4.2. Compute the frequency response for the filter matched to the signal
(a) $x(t) = \exp\left(\frac{-t^2}{2T}\right)$;
(b) $x(t) = u(t)\exp(-\alpha t)$ where α is a positive constant.

4.3. Repeat the example in Section 4.1 using $x(t) = u(t)\exp(-\alpha t)$.

4.4. Prove the properties of the radar ambiguity function.

4.5. A radar system uses LFM waveforms. The received signal is of the form $s_r(t) = As(t-\tau)+n(t)$, where τ is a time delay that depends on range, $s(t) = Rect(t/\tau')\cos(2\pi f_0 t - \phi(t))$, and $\phi(t) = -\pi Bt^2/\tau'$. Assume that the radar bandwidth is $B = 5MHz$, and the pulse width is $\tau' = 5\mu s$. (a) Give the quadrature components of the matched filter response that is matched to $s(t)$. (b) Write an expression for the output of the matched filter. (c) Compute the increase in SNR produced by the matched filter.

4.6. (a) Write an expression for the ambiguity function of an LFM waveform, where $\tau' = 6.4\mu s$ and the compression ratio is 32. (b) Give an expression for the matched filter impulse response.

4.7. (a) Write an expression for the ambiguity function of a LFM signal with bandwidth $B = 10MHz$, pulse width $\tau' = 1\mu s$, and wavelength $\lambda = 1cm$. (b) Plot the zero Doppler cut of the ambiguity function. (c) Assume a target moving toward the radar with radial velocity $v_r = 100m/s$. What is the Doppler shift associated with this target? (d) Plot the ambiguity function for the Doppler cut in part (c). (e) Assume that three pulses are transmitted with PRF $f_r = 2000Hz$. Repeat part (b).

4.8. (a) Give an expression for the ambiguity function for a pulse train consisting of 4 pulses, where the pulse width is $\tau' = 1\mu s$ and the pulse repetition interval is $T = 10\mu s$. Assume a wavelength of $\lambda = 1cm$. (b) Sketch the ambiguity function contour.

4.9. Hyperbolic frequency modulation (HFM) is better than LFM for high radial velocities. The HFM phase is

$$\phi_h(t) = \frac{\omega_0^2}{\mu_h}\ln\left(1 + \frac{\mu_h\alpha t}{\omega_0}\right)$$

where μ_h is an HFM coefficient and α is a constant. (a) Give an expression for the instantaneous frequency of an HFM pulse of duration τ'_h. (b) Show that HFM can be approximated by LFM. Express the LFM coefficient μ_l in terms of μ_h and in terms of B and τ'.

4.10. Consider a sonar system with range resolution $\Delta R = 4cm$. (a) A sinusoidal pulse at frequency $f_0 = 100KHz$ is transmitted. What is the pulse width, and what is the bandwidth? (b) By using an up-chirp LFM, centered at f_0, one can increase the pulse width for the same range resolution. If you want to increase the transmitted energy by a factor of 20, give an expression for the transmitted pulse. (c) Give an expression for the causal filter matched to the LFM pulse in part b.

4.11. A pulse train $y(t)$ is given by

$$y(t) = \sum_{n=0}^{2} w(n)x(t - n\tau')$$

where $x(t) = \exp(-t^2/2)$ is a single pulse of duration τ' and the weighting sequence is $\{w(n)\} = \{0.5, 1, 0.7\}$. Find and sketch the correlations R_{xx}, R_w, and R_y.

4.12. Repeat the previous problem for $x(t) = \exp(-t^2/2)\cos 2\pi f_0 t$.

4.13. Derive Eq. (4.29) and Eq. (4.30) when the input noise is not white.

4.14. Show that the zero Doppler cut for the ambiguity function of an arbitrary phase coded pulse with a pulse width τ_p is given by $Y(f) = |\sin c(f\tau_p)|^2$.

4.15. Show that

$$\int_{-\infty}^{\infty} tx^*(t)x'(t)\ dt = -\int_{-\infty}^{\infty} fX^*(f)X'(f)\ df$$

where $X(f)$, is the FT of $x(t)$ and $x'(t)$ is its derivative with respect to time. The function $X'(f)$ is the derivative of $X(f)$ with respect to frequency.

Chapter 5   The Ambiguity
               Function - Analog
               Waveforms

5.1. Introduction

The radar ambiguity function represents the output of the matched filter, and
it describes the interference caused by the range and/or Doppler shift of a target when compared to a reference target of equal RCS. The ambiguity function
evaluated at $(\tau, f_d) = (0, 0)$ is equal to the matched filter output that is perfectly matched to the signal reflected from the target of interest. In other
words, returns from the nominal target are located at the origin of the ambiguity function. Thus, the ambiguity function at nonzero τ and f_d represents
returns from some range and Doppler different from those for the nominal target.

The formula for the output of the matched filter was derived in Chapter 4, it
is, assuming a moving target with Doppler frequency f_d,

$$\chi(\tau, f_d) = \int_{-\infty}^{\infty} \tilde{x}(t)\tilde{x}^*(t-\tau)e^{j2\pi f_d t} dt \tag{5.1}$$

The modulus square of Eq. (5.1) is referred to as the ambiguity function. That
is,

$$|\chi(\tau, f_d)|^2 = \left| \int_{-\infty}^{\infty} \tilde{x}(t)\tilde{x}^*(t-\tau)e^{j2\pi f_d t} dt \right|^2 \tag{5.2}$$

The radar ambiguity function is normally used by radar designers as a means
of studying different waveforms. It can provide insight about how different
radar waveforms may be suitable for the various radar applications. It is also
used to determine the range and Doppler resolutions for a specific radar waveform. The three-dimensional (3-D) plot of the ambiguity function versus frequency and time delay is called the radar ambiguity diagram.

Denote E_x as the energy of the signal $\tilde{x}(t)$,

$$E_x = \int_{-\infty}^{\infty} |\tilde{x}(t)|^2 dt \tag{5.3}$$

The following list includes the properties for the radar ambiguity function:

1) The maximum value for the ambiguity function occurs at $(\tau, f_d) = (0, 0)$ and is equal to $4E_x^2$,

$$max\{|\chi(\tau;f_d)|^2\} = |\chi(0;0)|^2 = (2E_x)^2 \tag{5.4}$$

$$|\chi(\tau;f_d)|^2 \leq |\chi(0;0)|^2 \tag{5.5}$$

2) The ambiguity function is symmetric,

$$|\chi(\tau;f_d)|^2 = |\chi(-\tau;-f_d)|^2 \tag{5.6}$$

3) The total volume under the ambiguity function is constant,

$$\int\int |\chi(\tau;f_d)|^2 \, d\tau \, df_d = (2E_x)^2 \tag{5.7}$$

4) If the function $X(f)$ is the Fourier transform of the signal $x(t)$, then by using Parseval's theorem we get

$$|\chi(\tau;f_d)|^2 = \left| \int X^*(f)X(f-f_d)e^{-j2\pi f\tau} df \right|^2 \tag{5.8}$$

5) Suppose that $|\chi(\tau;f_d)|^2$ is the ambiguity function for the signal $\tilde{x}(t)$. Adding a quadratic phase modulation term to $\tilde{x}(t)$ yields

$$\tilde{x}_1(t) = \tilde{x}(t)e^{j\pi\mu t^2} \tag{5.9}$$

where μ is a constant. It follows that the ambiguity function for the signal $\tilde{x}_1(t)$ is given by

$$|\chi_1(\tau;f_d)|^2 = |\chi(\tau;(f_d + \mu\tau))|^2 \tag{5.10}$$

5.2. Examples of the Ambiguity Function

The ideal radar ambiguity function is represented by a spike of infinitesimally small width that peaks at the origin and is zero everywhere else, as illustrated in Fig. 5.1. An ideal ambiguity function provides perfect resolution between neighboring targets regardless of how close they may be to each other. Unfortunately, an ideal ambiguity function cannot physically exist because the ambiguity function must have finite peak value equal to $(2E_x)^2$ and a finite volume also equal to $(2E_x)^2$. Clearly, the ideal ambiguity function cannot meet those two requirements.

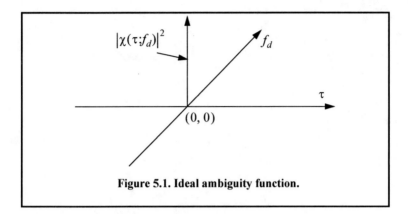

Figure 5.1. Ideal ambiguity function.

5.2.1. Single Pulse Ambiguity Function

The complex envelope of a single pulse is $\tilde{x}(t)$ defined by

$$\tilde{x}(t) = \frac{1}{\sqrt{\tau_0}} Rect\left(\frac{t}{\tau_0}\right)$$ (5.11)

From Eq. (5.1) we have

$$\chi(\tau;f_d) = \int_{-\infty}^{\infty} \tilde{x}(t)\tilde{x}^*(t-\tau)e^{j2\pi f_d t} dt$$ (5.12)

Substituting Eq. (5.11) into Eq. (5.12) and performing the integration yield

$$|\chi(\tau;f_d)|^2 = \left|\left(1 - \frac{|\tau|}{\tau_0}\right)\frac{\sin(\pi f_d(\tau_0 - |\tau|))}{\pi f_d(\tau_0 - |\tau|)}\right|^2 \qquad |\tau| \le \tau_0$$ (5.13)

Figures 5.2 a and b show 3-D and contour plots of single pulse ambiguity functions. This figure can be reproduced using the following MATLAB code

```
close all; clear all;
eps = 0.000001;
taup = 3;
[x] = single_pulse_ambg (taup);
taux = linspace(-taup,taup, size(x,1));
fdy = linspace(-5/taup+eps,5/taup-eps, size(x,1));
mesh(taux,fdy,x);
```

xlabel ('Delay in seconds');
ylabel ('Doppler in Hz');
zlabel ('Ambiguity function')
figure(2)
contour(taux,fdy,x);
xlabel ('Delay in seconds');
ylabel ('Doppler in Hz'); grid

The ambiguity function cut along the time-delay axis τ is obtained by setting $f_d = 0$. More precisely,

$$|\chi(\tau;0)| = \left(1 - \frac{|\tau|}{\tau_0}\right)^2 \qquad |\tau| \le \tau_0 \tag{5.14}$$

Note that the time autocorrelation function of the signal $\tilde{x}(t)$ is equal to $\chi(\tau;0)$. Similarly, the cut along the Doppler axis is

$$|\chi(0;f_d)|^2 = \left|\frac{\sin \pi \tau_0 f_d}{\pi \tau_0 f_d}\right|^2 \tag{5.15}$$

Figures 5.3 and 5.4, respectively, show the plots of the uncertainty function cuts defined by Eq. (5.14) and Eq. (5.15). Since the zero Doppler cut along the time-delay axis extends between $-\tau_0$ and τ_0, close targets will be unambiguous if they are at least τ_0 seconds apart.

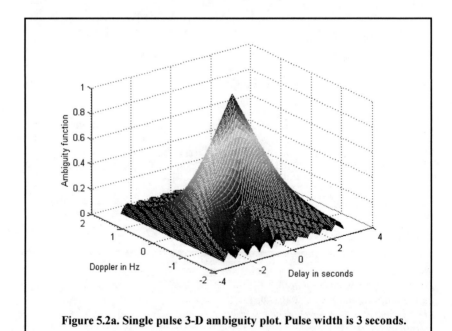

Figure 5.2a. Single pulse 3-D ambiguity plot. Pulse width is 3 seconds.

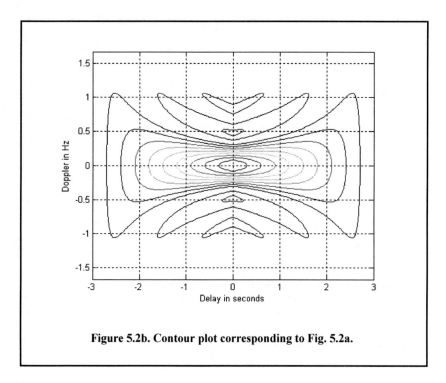

Figure 5.2b. Contour plot corresponding to Fig. 5.2a.

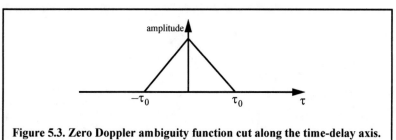

Figure 5.3. Zero Doppler ambiguity function cut along the time-delay axis.

The zero time cut along the Doppler frequency axis has a $(\sin x / x)^2$ shape. It extends from $-\infty$ to ∞. The first null occurs at $f_d = \pm 1/\tau_0$. Hence, it is possible to detect two targets that are shifted by $1/\tau_0$, without any ambiguity. Thus, a single pulse range and Doppler resolutions are limited by the pulse width τ_0. Fine range resolution requires that a very short pulse be used. Unfortunately, using very short pulses requires very large operating bandwidths and may limit the radar average transmitted power to impractical values.

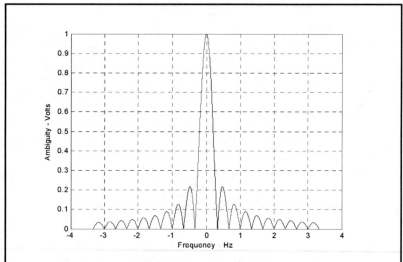

Figure 5.4. Ambiguity function of a single frequency pulse (zero delay). The pulse width is 3 seconds.

5.2.2. LFM Ambiguity Function

Consider the LFM complex envelope signal defined by

$$\tilde{x}(t) = \frac{1}{\sqrt{\tau_0}} Rect\left(\frac{t}{\tau_0}\right) e^{j\pi\mu t^2} \tag{5.16}$$

In order to compute the ambiguity function for the LFM complex envelope, we will first consider the case when $0 \leq \tau \leq \tau_0$. In this case the integration limits are from $-\tau_0/2$ to $(\tau_0/2) - \tau$. Substituting Eq. (5.16) into Eq. (5.1) yields

$$\chi(\tau;f_d) = \frac{1}{\tau_0} \int\limits_{-\infty}^{\infty} Rect\left(\frac{t}{\tau_0}\right) Rect\left(\frac{t-\tau}{\tau_0}\right) e^{j\pi\mu t^2} e^{-j\pi\mu(t-\tau)^2} e^{j2\pi f_d t} dt \tag{5.17}$$

It follows that

$$\chi(\tau;f_d) = \frac{e^{-j\pi\mu\tau^2}}{\tau_0} \int\limits_{\frac{-\tau_0}{2}}^{\frac{\tau_0}{2}-\tau} e^{j2\pi(\mu\tau + f_d)t} dt \tag{5.18}$$

Finishing the integration process in Eq. (5.18) yields

$$\chi(\tau;f_d) = e^{j\pi\tau f_d}\left(1 - \frac{\tau}{\tau_0}\right)\frac{\sin\left(\pi\tau_0(\mu\tau + f_d)\left(1 - \frac{\tau}{\tau_0}\right)\right)}{\pi\tau_0(\mu\tau + f_d)\left(1 - \frac{\tau}{\tau_0}\right)} \qquad 0 \le \tau \le \tau_0 \quad \textbf{(5.19)}$$

Similar analysis for the case when $-\tau_0 \le \tau \le 0$ can be carried out, where, in this case, the integration limits are from $(-\tau_0/2) - \tau$ to $\tau_0/2$. The same result can be obtained by using the symmetry property of the ambiguity function ($|\chi(-\tau, -f_d)| = |\chi(\tau, f_d)|$). It follows that an expression for $\chi(\tau;f_d)$ that is valid for any τ is given by

$$\chi(\tau;f_d) = e^{j\pi\tau f_d}\left(1 - \frac{|\tau|}{\tau_0}\right)\frac{\sin\left(\pi\tau_0(\mu\tau + f_d)\left(1 - \frac{|\tau|}{\tau_0}\right)\right)}{\pi\tau_0(\mu\tau + f_d)\left(1 - \frac{|\tau|}{\tau_0}\right)} \qquad |\tau| \le \tau_0 \quad \textbf{(5.20)}$$

and the LFM ambiguity function is

$$|\chi(\tau;f_d)|^2 = \left|\left(1 - \frac{|\tau|}{\tau_0}\right)\frac{\sin\left(\pi\tau_0(\mu\tau + f_d)\left(1 - \frac{|\tau|}{\tau_0}\right)\right)}{\pi\tau_0(\mu\tau + f_d)\left(1 - \frac{|\tau|}{\tau_0}\right)}\right|^2 \qquad |\tau| \le \tau_0 \quad \textbf{(5.21)}$$

Again the time autocorrelation function is equal to $\chi(\tau, 0)$. The reader can verify that the ambiguity function for a down-chirp LFM waveform is given by

$$|\chi(\tau;f_d)|^2 = \left|\left(1 - \frac{|\tau|}{\tau_0}\right)\frac{\sin\left(\pi\tau_0(\mu\tau - f_d)\left(1 - \frac{|\tau|}{\tau_0}\right)\right)}{\pi\tau_0(\mu\tau - f_d)\left(1 - \frac{|\tau|}{\tau_0}\right)}\right|^2 \qquad |\tau| \le \tau_0 \quad \textbf{(5.22)}$$

Incidentally, either Eq. (5.21) or (5.22) can be obtained from Eq. (5.13) by applying property 5 from Section 5.1. Figures 5.5 a and b show 3-D and contour plots for the LFM uncertainty and ambiguity functions for $\tau_0 = 1$ second and $B = 5Hz$ for a down-chirp pulse. This figure can be reproduced using the following MATLAB code.

```
% Use this program to reproduce Fig. 5.5 of text
close all;
clear all;
eps = 0.0001;
taup = 1.;
b = 5.;
up_down = -1.;
x = lfm_ambg(taup, b, up_down);
```

taux = linspace(-1.*taup,taup,size(x,1));
fdy = linspace(-1.5*b,1.5*b,size(x,1));
figure(1)
mesh(taux,fdy,sqrt(x))
xlabel ('Delay in seconds')
ylabel ('Doppler in Hz')
zlabel ('Ambiguity function')
axis tight
figure(2)
contour(taux,fdy,sqrt(x))
xlabel ('Delay in seconds')
ylabel ('Doppler in Hz')
grid

The up-chirp ambiguity function cut along the time delay axis τ is

$$|\chi(\tau;0)|^2 = \left| \left(1 - \frac{|\tau|}{\tau_0}\right) \frac{\sin\left(\pi\mu\tau\tau_0\left(1 - \frac{|\tau|}{\tau_0}\right)\right)}{\pi\mu\tau\tau_0\left(1 - \frac{|\tau|}{\tau_0}\right)} \right|^2 \qquad |\tau| \le \tau_0 \qquad (5.23)$$

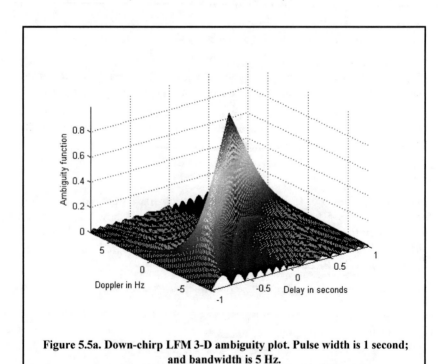

**Figure 5.5a. Down-chirp LFM 3-D ambiguity plot. Pulse width is 1 second;
and bandwidth is 5 Hz.**

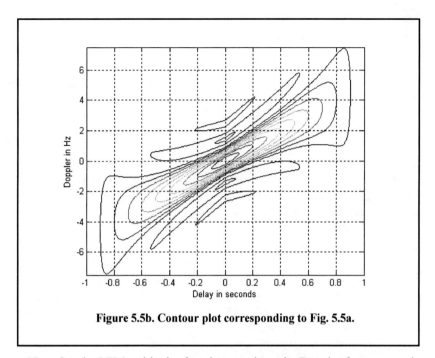

Figure 5.5b. Contour plot corresponding to Fig. 5.5a.

Note that the LFM ambiguity function cut along the Doppler frequency axis is similar to that of the single pulse. This should not be surprising since the pulse shape has not changed (only frequency modulation was added). However, the cut along the time-delay axis changes significantly. It is now much narrower compared to the unmodulated pulse cut. In this case, the first null occurs at

$$\tau_{n1} \approx 1/B \qquad (5.24)$$

Figure 5.6 shows a plot for a cut in the uncertainty function corresponding to Eq. (5.23). This figure can be reproduced using the following MATLAB code

```
close all; clear all;
taup = 1;
b =20.;
up_down = 1.;
taux = -1.5*taup:.01:1.5*taup;
mu = up_down * b / 2. / taup;
ii = 0.;
for tau = -1.5*taup:.01:1.5*taup
  ii = ii + 1;
  val1 = 1. - abs(tau) / taup;
  val2 = pi * taup * (1.0 - abs(tau) / taup);
```

```
val3 = (0 + mu * tau);
val = val2 * val3;
 x(ii) = abs( val1 * (sin(val+eps)/(val+eps)));
end
figure(1)
plot(taux,10*log10(x+0.001))
grid
xlabel ('Delay in seconds')
ylabel ('Ambiguity in dB')
axis tight
```

Equation (5.24) indicates that the effective pulse width (compressed pulse width) of the matched filter output is completely determined by the radar bandwidth. It follows that the LFM ambiguity function cut along the time-delay axis is narrower than that of the unmodulated pulse by a factor

$$\xi = \frac{\tau_0}{(1/B)} = \tau_0 B \qquad (5.25)$$

ξ is referred to as the compression ratio (also called time-bandwidth product and compression gain). All three names can be used interchangeably to mean the same thing. As indicated by Eq. (5.25) the compression ratio also increases as the radar bandwidth is increased.

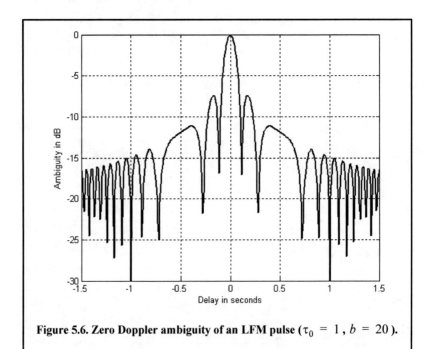

Figure 5.6. Zero Doppler ambiguity of an LFM pulse ($\tau_0 = 1$, $b = 20$).

Example:

Compute the range resolution before and after pulse compression corresponding to an LFM waveform with the following specifications: Bandwidth B = 1GHz and pulse width τ_0 = 10ms.

Solution:

The range resolution before pulse compression is

$$\Delta R_{uncomp} = \frac{c\tau_0}{2} = \frac{3 \times 10^8 \times 10 \times 10^{-3}}{2} = 1.5 \times 10^6 \ meters$$

Using Eq. (5.23) yields

$$\tau_{nl} = \frac{1}{1 \times 10^9} = 1 \ ns$$

$$\Delta R_{comp} = \frac{c\tau_{nl}}{2} = \frac{3 \times 10^8 \times 1 \times 10^{-9}}{2} = 15 \ cm$$

5.2.3. Coherent Pulse Train Ambiguity Function

Figure 5.7 shows a plot of a coherent pulse train. The pulse width is denoted as τ_0 and the PRI is T. The number of pulses in the train is N; hence, the train's length is $(N-1)T$ seconds. A normalized individual pulse $\tilde{x}(t)$ is defined by

$$\tilde{x}_1(t) = \frac{1}{\sqrt{\tau_0}} Rect\left(\frac{t}{\tau_0}\right) \tag{5.26}$$

When coherency is maintained between the consecutive pulses, then an expression for the normalized train is

$$\tilde{x}(t) = \frac{1}{\sqrt{N}} \sum_{i=0}^{N-1} \tilde{x}_1(t-iT) \tag{5.27}$$

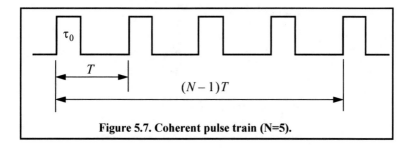

Figure 5.7. Coherent pulse train (N=5).

The output of the matched filter is

$$\chi(\tau;f_d) = \int_{-\infty}^{\infty} \tilde{x}(t)\tilde{x}^*(t-\tau)e^{j2\pi f_d t}\,dt \tag{5.28}$$

Substituting Eq. (5.27) into Eq. (5.28) and interchanging the summations and integration yield

$$\chi(\tau;f_d) = \frac{1}{N}\sum_{i=0}^{N-1}\sum_{j=0}^{N-1}\int_{-\infty}^{\infty}\tilde{x}_1(t-iT)\,\tilde{x}_1^*(t-jT-\tau)e^{j2\pi f_d t}\,dt \tag{5.29}$$

Making the change of variable $t_1 = t - iT$ yields

$$\chi(\tau;f_d) = \frac{1}{N}\sum_{i=0}^{N-1}e^{j2\pi f_d i T}\sum_{j=0}^{N-1}\int_{-\infty}^{\infty}\tilde{x}_1(t_1)\,\tilde{x}_1^*(t_1-[\tau-(i-j)T])e^{j2\pi f_d t_1}\,dt_1 \tag{5.30}$$

The integral inside Eq. (5.30) represents the output of the matched filter for a single pulse, and is denoted by χ_1. It follows that

$$\chi(\tau;f_d) = \frac{1}{N}\sum_{i=0}^{N-1}e^{j2\pi f_d i T}\sum_{j=0}^{N-1}\chi_1[\tau-(i-j)T;f_d] \tag{5.31}$$

When the relation $q = i-j$ is used, then the following relation is true:

$$\sum_{i=0}^{N}\sum_{m=0}^{N} = \left.\sum_{q=-(N-1)}^{0}\sum_{i=0}^{N-1-|q|}\right|_{for\ j=i-q} + \left.\sum_{q=1}^{N-1}\sum_{j=0}^{N-1-|q|}\right|_{for\ i=j+q} \tag{5.32}$$

Substituting Eq. (5.32) into Eq. (5.31) gives

$$\chi(\tau;f_d) = \frac{1}{N}\sum_{q=-(N-1)}^{0}\left\{\chi_1(\tau-qT;f_d)\sum_{i=0}^{N-1-|q|}e^{j2\pi f_d i T}\right\} \tag{5.33}$$

$$+\frac{1}{N}\sum_{q=1}^{N-1}\left\{e^{j2\pi f_d qT}\chi_1(\tau-qT;f_d)\sum_{j=0}^{N-1-|q|}e^{j2\pi f_d j T}\right\}$$

Setting $z = \exp(j2\pi f_d T)$, and using the relation

$$\sum_{j=0}^{N-1-|q|} z^j = \frac{1 - z^{N-|q|}}{1 - z} \tag{5.34}$$

yield

$$\sum_{i=0}^{N-1-|q|} e^{j2\pi f_d iT} = e^{[j\pi f_d(N-1-|q|T)]} \frac{\sin[\pi f_d(N-1-|q|)T]}{\sin(\pi f_d T)} \tag{5.35}$$

Using Eq. (5.35) in Eq. (5.31) yields two complementary sums for positive and negative q. Both sums can be combined as

$$\chi(\tau;f_d) = \frac{1}{N} \sum_{q=-(N-1)}^{N-1} \chi_1(\tau - qT;f_d) e^{[j\pi f_d(N-1+q)T]} \frac{\sin[\pi f_d(N-|q|)T]}{\sin(\pi f_d T)} \tag{5.36}$$

The second part of the right-hand side of Eq. (5.36) is the impact of the train on the ambiguity function; while the first part is primarily responsible for its shape details (according to the pulse type being used).

Finally, the ambiguity function associated with the coherent pulse train is computed as the modulus square of Eq. (5.36). For $\tau_0 < T/2$, the ambiguity function reduces to

$$|\chi(\tau;f_d)| = \frac{1}{N} \sum_{q=-(N-1)}^{N-1} |\chi_1(\tau - qT;f_d)| \left| \frac{\sin[\pi f_d(N-|q|)T]}{\sin(\pi f_d T)} \right| \quad ;|\tau| \leq NT \tag{5.37}$$

Within the region $|\tau| \leq \tau_0 \Rightarrow q = 0$, Eq. (5.37) can be written as

$$|\chi(\tau;f_d)| = |\chi_1(\tau;f_d)| \left| \frac{\sin[\pi f_d NT]}{N \sin(\pi f_d T)} \right| \quad ;|\tau| \leq \tau_0 \tag{5.38}$$

Thus, the ambiguity function for a coherent pulse train is the superposition of the individual pulse's ambiguity functions. The ambiguity function cuts along the time delay and Doppler axes are, respectively, given by

$$|\chi(\tau;0)|^2 = \left| \sum_{q=-(N-1)}^{N-1} \left(1 - \frac{|q|}{N}\right)\left(1 - \frac{|\tau - qT|}{\tau_0}\right) \right|^2 \quad ; |\tau - qT| < \tau_0 \tag{5.39}$$

$$|\chi(0;f_d)|^2 = \left| \frac{1}{N} \frac{\sin(\pi f_d \tau_0)}{\pi f_d \tau_0} \frac{\sin(\pi f_d NT)}{\sin(\pi f_d T)} \right|^2 \tag{5.40}$$

Figures 5.8a and 5.8b show the 3-D ambiguity plot and the corresponding contour plot for $N = 5$, $\tau_0 = 0.4$, and $T = 1$. This plot can be reproduced using the following MATLAB code.

```
clear all; close all;
taup = 0.4; pri = 1; n = 5;
x = train_ambg(taup, n, pri);
figure(1)
time = linspace(-(n-1)*pri-taup, n*pri-taup, size(x,2));
doppler = linspace(-1/taup, 1/taup, size(x,1));
surf(time, doppler, x); %mesh(time, doppler, x);
xlabel('Delay in seconds'); ylabel('Doppler in Hz');
zlabel('Ambiguity function'); axis tight;
figure(2)
contour(time, doppler, (x)); % surf(time, doppler, x);
xlabel('Delay in seconds'); ylabel('Doppler in Hz'); grid; axis tight;
```

Figures 5.8c and 5.8d, respectively shows sketches of the zero Doppler and zero delay cuts in the ambiguity function. The ambiguity function peaks along the frequency axis are located at multiple integers of the frequency $f = 1/T$. Alternatively, the peaks are at multiple integers of T along the delay axis. Width of the ambiguity function peaks along the delay axis is $2\tau_0$. The peak width along the Doppler axis is $1/(N-1)T$.

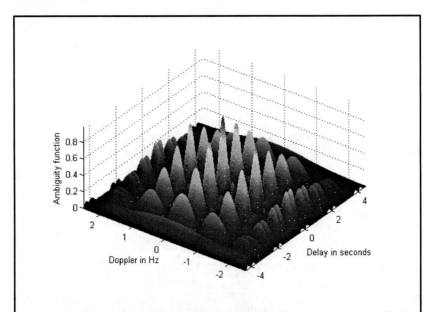

Figure 5.8a. Three-dimensional ambiguity plot for a five-pulse equal amplitude coherent train. Pulse width is 0.4 seconds; and PRI is 1 second, N=5.

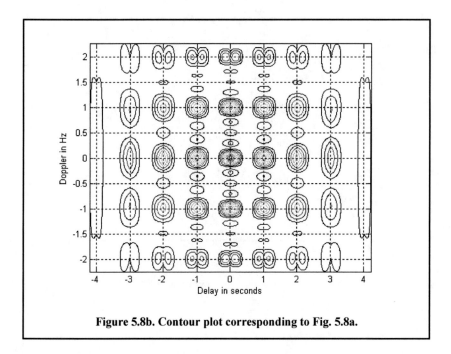

Figure 5.8b. Contour plot corresponding to Fig. 5.8a.

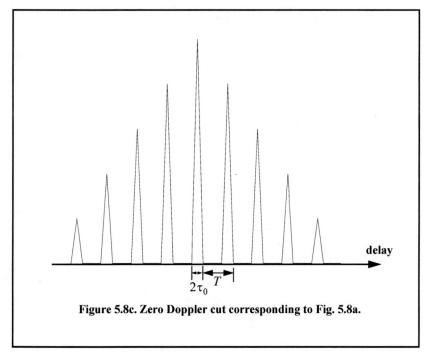

Figure 5.8c. Zero Doppler cut corresponding to Fig. 5.8a.

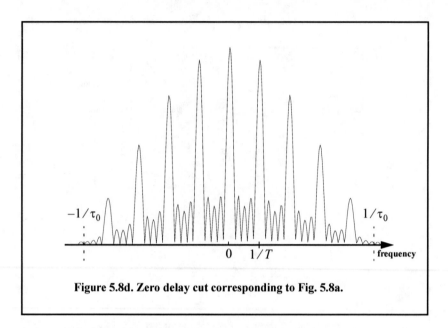

Figure 5.8d. Zero delay cut corresponding to Fig. 5.8a.

5.2.4. Pulse Train Ambiguity Function with LFM

In this case, the signal is as given in the previous section except for the LFM modulation within each pulse. This is illustrated in Fig. 5.9. Again let the pulse width be denoted by τ_0 and the PRI by T. The number of pulses in the train is N; hence, the train's length is $(N-1)T$ seconds. A normalized individual pulse $\tilde{x}_1(t)$ is defined by

$$\tilde{x}_1(t) = \frac{1}{\sqrt{\tau_0}} Rect\left(\frac{t}{\tau_0}\right) e^{j\pi\frac{B}{\tau_0}t^2} \tag{5.41}$$

where B is the LFM bandwidth.

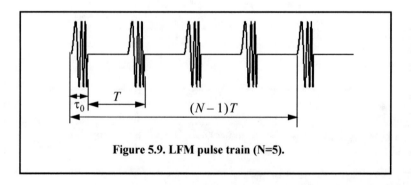

Figure 5.9. LFM pulse train (N=5).

The signal is now given by

$$\tilde{x}(t) = \frac{1}{\sqrt{N}} \sum_{i=0}^{N-1} \tilde{x}_1(t - iT) \tag{5.42}$$

Utilizing property 5 of Section 5.1 and Eq. (5.37) yields the following ambiguity function

$$|\chi(\tau;f_d)| = \sum_{q=-(N-1)}^{N-1} \left| \chi_1\left(\tau - qT;f_d + \frac{B}{\tau_0}\tau\right) \right| \left| \frac{\sin[\pi f_d(N - |q|)T]}{N\sin(\pi f_d T)} \right| \quad ; |\tau| \le NT \tag{5.43}$$

where χ_1 is the ambiguity function of the single pulse. Note that the shape of the ambiguity function is unchanged from the case of unmodulated train along the delay axis. This should be expected since only a phase modulation has been added which will impact the shape only along the frequency axis.

Figures 5.10 a and b show the ambiguity plot and its associated contour plot for the same example listed in the previous section except, in this case, LFM modulation is added and $N = 3$ pulses. This figure can be reproduced using the following MATLAB code.

```
% figure 5.10
clear all; close all;
taup = 0.4;
pri = 1;
n = 3;
bw = 10;
x = train_ambg_lfm(taup, n, pri, bw);
figure(1)
time = linspace(-(n-1)*pri-taup, n*pri-taup, size(x,2));
doppler = linspace(-bw,bw, size(x,1));
%mesh(time, doppler, x);
surf(time, doppler, x); shading interp;
xlabel('Delay in seconds');
ylabel('Doppler in Hz');
zlabel('Ambiguity function');
axis tight;
title('LFM pulse train, B\tau = 40, N = 3 pulses')
figure(2)
contour(time, doppler, (x));
%surf(time, doppler, x); shading interp; view(0,90);
xlabel('Delay in seconds');
ylabel('Doppler in Hz');
grid; axis tight;
title('LFM pulse train, B\tau = 40, N = 3 pulses')
```

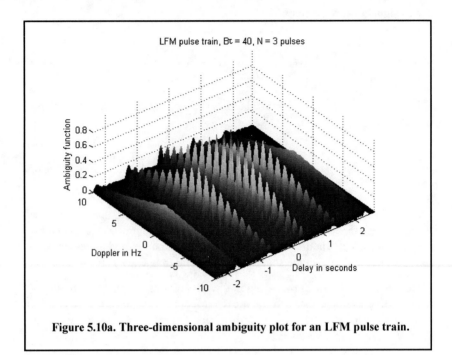

Figure 5.10a. Three-dimensional ambiguity plot for an LFM pulse train.

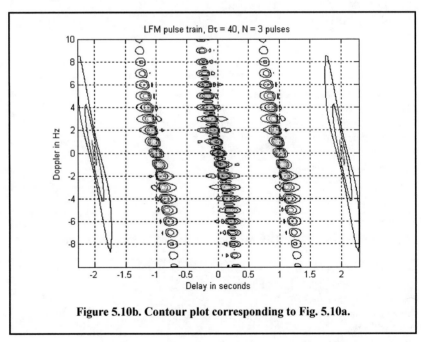

Figure 5.10b. Contour plot corresponding to Fig. 5.10a.

Understanding the difference between the ambiguity diagrams for a coherent pulse train and an LFM pulse train can be done with the help of Fig. 5.11a and Fig. 5.11b. In both figures a train of three pulses is used; in both cases the pulse width is $\tau_0 = 0.4 \, sec$ and the period is $T = 1 \, sec$. In the case, of LFM pulse train each pulse has LFM modulation with $B\tau_0 = 20$. Locations of the ambiguity peaks along the delay and Doppler axes are the same in both cases. This is true because peaks along the delay axis are T seconds apart and peaks along the Doppler axis are $1/T$ apart; in both cases T is unchanged. Additionally, the width of the ambiguity peaks along the Doppler axis are also the same in both cases, because this value depends only on the pulse train length which is the same in both cases (i.e., $(N-1)T$).

Width of the ambiguity peaks along the delay axis are significantly different, however. In the case of coherent pulse train, this width is approximately equal to twice the pulse width. Alternatively, this value is much smaller in the case of the LFM pulse train. The ratio between the two values is as given in Eq. (5.25). This clearly leads to the expected conclusion that the addition of LFM modulation significantly enhances the range resolution. Finally, the presence of the LFM modulation introduces a slope change in the ambiguity diagram; again a result that is also expected.

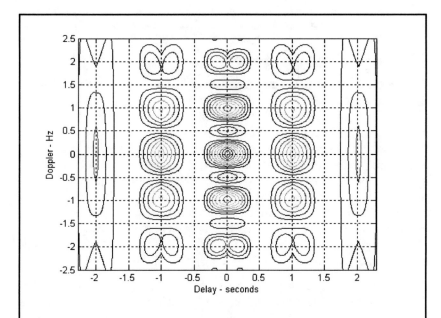

Figure 5.11a. Contour plot for the ambiguity function of a coherent pulse train.
$$N = 3; \tau_0 = 0.4; \quad T = 1$$

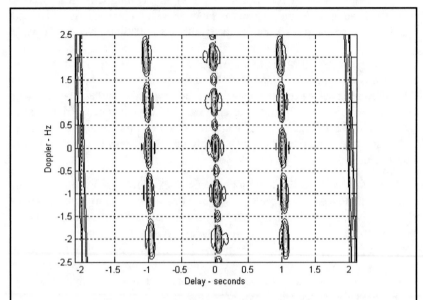

Figure 5.11b. Contour plot for the ambiguity function of a coherent pulse train.
$N = 3; \ B\tau_0 = 20; \ T = 1$

5.3. Stepped Frequency Waveforms

Stepped Frequency Waveforms (SFW) is a class of radar waveforms that are used in extremely wide bandwidth applications where very large time bandwidth product (or compression ratio as defined in Eq. (5.25)) is required. One may think of SFW as a special case of an extremely wide bandwidth LFM waveform. For this purpose, consider an LFM signal whose bandwidth is B_i and whose pulsewidth is T_i and refer to it as the primary LFM. Divide this long pulse into N subpulses each of width τ_0 to generate a sequence of pulses whose PRI is denoted by T. It follows that $T_i = (n-1)T$. One reason SFW is favored over an extremely wideband LFM is that it may be very difficult to maintain the LFM slope when the time bandwidth product is large. By using SFW, the same equivalent bandwidth can be achieved; however, phase errors are minimized since the LFM is chirped over a much shorter duration.

Define the beginning frequency for each subpulse as that value measured from the primary LFM at the leading edge of each subpulse, as illustrated in Fig. 5.12. That is

$$f_i = f_0 + i\Delta f; \ i = 0, N-1 \tag{5.44}$$

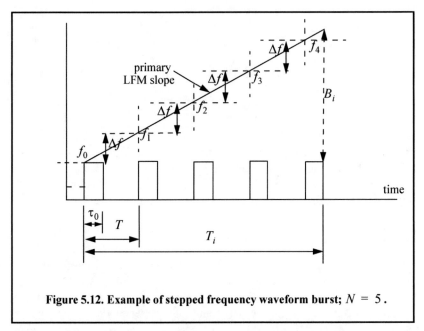

Figure 5.12. Example of stepped frequency waveform burst; $N = 5$.

where Δf is the frequency step from one subpulse to another. The set of n sub-pulses is often referred to as a burst. Each subpulse can have its own LFM modulation. To this end, assume that the subpulse LFM modulation corresponds to an LFM slope of $\mu = B/\tau_0$.

The complex envelope of a single subpluse with LFM modulation is

$$\tilde{x}_1 = \frac{1}{\sqrt{\tau_0}} Rect\left(\frac{t}{\tau_0}\right) e^{j\pi\mu t^2} \tag{5.45}$$

Of course if the subpulses do not have any LFM modulation, then the same equation holds true by setting $\mu = 0$. The overall complex envelope of the whole burst is

$$\tilde{x}(t) = \frac{1}{\sqrt{N}} \sum_{i=0}^{N-1} \tilde{x}_1(t - iT) \tag{5.46}$$

The ambiguity function of the matched filter corresponding to Eq. (5.46) can be obtained from that of the coherent pulse train developed in Section 5.2.3 along with property 5 of the ambiguity function. The details are fairly straightforward and are left to the reader as an exercise. The result is (see Problem 5.2)

$$|\chi(\tau;f_d)| = \sum_{q=-(N-1)}^{N-1} \left| \chi_1\left(\tau - qT; \left(f_d + \frac{B}{\tau_0}\tau\right)\right) \right| \times \qquad \text{(5.47)}$$

$$\left| \frac{\sin\left[\pi\left(f_d + \frac{\Delta f}{T}\tau\right)(N - |q|)T \right]}{N\sin\left(\pi\left(f_d + \frac{\Delta f}{T}\tau\right)T \right)} \right| \quad ; |\tau| \leq NT$$

where χ_1 is the ambiguity function of the single pulse. Unlike the case in Eq. (5.43), the second part of the right-hand side of Eq. (5.47) is now modified according to property 5 of Section 5.1. This is true since each subpulse has its own beginning frequency derived from the primary LFM slope.

5.4. Nonlinear FM

As clearly shown by Fig. 5.6 the output of the matched filter corresponding to an LFM pulse has sidelobe levels similar to those of the $\sin(x)/x$ signal, that is, 13.4 dB below the main beam peak. In many radar applications, these sidelobe levels are considered too high and may present serious problems for detection particularly in the presence of nearby interfering targets or other noise sources. Therefore, in most radar applications, sidelobe reduction of the output of the matched filter is always required. This sidelobe reduction can be accomplished using windowing techniques as described in Chapter 2. However, windowing techniques reduce the sidelobe levels at the expense of reducing of the SNR and widening the main beam (i.e., loss of resolution) which are considered to be undesirable features in many radar applications.

These effects can be mitigated by using non-linear FM (NLFM) instead of LFM waveforms. In this case, the LFM waveform spectrum is shaped according to a specific predetermined frequency function. Effectively, in NLFM, the rate of change of the LFM waveform phase is varied so that less time is spent on the edges of the bandwidth, as illustrated in Fig. 5.13. The concept of NLFM can be better analyzed and understood in the context of the stationary phase.

5.4.1. The Concept of Stationary Phase

Consider the following bandpass signal

$$x(t) = x_I(t)\cos(2\pi f_0 t + \phi(t)) - x_Q(t)\sin(2\pi f_0 t + \phi(t)) \qquad \text{(5.48)}$$

where $\phi(t)$ is the frequency modulation. The corresponding analytic signal (pre-envelope) is

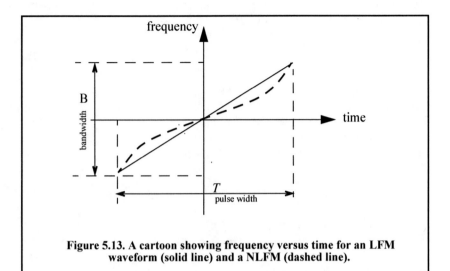

Figure 5.13. A cartoon showing frequency versus time for an LFM waveform (solid line) and a NLFM (dashed line).

$$\psi(t) = \tilde{x}(t)e^{j2\pi f_0 t} = r(t)e^{j\phi(t)}e^{j2\pi f_0 t} \tag{5.49}$$

where $\tilde{x}(t)$ is the complex envelope and is given by

$$\tilde{x}(t) = r(t)e^{j\phi(t)} \tag{5.50}$$

The lowpass signal $r(t)$ represents the envelope of the transmitted signal; it is given by

$$r(t) = \sqrt{x_I^2(t) + x_Q^2(t)} \tag{5.51}$$

It follows that the FT of the signal $\tilde{x}(t)$ can then be written as

$$X(\omega) = \int_{-\infty}^{\infty} r(t)e^{j(-\omega t + \phi(t))} \, dt \tag{5.52}$$

$$X(\omega) = |X(\omega)|e^{j\Phi(\omega)} \tag{5.53}$$

where $|X(\omega)|$ is the modulus of the FT and $\Phi(\omega)$ is the corresponding phase frequency response. It is clear that the integrand is an oscillating function of time varying at a rate

$$\frac{d}{dt}[\omega t - \phi(t)] \tag{5.54}$$

Most contribution to the FT spectrum occurs when this rate of change is minimal. More specifically, it occurs when

$$\frac{d}{dt}[\omega t - \phi(t)] = 0 \Rightarrow \omega - \phi'(t) = 0 \tag{5.55}$$

The expression in Eq. (5.55) is parametric since it relates two independent variables. Thus, for each value ω_n there is only one specific $\phi'(t_n)$ that satisfies Eq. (5.55). Thus, the time when this phase term is stationary will be different for different values of ω_n. Expanding the phase term in Eq. (5.55) about an incremental value t_n using Taylor series expansion yields

$$\omega_n t - \phi(t) = \omega_n t_n - \phi(t_n) + (\omega_n - \phi'(t_n))(t - t_n) - \frac{\phi''(t_n)}{2!}(t - t_n)^2 + \dots \tag{5.56}$$

An acceptable approximation of Eq. (5.56) is obtained by using the first three terms, provided that the difference $(t - t_n)$ is very small. Now, using the right-hand side of Eq. (5.55) into Eq. (5.56) and terminating the expansion to the first three terms yield

$$\omega_n t - \phi(t) = \omega_n t_n - \phi(t_n) - \frac{\phi''(t_n)}{2!}(t - t_n)^2 \tag{5.57}$$

By substituting Eq. (5.57) into Eq. (5.52) and using the fact that $r(t)$ is relatively constant (slow varying) when compared to the rate at which the carrier signal is varying, gives

$$X(\omega_n) = r(t_n) \int_{t_n^-}^{t_n^+} e^{-j\left(\omega_n t_n - \phi(t_n) - \frac{\phi''(t_n)}{2}(t - t_n)^2\right)} dt \tag{5.58}$$

where t_n^+ and t_n^- represent infinitesimal changes about t_n. Equation (5.58) can be written as

$$X(\omega_n) = r(t_n) e^{j(-\omega_n t_n - \phi(t_n))} \int_{t_n^-}^{t_n^+} e^{j\left(\frac{\phi''(t_n)}{2}(t - t_n)^2\right)} dt \tag{5.59}$$

Consider the changes of variables

$$t - t_n = \lambda \Rightarrow dt = d\lambda \tag{5.60}$$

$$\sqrt{\phi''(t_n)}\lambda = \sqrt{\pi}\, y \Rightarrow d\lambda = \frac{\sqrt{\pi}}{\sqrt{\phi''(t_n)}} dy \tag{5.61}$$

Using these changes of variables leads to

$$X(\omega_n) = \frac{2\sqrt{\pi}\ r(t_n)}{\sqrt{\phi''(t_n)}}e^{j(-\omega_n t_n - \phi(t_n))}\int_0^{y_0} e^{j\left(\frac{\pi y^2}{2}\right)}\ dy \qquad (5.62)$$

where

$$y_0 = \sqrt{\frac{|\phi''(t_n)|}{\pi}} \qquad (5.63)$$

The integral in Eq. (5.62) is that of the form of a Fresnel integral, which has an upper limit approximated by

$$\frac{\exp\left(j\frac{\pi}{4}\right)}{\sqrt{2}} \qquad (5.64)$$

Substituting Eq. (5.64) into Eq. (5.62) yields

$$X(\omega_n) = \frac{\sqrt{2\pi}\ r(t_n)}{\sqrt{\phi''(t_n)}}e^{j\left(-\omega_n t_n - \phi(t_n) + \frac{\pi}{4}\right)} \qquad (5.65)$$

Thus, for all possible values of ω

$$|X(\omega_t)|^2 \approx 2\pi\frac{r^2(t)}{|\phi''(t)|} \Rightarrow |X(\omega)| = \frac{\sqrt{2\pi}}{\sqrt{|\phi''(t)|}}\ r(t) \qquad (5.66)$$

The subscript t was used to indicate the dependency of ω on time.

Using a similar approach that led to Eq. (5.66), an expression for $\tilde{x}(t_n)$ can be obtained. From Eq. (5.53), the signal $\tilde{x}(t)$

$$\tilde{x}(t) = \frac{1}{2\pi}\int_{-\infty}^{\infty}|X(\omega)|\ e^{j(\Phi(\omega)+\omega t)}\ d\omega \qquad (5.67)$$

The phase term $\Phi(\omega)$ is (using Eq. (5.65))

$$\Phi(\omega) = -\omega t - \phi(t) + \frac{\pi}{4} \qquad (5.68)$$

Differentiating with respect to ω yields

$$\frac{d}{d\omega}\Phi(\omega) = -t - \left(\frac{dt}{d\omega}\right)\left[\omega - \frac{d}{dt}\phi(t)\right] = \Phi'(\omega) \qquad (5.69)$$

Using the stationary phase relation in Eq. (5.55) (i.e., $\omega - \phi'(t) = 0$) yields

$$\Phi'(\omega) = -t \qquad (5.70)$$

and

$$\Phi''(\omega) = -\frac{dt}{d\omega} \qquad (5.71)$$

Define the signal group time delay function as

$$T_g(\omega) = -\Phi'(\omega) \qquad (5.72)$$

then the signal instantaneous frequency is the inverse of the $T_g(\omega)$. Figure 5.14 shows a drawing illustrating this inverse relationship between the NLFM frequency modulation and the corresponding group time delay function.

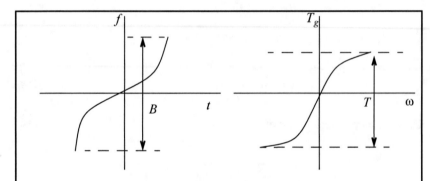

Figure 5.14. Matched filter time delay and frequency modulation for a NLFM waveform.

Comparison of Eq. (5.67) and Eq. (5.52) indicates that both equation have similar form. Thus, if one substitutes $X(\omega)/2\pi$ for $r(t)$, $\Phi(\omega)$ for $\phi(t)$, ω for t, and $-t$ for ω in Eq. (5.52), a similar expression to that in Eq. (5.65) can be derived. That is,

$$|\tilde{x}(t_\omega)|^2 \approx \frac{1}{2\pi} \frac{|X(\omega)|^2}{|\Phi''(\omega)|} \qquad (5.73)$$

the subscript ω was used to indicate the dependency of t on frequency. However, from Eq. (5.60)

$$|\tilde{x}(t)|^2 = |r(t)e^{j\phi(t)}|^2 = r^2(t) \qquad (5.74)$$

It follows that Eq. (5.73) can be rewritten as

$$r^2(t_\omega) \approx \frac{1}{2\pi} \frac{|X(\omega)|^2}{|\Phi''(\omega)|} \Rightarrow r(t) = \frac{|X(\omega)|}{\sqrt{2\pi|\Phi''(\omega)|}} \quad (5.75)$$

substituting Eq. (5.71) into Eq. (5.75) yields a general relationship for any t

$$r^2(t) \; dt = \frac{1}{2\pi}|X(\omega)|^2 d\omega \quad (5.76)$$

Clearly, the functions $r(t)$, $\phi(t)$, $X(\omega)$, and $\Phi(\omega)$ are related to each other as Fourier transform pairs, as given by

$$r(t)e^{j\phi(t)} = \frac{1}{2\pi} \int_{-\infty}^{\infty} |X(\omega)| \; e^{j(\Phi(\omega) + \omega t)} \; d\omega \quad (5.77)$$

$$|X(\omega)| \; e^{j\Phi(\omega)} = \int_{-\infty}^{\infty} r(t) \; e^{-j(\omega t - \phi(t))} \; d\omega \quad (5.78)$$

They are also related using the Parseval's theorem by

$$\int_{-\infty}^{t} r^2(\zeta) \; d\zeta = \frac{1}{2\pi} \int_{\omega}^{\infty} |X(\lambda)|^2 \; d\lambda \quad (5.79)$$

or

$$\int_{-\infty}^{t} r^2(\zeta) \; d\zeta = \frac{1}{2\pi} \int_{-\infty}^{\omega} |X(\lambda)|^2 \; d\lambda \quad (5.80)$$

The formula for the output of the matched filter was derived earlier and is repeated here as Eq. (5.81)

$$\chi(\tau, f_d) = \int_{-\infty}^{\infty} \tilde{x}(t)\tilde{x}^*(t - \tau)e^{j2\pi f_d t} dt \quad (5.81)$$

Substituting the right-hand side of Eq. (5.50) into Eq. (5.89) yields

$$\chi(\tau, f_d) = \int_{-\infty}^{\infty} r(t)r^*(t - \tau)e^{j2\pi f_d t} dt \quad (5.82)$$

It follows that the zero Doppler and zero delay cuts of the ambiguity function can be written as

$$\chi(\tau, 0) = \frac{1}{2\pi} \int_{-\infty}^{\infty} |X(\omega)|^2 \, e^{j\omega\tau} d\omega \qquad (5.83)$$

$$\chi(0, f_d) = \int_{-\infty}^{\infty} |r(t)|^2 \, e^{j2\pi f_d t} dt \qquad (5.84)$$

These two equations, imply that the shape of the ambiguity function cuts are controlled by selecting different functions X and r (related as defined in Eq. (5.76)). In other words, the ambiguity function main beam and its delay axis sidelobes can be controlled (shaped) by the specific choices of these two functions; and hence, the term *spectrum shaping* is used. Using this concept of spectrum shaping, one can control the frequency modulation of an LFM (see Fig. 5.13) to produce an ambiguity function with the desired sidelobe levels.

5.4.2. Frequency Modulated Waveform Spectrum Shaping

One class of FM waveforms which takes advantage of the stationary phase principles to control (shape) the spectrum is

$$|X(\omega;n)|^2 = \left(\cos\pi\left(\frac{\pi\omega}{B_n}\right) \right)^n \quad ; \ |\omega| \le \frac{B_n}{2} \qquad (5.85)$$

where the value n is an integer greater than zero. It can be easily shown using direct integration and by utilizing Eq. (5.85) that

$$n = 1 \Rightarrow T_{g1}(\omega) = \frac{T}{2}\sin\left(\frac{\pi\omega}{B_1}\right) \qquad (5.86)$$

$$n = 2 \Rightarrow T_{g2}(\omega) = T\left[\frac{\omega}{B_2} + \frac{1}{2\pi}\sin\left(\frac{2\pi\omega}{B_2}\right)\right] \qquad (5.87)$$

$$n = 3 \Rightarrow T_{g3}(\omega) = \frac{T}{4}\left\{ \sin\left(\frac{\pi\omega}{B_3}\right)\left[\left(\cos\frac{\pi\omega}{B_3}\right)^2 + 2\right] \right\} \qquad (5.88)$$

$$n = 4 \Rightarrow T_{g4}(\omega) = T\left\{ \frac{\omega}{B_4} + \frac{1}{2\pi}\sin\frac{2\pi\omega}{B_4} + \frac{2}{3\pi}\left(\cos\frac{\pi\omega}{B_4}\right)^3\sin\frac{\pi\omega}{B_4} \right\} \qquad (5.89)$$

Figure 5.15 shows a plot for Eq. (5.86) through Eq. (5.89). These plots assume $T = 1$ and the x-axis is normalized, with respect to B. This figure can be reproduced using the following MATLAB code:

```
% Figure 5.15
clear all; close all;
delw = linspace(-.5,.5,75);
T1 = .5 .* sin(pi.*delw);
T2 = delw + (1/2/pi) .* sin(2*pi.*delw);
T3 = .25 .* (sin(pi.*delw)) .* ((cos(pi.*delw)).^2 + 2);
T4 = delw + (1/2/pi) .* sin(2*pi.*delw) + (2/3/pi) .* (cos(pi.*delw)).^3 .* sin(delw);
figure (1)
plot(delw,T1,'k*',delw,T2,'k:',delw,T3,'k.',delw,T4,'k');
grid
ylabel('Group delay function'); xlabel('\omega/B')
legend('n=1','n=2','n=3','n=4')
```

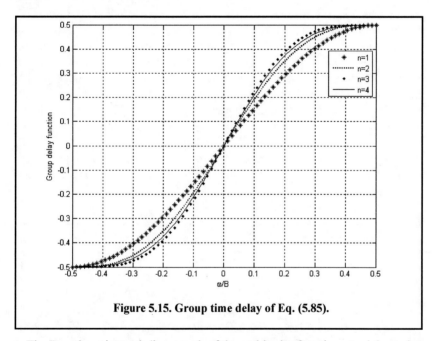

Figure 5.15. Group time delay of Eq. (5.85).

The Doppler mismatch (i.e, a peak of the ambiguity function at a delay value other than zero) is proportional to the amount of Doppler frequency f_d. Hence, an error in measuring target range is always expected when LFM waveforms are used. To achieve sidelobe levels for the output of the matched filter that do not exceed a predetermined level use this class of NLFM waveforms

$$|X(\omega;n;k)|^2 = k + (1-k)\left(\cos\pi\left(\frac{\pi\omega}{B_n}\right)\right)^n \qquad ;|\omega| \leq \frac{B_n}{2} \qquad \textbf{(5.90)}$$

For example, using the combination $n = 2$, $k = 0.08$ yields sidelobe levels less than $-40dB$.

5.5. *Ambiguity Diagram Contours*

Plots of the ambiguity function are called ambiguity diagrams. For a given waveform, the corresponding ambiguity diagram is normally used to determine the waveform properties such as the target resolution capability, measurements (time and frequency) accuracy, and its response to clutter. The ambiguity diagram contours are cuts in the 3-D ambiguity plot at some value, Q, such that $Q < |\chi(0, 0)|^2$. The resulting plots are ellipses (see Problem 5.11). The width of a given ellipse along the delay axis is proportional to the signal effective duration, τ_e, defined in Chapter 2. Alternatively, the width of an ellipse along the Doppler axis is proportional to the signal effective bandwidth, B_e.

Figure 5.16 shows a sketch of typical ambiguity contour plots associated with a single unmodulated pulse. As illustrated in Fig. 5.16, narrow pulses provide better range accuracy than long pulses. Alternatively, the Doppler accuracy is better for a wider pulse than it is for a short one. This trade-off between range and Doppler measurements comes from the uncertainty associated with the time-bandwidth product of a single sinusoidal pulse, where the product of uncertainty in time (range) and uncertainty in frequency (Doppler) cannot be much smaller than unity (see Problem 5.12). Figure 5.17 shows the ambiguity contour plot associated with an LFM waveform. The slope is an indication of the LFM modulation. The values σ_τ, σ_{f_d}, $\sigma_{\tau RDC}$, and $\sigma_{f_d RDC}$ were derived in Chapter 4 and were, respectively, given in Eq. (4.107), Eq. (4.111), Eq. (4.136), and Eq. (4.137).

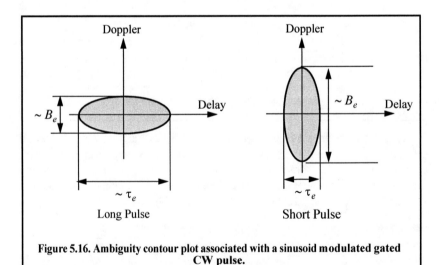

Figure 5.16. Ambiguity contour plot associated with a sinusoid modulated gated CW pulse.

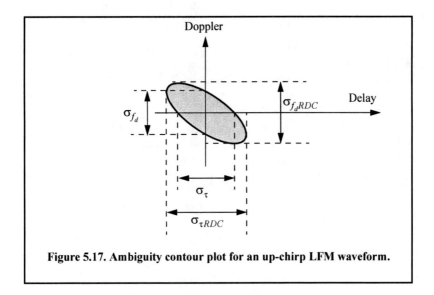

Figure 5.17. Ambiguity contour plot for an up-chirp LFM waveform.

5.6. Interpretation of Range-Doppler Coupling in LFM Signals

An expression of the range-Doppler for LFM signals was derived in Chapter 4. Range-Doppler coupling affects the radar's ability to compute target range and Doppler estimates. An interpretation of this term in the context of the ambiguity function can be explained further with the help of Eq. (5.20). Observation of this equation indicates that ambiguity function for the LFM pulse has a peak value not at $\tau = 0$ but rather at

$$(B/\tau_0)\tau - f_d = 0 \Rightarrow \tau = f_d - \tau_0/B \qquad \text{(5.91)}$$

This Doppler mismatch (i.e, a peak of the ambiguity function at a delay value other than zero) is proportional to the amount of Doppler frequency f_d. Hence, an error in measuring target range is always expected when LFM waveforms are used.

Most radar systems using LFM waveforms will correct for the effect of range-Doppler coupling by repeating the measurement with an LFM waveform of the opposite slope and averaging the two measurements. This way, the range measurement error is negated and the true target range is extracted from the averaged value. However, some radar systems, particularly those used for long range surveillance applications, may actually take advantage of range-Doppler coupling effect; and here is how it works: Typically radars during the search mode utilize very wide range bins which may contain many targets with different distinct Doppler frequencies. It follows that the output of the matched filter has several targets that have equal delay but different Doppler mismatches.

All targets with Doppler mismatches greater than $1/\tau_0$ are significantly attenuated by the ambiguity function (because of the sharp decaying slope of the ambiguity function along the Doppler axis) and thus will most likely go undetected along the Doppler axis. The combined target complex within that range bin is then detected by the LFM as if all targets had Doppler mismatch corresponding to the target whose Doppler mismatch is less or equal to $1/\tau_0$. Thus, all targets within that wide range bin are detected as one narrowband target. Because of this range-Doppler coupling LFM waveforms are often referred to as Doppler intolerant (insensitive) waveforms.

5.7. MATLAB Programs and Functions

This section presents listings for all the MATLAB programs used to produce all of the MATLAB-generated figures in this chapter. They are listed in the same order in which they appear in the text.

5.7.1. Single Pulse Ambiguity Function

The MATLAB function *"single_pulse_ambg.m"* implements Eq. (5.11). The syntax is as follows:

$$single\_pulse\_ambg \ [taup]$$

taup is the pulse width.

MATLAB Function *"single_pulse_ambg.m"* **Listing**

```
function [x] = single_pulse_ambg (taup)
eps = 0.000001;
i = 0;
del = 2*taup/150;
for tau = -taup:del:taup
  i = i + 1;
  j = 0;
  fd = linspace(-5/taup,5/taup,151);
  val1 = 1. - abs(tau) / taup;
  val2 = pi * taup .* (1.0 - abs(tau) / taup) .* fd;
  x(:,i) = abs( val1 .* sin(val2+eps)./(val2+eps));
end
```

5.7.2. LFM Ambiguity Function

The function *"lfm_ambg.m"* implements Eq. (5.20). The syntax is as follows:

lfm_ambg [taup, b, up_down]

where

Symbol	Description	Units	Status
taup	*pulse width*	*seconds*	*input*
b	*bandwidth*	*Hz*	*input*
up_down	*up_down = 1 for up-chirp* *up_down = -1 for down-chirp*	*none*	*input*

MATLAB Function *"lfm_ambg.m"* **Listing**

```
function [x] = single_pulse_ambg (taup)
% Single umodulated pulse
eps = 0.000001;
i = 0;
del = 2*taup/150;
for tau = -taup:del:taup
  i = i + 1;
  j = 0;
  fd = linspace(-5/taup,5/taup,151);
  val1 = 1. - abs(tau) / taup;
  val2 = pi * taup .* (1.0 - abs(tau) / taup) .* fd;
  x(:,i) = abs( val1 .* sin(val2+eps)./(val2+eps));
end
```

5.7.3. Pulse Train Ambiguity Function

The function *"train_ambg.m"* implements Eq. (5.35). The syntax is as follows:

train_ambg [taup, n, pri]

where

Symbol	Description	Units	Status
taup	*pulse width*	*seconds*	*input*
n	*number of pulses in train*	*none*	*input*
pri	*pulse repetition interval*	*seconds*	*input*

MATLAB Function *"train_ambg.m"* **Listing**

```
function x = train_ambg(taup, n, pri)
% This code was developed by Stephen Robinson, a senior radar engineer at
% deciBel Research in Hunstville AL
if (taup >= pri/2)
  'ERROR. Pulse width must be less than the PRI/2.'
```

```
    return
end
eps = 1.0e-6;
bw = 1/taup;
q = -(n-1):1:n-1;
offset = 0:0.0533:pri;
[Q, S] = meshgrid(q, offset);
Q = reshape(Q, 1, length(q)*length(offset));
S = reshape(S, 1, length(q)*length(offset));
tau = (-taup * ones(1,length(S))) + S;
fd = -bw:0.033:bw;
[T, F] = meshgrid(tau, fd);
Q = repmat(Q, length(fd), 1);
S = repmat(S, length(fd), 1);
N = n * ones(size(T));
val1 = 1.0-(abs(T))/taup;
val2 = pi*taup*F.*val1;
val3 = abs(val1.*sin(val2+eps)./(val2+eps));
val4 = abs(sin(pi*F.*(N-abs(Q))*pri+eps)./sin(pi*F*pri+eps));
x = val3.*val4./N;
[rows, cols] = size(x);
x = reshape(x, 1, rows*cols);
T = reshape(T, 1, rows*cols);
indx = find(abs(T) > taup);
x(indx) = 0.0;
x = reshape(x, rows, cols);
return
```

5.7.4. Pulse Train Ambiguity Function with LFM

The function *"train_ambg_lfm.m"* implements Eq. (5.43). The syntax is as follows:

$$x = train\_ambg\_lfm(taup, n, pri, bw)$$

where

Symbol	Description	Units	Status
taup	pulse width	seconds	input
n	number of pulses in train	none	input
pri	pulse repetition interval	seconds	input
bw	the LFM bandwidth	Hz	input
x	array of bimodality function	none	output

Note this function will generate identical results to the function *"train_ambg.m"* when the value of *bw* is set to zero. In this case, Eq. (4.43) and (4.35) are identical.

MATLAB Function *"train_ambg_lfm.m"* **Listing**

```
function x = train_ambg_lfm(taup, n, pri, bw)
% This code was developed by Stephen Robinson, a senior radar engineer at
% deciBel Research in Hunstville AL
if (taup >= pri/2)
    'ERROR. Pulse width must be less than the PRI/2.'
    return
end
eps = 1.0e-6;
q = -(n-1):1:n-1;
offset = 0:0.0533:pri;
[Q, S] = meshgrid(q, offset);
Q = reshape(Q, 1, length(q)*length(offset));
S = reshape(S, 1, length(q)*length(offset));
tau = (-taup * ones(1,length(S))) + S;
fd = -bw:0.033:bw;
[T, F] = meshgrid(tau, fd);
Q = repmat(Q, length(fd), 1);
S = repmat(S, length(fd), 1);
N = n * ones(size(T));
val1 = 1.0-(abs(T))/taup;
val2 = pi*taup*(F+T*(bw/taup)).*val1;
val3 = abs(val1.*sin(val2+eps)./(val2+eps));
val4 = abs(sin(pi*F.*(N-abs(Q))*pri+eps)./sin(pi*F*pri+eps));
x = val3.*val4./N;
[rows, cols] = size(x);
x = reshape(x, 1, rows*cols);
T = reshape(T, 1, rows*cols);
indx = find(abs(T) > taup);
x(indx) = 0.0;
x = reshape(x, rows, cols);
return
```

Problems

5.1. Derive Eq. (5.47).

5.2. Show that Eq. (5.79) and Eq. (5.80) are equivalent.

5.3. Derive an expression for the ambiguity function of a Gaussian pulse defined by

$$x(t) = \frac{1}{\sqrt{\sigma}\,\sqrt[4]{\pi}}\exp\left[\frac{-t^2}{2\sigma^2}\right] \qquad ;0 < t < T$$

where T is the pulsewidth and σ is a constant.

5.4. Write a MATLAB code to plot the 3-D and the contour plots for the results in Problem 5.3.

5.5. Derive an expression for the ambiguity function of a V-LFM waveform, illustrated in figure below. In this case, the overall complex envelope is

$$\tilde{x}(t) = \tilde{x}_1(t) + \tilde{x}_2(t) \qquad ;-(T < t < T)$$

where

$$\tilde{x}_1(t) = \frac{1}{\sqrt{2T}}\exp[-\mu t^2] \qquad ;-T < t < 0$$

and

$$\tilde{x}_2(t) = \frac{1}{\sqrt{2T}}\exp[\mu t^2] \qquad ;0 < t < T$$

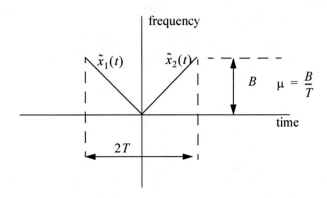

5.6. Using the stationary phase concept, find the instantaneous frequency for the waveform whose envelope and complex spectrum are, respectively, given by

$$r(t) = \frac{1}{\sqrt{T}}\exp\left[-\left(\frac{2t}{T}\right)^2\right] \qquad ;0 < t < T$$

and

$$|X(f)| = \frac{1}{\sqrt{B}}\exp\left[-\left(\frac{2f}{B}\right)^2\right]$$

5.7. Using the stationary phase concept find the instantaneous frequency for the waveform whose envelope and complex spectrum are respectively given by

$$r(t) = \frac{1}{\sqrt{\tau_0}} Rect\left(\frac{t}{\tau_0}\right) \quad ; \ 0 < t < \tau_0$$

and

$$|X(\omega)| = \frac{2}{\sqrt{B}} \frac{1}{\sqrt{1 + (2\omega/B)^2}}$$

5.8. Write detailed MATLAB code to compute the ambiguity function for an NLFM waveform. Your code must be able to produce 3-D and contour plots of the resulting ambiguity function. Hint: Use Eq. (5.90).

5.9. Revisit the analyses performed in Chapter 2 for the effective bandwidth and effective duration of the LFM waveform. Write a short discussion to outline how do the range and Doppler resolution are different from the theoretical limits used in this chapter.

5.10. Write a detailed MATLAB code to compute the ambiguity function for an SFW waveform. Your code must be able to produce 3-D and contour plots of the resulting ambiguity function. Hint: use Eq. (5.43).

5.11. Prove that cuts in the ambiguity function are always defined by an ellipse. Hint: Approximate the ambiguity function using a Taylor series expansion about the values $(\tau, f_d) = (0, 0)$; use only the first three terms in the Taylor series expansion.

5.12. The radar uncertainty principle establishes a lower bound for the time bandwidth product. More specifically, if the radar effective duration is τ_e and its effective bandwidth is B_e; show that $B_e^2 \tau_e^2 - \rho_{RDC}^2 \geq \pi^2$, where ρ_{RDC} is the range-Doppler coupling coefficient defined in Chapter 4. Hint: Assume a signal $x(t)$, write down the definition of ρ_{RDC}, and use Shwarz inequality on the integral

$$(-j2\pi) \int_{-\infty}^{\infty} t x^*(t) x'(t) dt.$$

Chapter 6

The Ambiguity Function - Discrete Coded Waveforms

The concepts of resolution and ambiguity were introduced in Chapter 4. The relationship between the waveform resolution (range and Doppler) and its corresponding ambiguity function was discussed and analyzed. It was determined that the *goodness* of a given waveform is based on its range and Doppler resolutions, which can be analyzed in the context of the ambiguity function. For this purpose, a few common analog radar waveforms were analyzed in Chapter 5. In this chapter, another type of radar waveform based on discrete codes is introduced. This topic has been and continues to be a major research thrust area for many radar scientist, designers, and engineers. Discrete coded waveforms are more effective in improving range characteristics than Doppler (velocity) characteristics. Furthermore, in some radar applications, discrete coded waveforms are heavily favored because of their inherent anti-jamming capabilities. In this chapter, a quick overview of discrete coded waveforms is presented. Three classes of discrete codes are analyzed. They are unmodulated pulse-train codes (uniform and staggered), phase-modulated (binary or polyphase) codes, and frequency modulated codes.

6.1. Discrete Code Signal Representation

The general form for a discrete coded signal can be written as

$$x(t) = e^{j\omega_0 t} \sum_{n=1}^{N} u_n(t) = e^{j\omega_0 t} \sum_{n=1}^{N} P_n(t) e^{j(\omega_n t + \theta_n)} \tag{6.1}$$

where ω_0 is the carrier frequency in radians, (ω_n, θ_n) are constants, N is the code length (number of bits in the code), and the signal $P_n(t)$ is given by

$$P_n(t) = a_n Rect\left(\frac{t}{\tau_0}\right) \tag{6.2}$$

the constant a_n is either (1) or (0), and

$$Rect\left(\frac{t}{\tau_0}\right) = \begin{cases} 1 & ; \quad 0 < t < \tau_0 \\ 0 & ; \quad elsewhere \end{cases} \tag{6.3}$$

Using this notation the discrete code can be described through the sequence

$$U[n] = \{u_n, n = 1, 2, ..., N\} \tag{6.4}$$

which, in general, is a complex sequence depending on the values of ω_n and θ_n. The sequence $U[n]$ is called the code and for convenience it will be denoted by U.

In general, the output of the matched filter is

$$\chi(\tau, f_d) = \int_{-\infty}^{\infty} x^*(t)x(t+\tau)e^{-j2\pi f_d t} dt \tag{6.5}$$

Substituting Eq. (6.1) into Eq. (6.5) yields

$$\chi(\tau, f_d) = \sum_{n=1}^{N} \sum_{k=1}^{N} \int_{-\infty}^{\infty} u_n^*(t)u_k(t+\tau)e^{-j2\pi f_d t} dt \tag{6.6}$$

Depending on the choice of combination for a_n, ω_n, and θ_n, different class of codes can be generated. More precisely, pulse-train codes are generated when

$$\theta_n = \omega_n = 0 \quad ; \quad and \ a_n = 1, or \ 0 \tag{6.7}$$

Binary phase codes and polyphase codes are generated when

$$\omega_n = 0 \quad ; \quad and \ a_n = 1 \tag{6.8}$$

Finally, frequency codes are generated when

$$\theta_n = 0 \quad ; \quad and \ a_n = 1, or \ 0 \tag{6.9}$$

6.2. Pulse-Train Codes

The idea behind this class of code is to divide a relatively long pulse of length T_P into N subpulses, each being a rectangular pulse with pulsewidth τ_0 and amplitude of 1 or 0. It follows that the code U is the sequence of 1's and 0's. More precisely, the signal representing this class of code can written as

$$x(t) = e^{j\omega_0 t} \sum_{n=1}^{N} P_n(t) = e^{j\omega_0 t} \sum_{n=1}^{N} a_n Rect\left(\frac{t}{\tau_0}\right) \tag{6.10}$$

One way to generate a train-pulse class code can be by setting

$$a_n = \begin{cases} 1 & n-1 = 0 \ modulu \ q \\ 0 & n-1 \neq 0 \ modulu \ q \end{cases} \tag{6.11}$$

where q is a positive integer that divides evenly into $N-1$. That is,

$$M-1 = (N-1)/q \tag{6.12}$$

where M is the number of 1's in the code. For example, when $N = 21$ and $q = 5$, then $M = 5$, and the resulting code is

$$\{U\} = \{10000 \ 10000 \ 10000 \ 10000 \ 1\} \tag{6.13}$$

This is illustrated in Fig. 6.1. In previous chapters this code would have been represented by the following continuous time domain signal

$$x_1(t) = e^{j\omega_0 t} \sum_{m=0}^{4} Rect\left(\frac{t - mT}{\tau_0}\right) \tag{6.14}$$

where the period is $T = 5\tau_0$. Using this analogy yields

$$\frac{T_p}{M-1} \equiv T \tag{6.15}$$

and Eq. (6.10) can now be written as

$$x(t) = e^{j\omega_0 t} \sum_{m=1}^{M-1} Rect\left(\frac{t - m\left(\frac{T_p}{M-1}\right)}{\tau_0}\right) \tag{6.16}$$

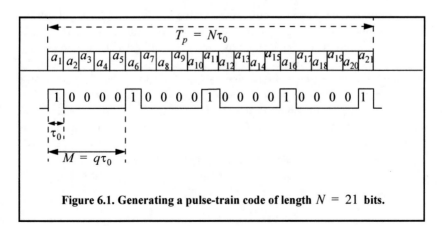

Figure 6.1. Generating a pulse-train code of length $N = 21$ **bits.**

In Chapter 5 (Section 5.2.3) an expression for the ambiguity function for a coherent train of pulses was derived. Comparison of Eq. (6.16) and Eq. (5.37) show that the two equations are equivalent when the condition in Eq. (6.15) is true except for the ratio $(1/\sqrt{N})$. It follows that the ambiguity function for the signal defined in Eq. (6.16) is

$$|\chi(\tau;f_d)| = \sum_{k=-M}^{M} \left| \frac{\sin\left[\pi f_d\left([M-|k|]\frac{T_p}{M-1}\right)\right]}{\sin\left(\pi f_d\frac{T_p}{M-1}\right)} \right| \left| \frac{\sin\left[\pi f_d\left(\tau_0 - \left|\tau - \frac{kT_p}{M-1}\right|\right)\right]}{\pi f_d} \right| \tag{6.17}$$

The zero Doppler and zero delay cuts of the ambiguity function are derived from Eq. (6.17). They are given by

$$|\chi(\tau;0)| = M\tau_0 \sum_{k=-M}^{M} \left[1 - \frac{|k|}{M}\right]\left(1 - \frac{\left|\tau - \frac{kT_p}{M-1}\right|}{\tau_0}\right) \tag{6.18}$$

$$|\chi(0;f_d)| = \sum_{k=-M}^{M} \left| \frac{\sin\left[\pi M f_d\left(\frac{T_p}{M-1}\right)\right]}{\sin\left(\pi f_d\frac{T_p}{M-1}\right)} \right| \left| \frac{\sin(\pi f_d\tau_0)}{\pi f_d\tau_0} \right| \tag{6.19}$$

Figure 6.2a shows the three-dimensional ambiguity plot for the code shown in Fig. 6.1, while Fig. 6.2b shows the corresponding contour plot. This figure can be reproduced using the following MATLAB code.

```
close all; clear all;
U = [1 0 0 0 0 1 0 0 0 0 1 0 0 0 0 1];
ambiguity = ambiguity_code(U);
```

A cartoon showing contour cuts of the ambiguity function for a pulse-train code is shown in Fig. 6.2c. Clearly, the width of the ambiguity function main lobe (i.e., resolution) is directly tied to the code length. As one would expect, longer codes will produce narrower main lobe and thus have better resolution than shorter ones. Further observation of Fig. 6.2 shows that this ambiguity function has strong grating lobe structure along with high sidelobe levels. The presence of such strong lobing structure limits the effectiveness of the code and will cause detection ambiguities. These lobes are a direct result from the uniform equal spacing between the 1's within a code (i.e., periodicity of the code). These lobes can be significantly reduced by getting rid of the periodic structure of the code, i.e., placing the pulses at nonuniform spacing. This is called code staggering (PRF staggering).

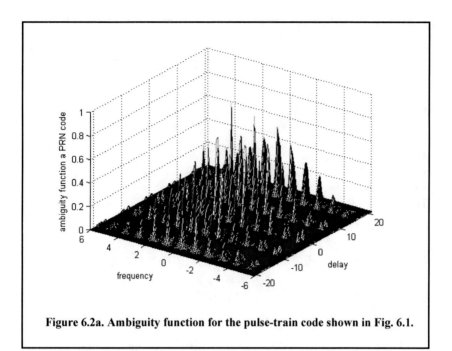

Figure 6.2a. Ambiguity function for the pulse-train code shown in Fig. 6.1.

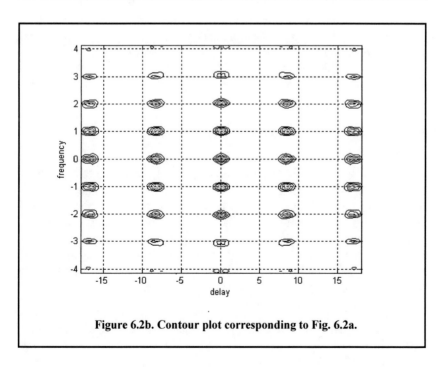

Figure 6.2b. Contour plot corresponding to Fig. 6.2a.

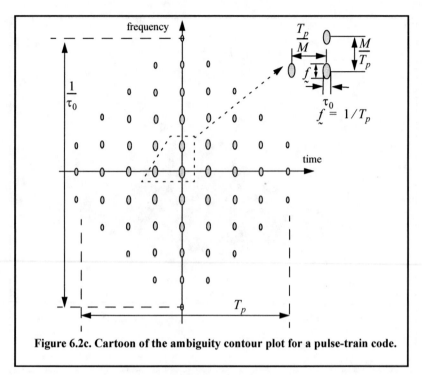

Figure 6.2c. Cartoon of the ambiguity contour plot for a pulse-train code.

For example, consider a pulse-train code of length $N = 21$. A staggered train-pulse code can then be obtained by using the following sequence a_n

$$\{a_n\} = 1 \qquad n = 1, 4, 6, 12, 15, 21 \qquad \textbf{(6.20)}$$

Thus, the resulting code is

$$\{U\} = \{100101000001001000001\} \qquad \textbf{(6.21)}$$

Figure 6.3 shows the ambiguity plot corresponding to this code. As indicated by Fig. 6.3 the ambiguity function corresponding to a staggered pulse-train code approaches a thumb-tack shape. The choice of the optimum staggered code has been researched extensively by numerous people. Resnick[1] defined the optimum staggered pulse-train code as that whose ambiguity function has absolutely uniform sidelobe levels that are equal to unity. Other researchers, have introduced different definitions for optimum staggering, none of which is necessarily better than the others, except when considered for the particular application being analyzed by the respective researcher.

---

1. Resnick, J. B., *High Resolution Waveforms Suitable for a Multiple Target Environment*, MS Thesis, MIT, Cambridge, MA, June 1962.

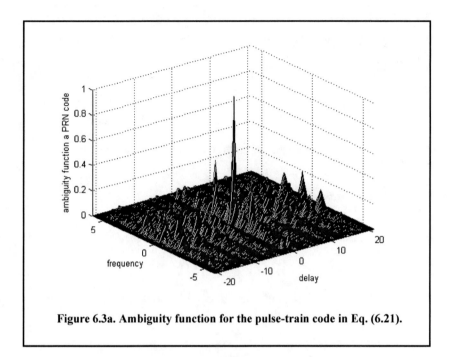

Figure 6.3a. Ambiguity function for the pulse-train code in Eq. (6.21).

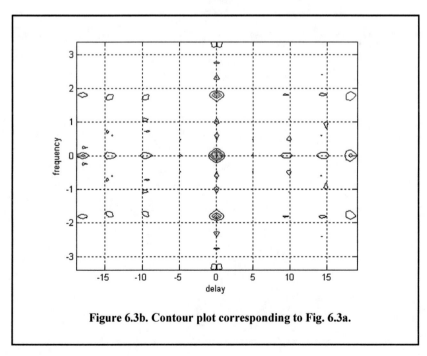

Figure 6.3b. Contour plot corresponding to Fig. 6.3a.

6.3. Phase Coding

The signal corresponding to this class of code is obtained from Eq. (6.1) by letting $\omega_n = 0$. It follows that

$$x(t) = e^{j\omega_0 t} \sum_{n=1}^{N} u_n(t) = e^{j\omega_0 t} \sum_{n=1}^{N} P_n(t) e^{j\theta_n} \tag{6.22}$$

Two subclasses of phase codes are analyzed. They are binary phase codes and polyphase codes.

6.3.1. Binary Phase Codes

In this case, the phase θ_n is set equal to either (0) or (π), and hence, the term *binary* is used. For this purpose, define the coefficient D_n as

$$D_n = e^{j\theta_n} = \pm 1 \tag{6.23}$$

The ambiguity function for this class of code is derived by substituting Eq. (6.22) into Eq. (6.5). The resulting ambiguity function is given by

$$\chi(\tau; f_d) = \begin{cases} \chi_0(\tau', f_d) \displaystyle\sum_{n=1}^{N-k} D_n D_{n+k} e^{-j2\pi f_d(n-1)\tau_0} + \\[2mm] \chi_0(\tau_0 - \tau', f_d) \displaystyle\sum_{n=1}^{N-(k+1)} D_n D_{n+k+1} e^{-j2\pi f_d n\tau_0} \end{cases} \quad 0 < \tau < N\tau_0 \tag{6.24}$$

where

$$\tau = k\tau_0 + \tau' \qquad \begin{cases} 0 < \tau' < \tau_0 \\ k = 0, 1, 2, \dots, N \end{cases} \tag{6.25}$$

$$\chi_0(\tau', f_d) = \int_0^{\tau_0 - \tau'} \exp(-j2\pi f_d t) dt \qquad 0 < \tau' < \tau_0 \tag{6.26}$$

The corresponding zero Doppler cut is then given by

$$\chi(\tau; 0) = \tau_0 \left(1 - \frac{|\tau'|}{\tau_0}\right) \sum_{n=1}^{N-|k|} D_n D_{n+k} + |\tau'| \sum_{n=1}^{N-|k+1|} D_n D_{n+k+1} \tag{6.27}$$

and when $\tau' = 0$ then

$$\chi(k;0) = \tau_0 \sum_{n=1}^{N-|k|} D_n D_{n+k} \tag{6.28}$$

Barker Codes

In this case, a long pulse of width T_p is divided into N smaller pulses; each is of width $\tau_0 = T_p/N$. Then, the phase of each subpulse is chosen as either 0 or π radians relative to some code. It is customary to characterize a subpulse that has 0 phase (amplitude of +1 Volt) as either "1" or "+." Alternatively, a subpulse with phase equal to π (amplitude of -1 Volt) is characterized by either "0" or "-." Barker code is optimum in accordance with the definition set by Resnick. Figure 6.4 illustrates this concept for a Barker code of length seven. A Barker code of length N is denoted as B_N. There are only seven known Barker codes that share this unique property; they are listed in Table 6.1. Note that B_2 and B_4 have complementary forms that have the same characteristics.

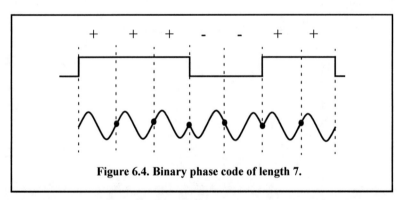

Figure 6.4. Binary phase code of length 7.

In general, the autocorrelation function (which is an approximation for the matched filter output) for a B_N Barker code will be $2N\tau_0$ wide. The main lobe is $2\tau_0$ wide; the peak value is equal to N. There are $(N-1)/2$ side-lobes on either side of the main lobe; this is illustrated in Fig. 6.5 for a B_{13}. Notice that the main lobe is equal to 13, while all side-lobes are unity.

The most side-lobe reduction offered by a Barker code is $-22.3\,dB$, which may not be sufficient for the desired radar application. However, Barker codes can be combined to generate much longer codes. In this case, a B_M code can be used within a B_N code (M within N) to generate a code of length MN. The compression ratio for the combined B_{MN} code is equal to MN. As an example, a combined B_{54} is given by

$$B_{54} = \{11101, 11101, 00010, 11101\} \tag{6.29}$$

and is illustrated in Fig. 6.6. Unfortunately, the side-lobes of a combined Barker code autocorrelation function are no longer equal to unity. Some side-lobes of a combined Barker code autocorrelation function can be reduced to zero if the matched filter is followed by a linear transversal filter with impulse response given by

$$h(t) = \sum_{k=-N}^{N} \beta_k \delta(t - 2k\tau_0) \tag{6.30}$$

where N is the filter's order, the coefficients β_k ($\beta_k = \beta_{-k}$) are to be determined, $\delta(\cdot)$ is the delta function, and τ_0 is the Barker code subpulse width. A filter of order N produces N zero side-lobes on either side of the main lobe. The main lobe amplitude and width do not change, as illustrated in Fig. 6.7.

TABLE 6.1. Barker codes

Code Symbol	Code Length	Code Elements	Side Lode Reduction (dB)
B_2	2	+- ++	6.0
B_3	3	++-	9.5
B_4	4	++-+ +++-	12.0
B_5	5	+++-+	14.0
B_7	7	+++--+-	16.9
B_{11}	11	+++---+--+-	20.8
B_{13}	13	+++++--++-+-+	22.3

In order to illustrate this approach, consider the case where the input to the matched filter is B_{11}, and assume $N = 4$. The autocorrelation for a B_{11} is

$$\phi_{11} = \{-1, 0, -1, 0, -1, 0, -1, 0, -1, 0, 11, \\ 0, -1, 0, -1, 0, -1, 0, -1, 0, -1\} \tag{6.31}$$

The output of the transversal filter is the discrete convolution between its impulse response and the sequence ϕ_{11}. At this point we need to compute the coefficients β_k that guarantee the desired filter output (i.e., unchanged main lobe and four zero side-lobe levels).

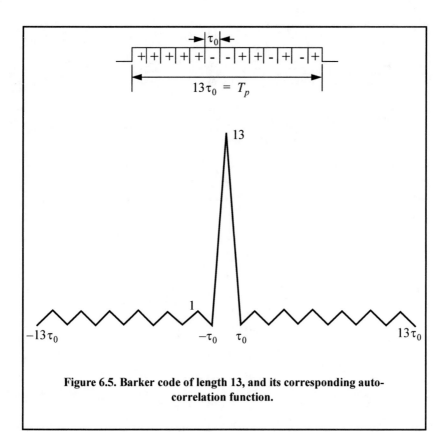

Figure 6.5. Barker code of length 13, and its corresponding auto-correlation function.

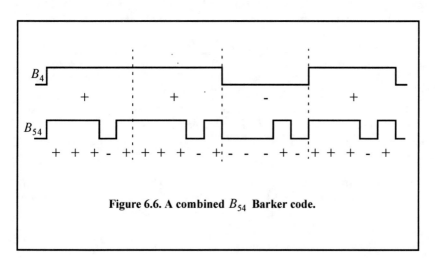

Figure 6.6. A combined B_{54} Barker code.

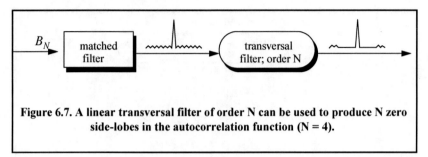

Figure 6.7. A linear transversal filter of order N can be used to produce N zero side-lobes in the autocorrelation function (N = 4).

Performing the discrete convolution as defined in Eq. (6.30) and collecting equal terms ($\beta_k = \beta_{-k}$) yield the following set of five linearly independent equations:

$$
\begin{bmatrix}
11 & -2 & -2 & -2 & -2 \\
-1 & 10 & -2 & -2 & -1 \\
-1 & -2 & 10 & -2 & -1 \\
-1 & -2 & -1 & 11 & -1 \\
-1 & -1 & -1 & -1 & 11
\end{bmatrix}
\begin{bmatrix}
\beta_0 \\ \beta_1 \\ \beta_2 \\ \beta_3 \\ \beta_4
\end{bmatrix}
=
\begin{bmatrix}
11 \\ 0 \\ 0 \\ 0 \\ 0
\end{bmatrix}
\qquad (6.32)
$$

Solving Eq. (6.32) yields

$$
\begin{bmatrix}
\beta_0 \\ \beta_1 \\ \beta_2 \\ \beta_3 \\ \beta_4
\end{bmatrix}
=
\begin{bmatrix}
1.1342 \\ 0.2046 \\ 0.2046 \\ 0.1731 \\ 0.1560
\end{bmatrix}
\qquad (6.33)
$$

Note that setting the first equation equal to 11 and all other equations to 0 and then solving for β_k guarantees that the main peak remains unchanged, and that the next four side-lobes are zeros. So far we have assumed that coded pulses have rectangular shapes. Using other pulses of other shapes, such as Gaussian, may produce better side-lobe reduction and a larger compression ratio.

Figure 6.8 shows the output of this function when B_{13} is used as an input. Figure 6.9 is similar to Fig. 6.8, except in this case B_7 is used as an input. Figure 6.10 shows the ambiguity function, the zero Doppler cut, and the contour plot for the combined Barker code defined in Fig. 6.6.

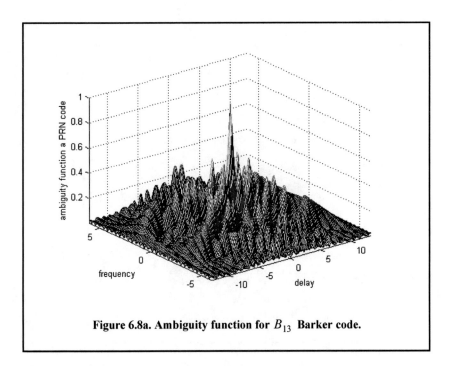

Figure 6.8a. Ambiguity function for B_{13} Barker code.

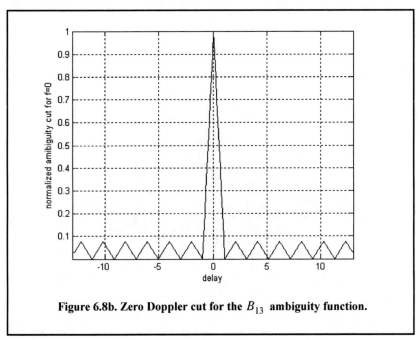

Figure 6.8b. Zero Doppler cut for the B_{13} ambiguity function.

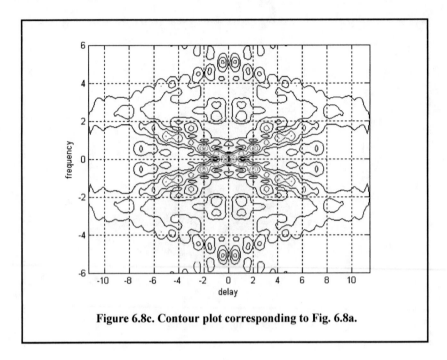

Figure 6.8c. Contour plot corresponding to Fig. 6.8a.

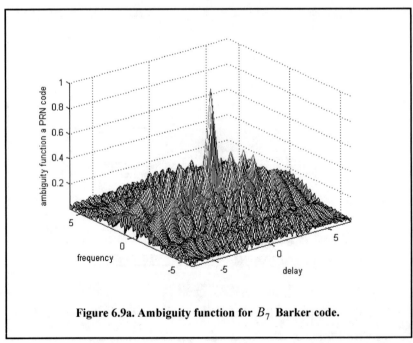

Figure 6.9a. Ambiguity function for B_7 Barker code.

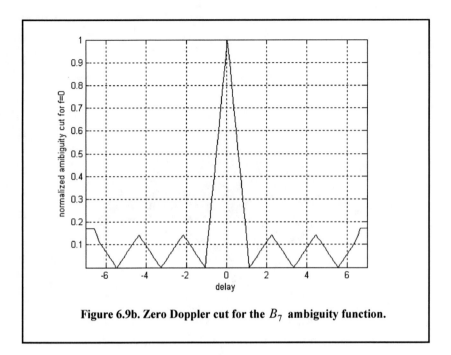

Figure 6.9b. Zero Doppler cut for the B_7 ambiguity function.

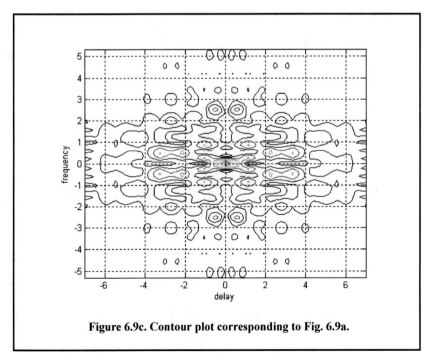

Figure 6.9c. Contour plot corresponding to Fig. 6.9a.

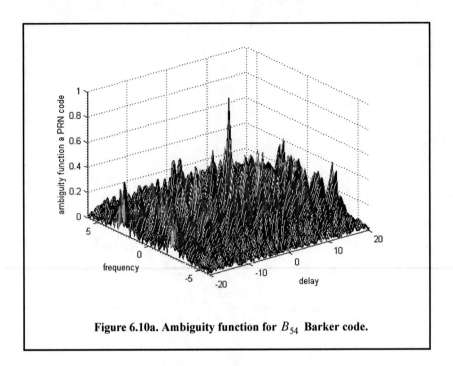

Figure 6.10a. Ambiguity function for B_{54} Barker code.

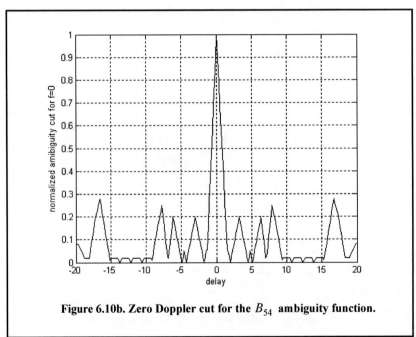

Figure 6.10b. Zero Doppler cut for the B_{54} ambiguity function.

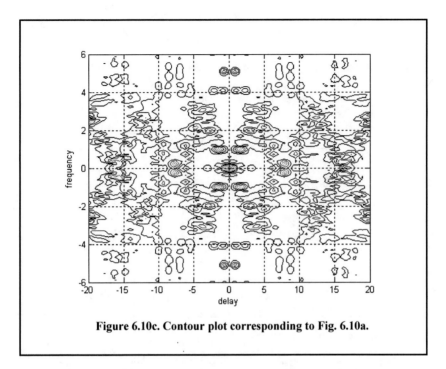

Figure 6.10c. Contour plot corresponding to Fig. 6.10a.

Pseudo-Random Number (PRN) Codes

Pseudo-Random Number (PRN) codes are also known as Maximal Length Sequences (MLS) codes. These codes are called pseudo-random because the statistics associated with their occurrence are similar to those associated with the coin-toss sequences. Maximum length sequences are periodic. The MLS codes have the following distinctive properties:

1. The number of ones per period is one more than the number of minus ones.

2. Half the runs (consecutive states of the same kind) are of length one and one fourth are of length two.

3. Every maximal length sequence has the "shift and add" property. This means that, if a maximal length sequence is added (modulo 2) to a shifted version of itself, then the resulting sequence is a shifted version of the original sequence.

4. Every n-tuple of the code appears once and only once in one period of the sequence.

5. The correlation function is periodic and is given by

$$\phi(n) = \begin{cases} L & n = 0, \pm L, \pm 2L, ... \\ -1 & elsewhere \end{cases} \qquad \text{(6.34)}$$

Figure 6.11 shows a typical sketch for an MLS autocorrelation function. Clearly these codes have the advantage that the compression ratio becomes very large as the period is increased. Additionally, adjacent peaks (grating lobes) become farther apart.

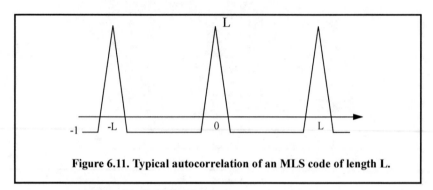

Figure 6.11. Typical autocorrelation of an MLS code of length L.

Linear Shift Register Generators

There are numerous ways to generate MLS codes. The most common is to use linear shift registers. When the binary sequence generated using a shift register implementation is periodic and has maximal length, it is referred to as an MLS binary sequence with period L, where

$$L = 2^n - 1 \qquad \text{(6.35)}$$

n is the number of stages in the shift register generator. A linear shift register generator basically consists of a shift register with modulo-two adders added to it. The adders can be connected to various stages of the register, as illustrated in Fig. 6.12 for $n = 4$ (i.e., $L = 15$). Note that the shift register initial state cannot be 0.

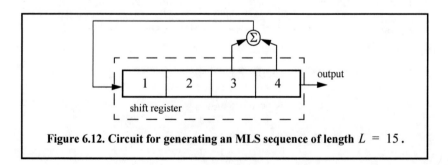

Figure 6.12. Circuit for generating an MLS sequence of length $L = 15$.

The feedback connections associated with a shift register generator determine whether the output sequence will be maximal. For a given size shift register, only a few feedback connections lead to maximal sequence outputs. In order to illustrate this concept, consider the two 5-stage shift register generators shown in Fig. 6.13. The shift register generator shown in Fig. 6.13 a generates a maximal length sequence, as clearly depicted by its state diagram. However, the shift register generator shown in Fig. 6.13 b produces three non-maximal length sequences (depending on the initial state).

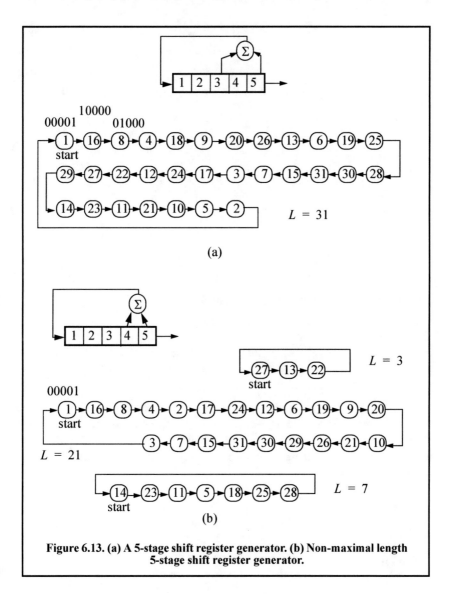

Figure 6.13. (a) A 5-stage shift register generator. (b) Non-maximal length 5-stage shift register generator.

Given an n-stage shift register generator, one would be interested in knowing how many feedback connections will yield maximal length sequences. Zierler[1] showed that the number of maximal length sequences possible for a given n-stage linear shift register generator is given by

$$N_L = \frac{\varphi(2^n - 1)}{n} \tag{6.36}$$

φ is the Euler's totient (Euler's phi) function and is defined by

$$\varphi(k) = k \prod_i \frac{(p_i - 1)}{p_i} \tag{6.37}$$

where p_i are the prime factors of k. Note that when p_i has multiples, only one of them is used. Also note that when k is a prime number, the Euler's phi function is

$$\varphi(k) = k - 1 \tag{6.38}$$

For example, a 3-stage shift register generator will produce

$$N_L = \frac{\varphi(2^3 - 1)}{3} = \frac{\varphi(7)}{3} = \frac{7 - 1}{3} = 2 \tag{6.39}$$

and a 6-stage shift register,

$$N_L = \frac{\varphi(2^6 - 1)}{6} = \frac{\varphi(63)}{6} = \frac{63}{6} \times \frac{(3 - 1)}{3} \times \frac{(7 - 1)}{7} = 6 \tag{6.40}$$

Maximal Length Sequence Characteristic Polynomial

Consider an n-stage maximal length linear shift register whose feedback connections correspond to n, k, m, etc. This maximal length shift register can be described using its characteristic polynomial defined by

$$x^n + x^k + x^m + \ldots + 1 \tag{6.41}$$

where the additions are modulo 2. Therefore, if the characteristic polynomial for an n-stage shift register is known, one can easily determine the register feedback connections and consequently deduce the corresponding maximal length sequence. For example, consider a 6-stage shift register whose characteristic polynomial is

$$x^6 + x^5 + 1 \tag{6.42}$$

1. Zierler, N., *Several Binary-Sequence Generators*, MIT Technical Report No. 95, Sept. 1955.

It follows that the shift register which generates a maximal length sequence is shown in Fig. 6.14.

One of the most important issues associated with generating a maximal length sequence using a linear shift register is determining the characteristic polynomial. This has been and continues to be a subject of research for many radar engineers and designers. It has been shown that polynomials which are both irreducible (not factorable) and primitive will produce maximal length shift register generators.

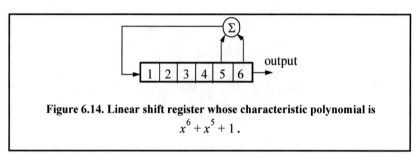

Figure 6.14. Linear shift register whose characteristic polynomial is
$$x^6 + x^5 + 1 .$$

A polynomial of degree n is irreducible if it is not divisible by any polynomial of degree less than n. It follows that all irreducible polynomials must have an odd number of terms. Consequently, only linear shift register generators with an even number of feedback connections can produce maximal length sequences. An irreducible polynomial is primitive if and only if it divides $x^n - 1$ for no value of n less than $2^n - 1$.

The MATLAB function *"prn_ambig.m"* calculates and plots the ambiguity function associated with a given PRN code. Figure 6.15 shows the output of this function for

$u31 = [1 -1 -1 -1 -1 1 -1 1 -1 1 1 1 -1 1 1 -1 -1 -1 -1 1 1 1 1 1 -1 -1 1 1 -1 1 -1 -1]$

Figure 6.16 is similar to Fig. 6.15, except in this case the input maximal length sequence is

$u15=[1 -1 -1 -1 1 1 1 1 -1 1 -1 1 1 -1 -1]$

6.3.2. Polyphase Codes

The signal corresponding to polyphase codes is as that given in Eq. (6.22) and the corresponding ambiguity function was given in Eq. (6.24). The only exception being that the phase θ_n is no longer restricted to $(0, \pi)$. Hence, the coefficient D_n are no longer equal to ± 1 but can be complex depending on the value of θ_n. Polyphase Barker codes have been investigated by many scientists and much is well documented in the literature. In this chapter the discussion will be limited to Frank codes.

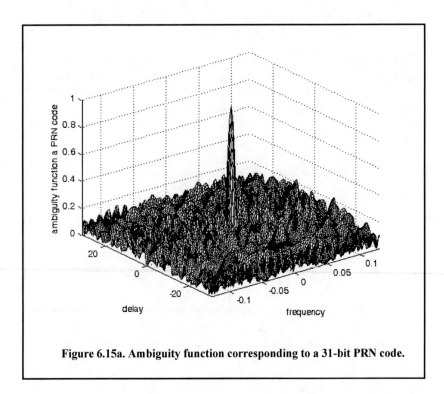

Figure 6.15a. Ambiguity function corresponding to a 31-bit PRN code.

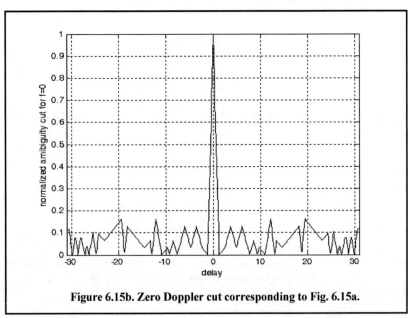

Figure 6.15b. Zero Doppler cut corresponding to Fig. 6.15a.

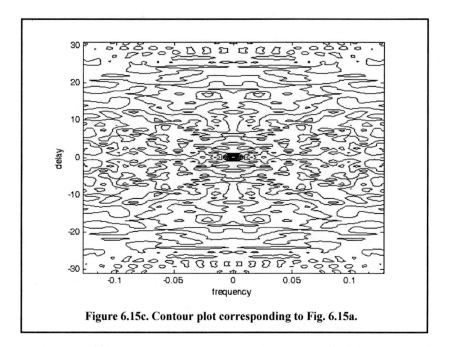

Figure 6.15c. Contour plot corresponding to Fig. 6.15a.

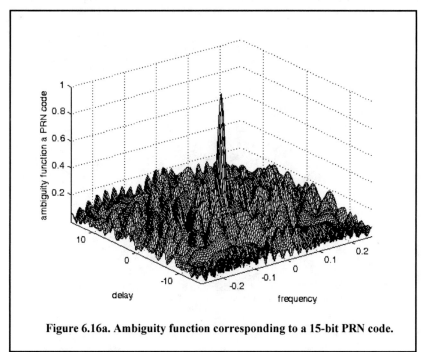

Figure 6.16a. Ambiguity function corresponding to a 15-bit PRN code.

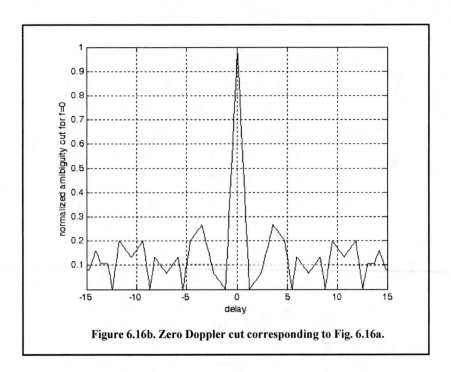

Figure 6.16b. Zero Doppler cut corresponding to Fig. 6.16a.

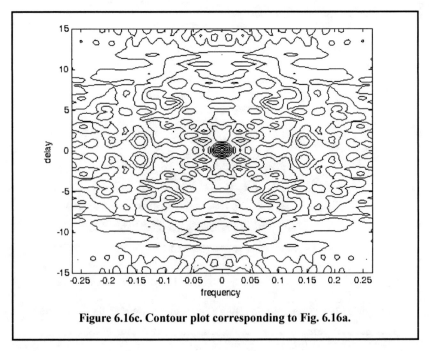

Figure 6.16c. Contour plot corresponding to Fig. 6.16a.

Frank codes

In this case, a single pulse of width T_p is divided into N equal groups; each group is subsequently divided into other N subpulses each of width τ_0. Therefore, the total number of subpulses within each pulse is N^2, and the compression ratio is $\xi = N^2$. As previously, the phase within each subpulse is held constant with respect to some CW reference signal.

A Frank code of N^2 subpulses is referred to as an N-phase Frank code. The first step in computing a Frank code is to divide $360°$ by N and define the result as the fundamental phase increment $\Delta\varphi$. More precisely,

$$\Delta\varphi = 360°/N \tag{6.43}$$

Note that the size of the fundamental phase increment decreases as the number of groups is increased, and because of phase stability, this may degrade the performance of very long Frank codes. For N-phase Frank code the phase of each subpulse is computed from

$$\begin{pmatrix} 0 & 0 & 0 & 0 & \dots & 0 \\ 0 & 1 & 2 & 3 & \dots & N-1 \\ 0 & 2 & 4 & 6 & \dots & 2(N-1) \\ \dots & \dots & \dots & \dots & \dots & \dots \\ \dots & \dots & \dots & \dots & \dots & \dots \\ 0 & (N-1) & 2(N-1) & 3(N-1) & \dots & (N-1)^2 \end{pmatrix} \Delta\varphi \tag{6.44}$$

where each row represents a group, and a column represents the subpulses for that group. For example, a 4-phase Frank code has $N = 4$, and the fundamental phase increment is $\Delta\varphi = (360°/4) = 90°$. It follows that

$$\begin{pmatrix} 0 & 0 & 0 & 0 \\ 0 & 90° & 180° & 270° \\ 0 & 180° & 0 & 180° \\ 0 & 270° & 180° & 90° \end{pmatrix} \Rightarrow \begin{pmatrix} 1 & 1 & 1 & 1 \\ 1 & j & -1 & -j \\ 1 & -1 & 1 & -1 \\ 1 & -j & -1 & j \end{pmatrix} \tag{6.45}$$

Therefore, a Frank code of 16 elements is given by

$$F_{16} = \{1 \ 1 \ 1 \ 1 \ 1 \ j \ -1 \ -j \ 1 \ -1 \ 1 \ -1 \ 1 \ -j \ -1 \ j\} \tag{6.46}$$

A plot of the ambiguity function for F_{16} is shown in Fig. 6.17. Note the thumb-tack shape of the ambiguity function. This plot can be reproduced using the following MATLAB code. The phase increments within each row represent a step-wise approximation of an up-chirp LFM waveform. The phase increments for subsequent rows increase linearly versus time. Thus, the corresponding LFM chirp slopes also increase linearly for subsequent rows. This is illustrated in Fig. 6.18, for F_{16}.

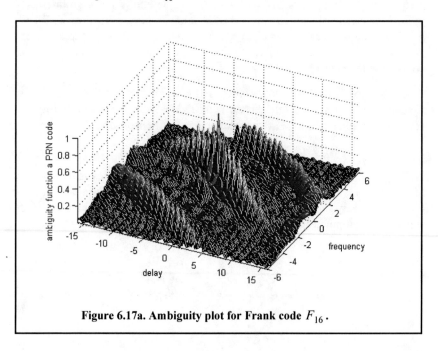

Figure 6.17a. Ambiguity plot for Frank code F_{16}.

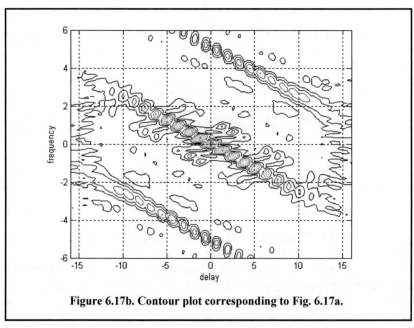

Figure 6.17b. Contour plot corresponding to Fig. 6.17a.

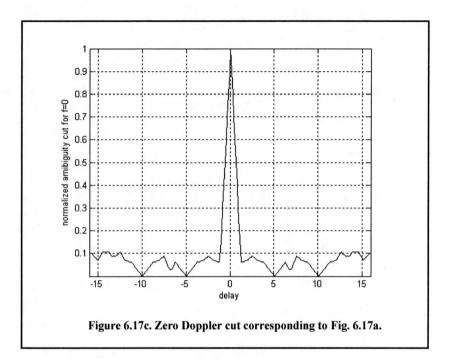

Figure 6.17c. Zero Doppler cut corresponding to Fig. 6.17a.

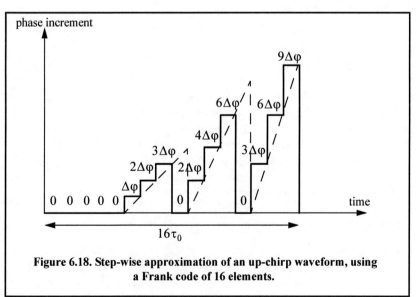

Figure 6.18. Step-wise approximation of an up-chirp waveform, using a Frank code of 16 elements.

6.4. Frequency Codes

Frequency codes are derived from Eq. (6.1) under the condition stated in Eq. (6.9) (i.e., $\theta_n = 0$;*and* $a_n = 1, or\ 0$). The Stepped Frequency Waveform (SFW) discussed in the previous chapter is considered to be a code under this class of discrete coded waveforms. The ambiguity function was derived in Chapter 5 for SFW. In this chapter the focus is on another type of frequency codes that is called the Costas frequency code.

6.4.1. Costas Codes

Construction of Costas codes can be understood in the context of SFW. In SFW, a relatively long pulse of length T_p is divided into N subpulses, each of width τ_0 ($T_p = N\tau_0$). Each group of N subpulses is called a burst. Within each burst the frequency is increased by Δf from one subpulse to the next. The overall burst bandwidth is $N\Delta f$. More precisely,

$$\tau_0 = T_p/N \tag{6.47}$$

and the frequency for the *ith* subpulse is

$$f_i = f_0 + i\Delta f;\ \ i = 1, N \tag{6.48}$$

where f_0 is a constant frequency and $f_0 \gg \Delta f$. It follows that the time-band-width product of this waveform is

$$\Delta f T_p = N^2 \tag{6.49}$$

Costas[1] signals (or codes) are similar to SFW, except that the frequencies for the subpulses are selected in a random fashion, according to some predeter-mined rule or logic. For this purpose, consider the $N \times N$ matrix shown in Fig. 6.19 b. In this case, the rows are indexed from $i = 1, 2, ..., N$ and the columns are indexed from $j = 0, 1, 2, ..., (N-1)$. The rows are used to denote the subpulses and the columns are used to denote the frequency. A *dot* indicates the frequency value assigned to the associated subpulse. In this fashion, Fig. 6.19 a shows the frequency assignment associated with an SFW. Alternatively, the frequency assignments in Fig. 6.19b are chosen randomly. For a matrix of size $N \times N$, there are a total of $N!$ possible ways of assigning the dots (i.e., $N!$ possible codes).

The sequences of dot assignments for which the corresponding ambiguity function approaches an ideal or a *thumb-tack* response are called Costas codes.

1. Costas, J. P., A Study of a Class of Detection Waveforms Having Nearly Ideal Range-Doppler Ambiguity Properties, *Proc. IEEE* 72, 1984, pp. 996-1009.

A near thumb-tack response was obtained by Costas using the following logic: There is only one frequency per time slot (row) and per frequency slot (column). Therefore, for an $N \times N$ matrix the number of possible Costas codes is drastically less than $N!$. For example, there are $N_c = 4$ possible Costas codes for $N = 3$, and $N_c = 40$ possible codes for $N = 5$. It can be shown that the code density, defined as the ratio $N_c/N!$, gets significantly smaller as N becomes larger

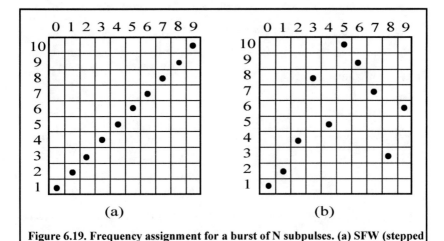

Figure 6.19. Frequency assignment for a burst of N subpulses. (a) SFW (stepped LFM); (b) Costas code of length Nc = 10.

There are numerous analytical ways to generate Costas codes. In this section we will describe two of these methods. First, let q be an odd prime number, and choose the number of subpulses as

$$N = q - 1 \qquad \text{(6.50)}$$

Define γ as the primitive root of q. A primitive root of q (an odd prime number) is defined as γ such that the powers $\gamma, \gamma^2, \gamma^3, ..., \gamma^{q-1}$ modulo q generate every integer from 1 to $q-1$.

In the first method, for an $N \times N$ matrix, label the rows and columns, respectively, as

$$i = 0, 1, 2, ..., (q-2)$$
$$j = 1, 2, 3, ..., (q-1) \qquad \text{(6.51)}$$

Place a dot in the location (i, j) corresponding to f_i if and only if

$$i = (\gamma)^j \ (modulo \ q) \qquad \text{(6.52)}$$

In the next method, Costas code is first obtained from the logic described above; then by deleting the first row and first column from the matrix a new code is generated. This method produces a Costas code of length $N = q-2$.

Define the normalized complex envelope of the Costas signal as

$$x(t) = \frac{1}{\sqrt{N\tau_0}} \sum_{l=0}^{N-1} x_l(t - l\tau_0) \tag{6.53}$$

$$x_l(t) = \begin{pmatrix} \exp(j2\pi f_l t) & 0 \le t \le \tau_0 \\ 0 & elsewhere \end{pmatrix} \tag{6.54}$$

Costas showed that the output of the matched filter is

$$\chi(\tau, f_d) = \frac{1}{N} \sum_{l=0}^{N-1} \exp(j2\pi l f_d \tau) \left\{ \Phi_{ll}(\tau, f_d) + \sum_{\substack{q=0 \\ q \ne l}}^{N-1} \Phi_{lq}(\tau - (l-q)\tau_0, f_d) \right\} \tag{6.55}$$

$$\Phi_{lq}(\tau, f_d) = \left(\tau_0 - \frac{|\tau|}{\tau_0}\right) \frac{\sin\alpha}{\alpha} \exp(-j\beta - j2\pi f_q \tau) \quad , \quad |\tau| \le \tau_1 \tag{6.56}$$

$$\alpha = \pi(f_l - f_q - f_d)(\tau_0 - |\tau|) \tag{6.57}$$

$$\beta = \pi(f_l - f_q - f_d)(\tau_0 + |\tau|) \tag{6.58}$$

Three-dimensional plots of the ambiguity function of Costas signals show the near thumb-tack response of the ambiguity function. All side-lobes, except for a few around the origin, have amplitude $1/N$. Few sidelobes close to the origin have amplitude $2/N$, which is typical of Costas codes. The compression ratio of a Costas code is approximately N.

6.5. Ambiguity Plots for Discrete Coded Waveforms

Plots of the ambiguity function for a given code and the corresponding cuts along zero delay and zero Doppler provide strong indication about the code's characteristics in range and Doppler. Earlier, it was stated that the *goodness* of a given code is measured by its range and Doppler resolution characteristics. Therefore, plotting the ambiguity function of a given code is a key part of the design and analysis of radar waveforms. Unfortunately, some of the formulas for the ambiguity function are rather complicated and fairly difficult to code by the nonexpert programmer. In this section, a numerical technique for plotting the ambiguity function of any code is presented. This technique takes advantage of the computation power of MATLAB by exploiting one of the properties of the ambiguity function. Three-dimensional plots are built successively from cuts of the ambiguity function as different Doppler mismatches.

For this purpose, consider the ambiguity function property given in Eq. (5.8) and repeated here as Eq. (6.59)

$$|\chi(\tau;f_d)|^2 = \left| \int X^*(f)X(f-f_d)e^{-j2\pi f\tau} df \right|^2 \qquad \text{(6.59)}$$

where $X(f)$ is the Fourier transform of the signal $x(t)$. Using Eq. (6.59), one can compute the ambiguity function by first computing the FT of the signal under consideration, delaying it by some value f_d, and then taking the inverse FT. When the signal under consideration is a discrete coded waveform then the Fast Fourier transform is utilized. From this one can compute plots of the ambiguity function using the following technique:

1. Determine the code U under consideration. Note that U may have complex values in accordance with the class of code being considered.
2. Extend the length of the code to the next power of 2 by zero padding (see Chapter 2 for details on interpolation).
3. For better display utilize an FFT whose size is 8 times or higher than the power integer of 2 computed in step 2.
4. Compute the FFT of the extended sequence.
5. Generate vectors of frequency mismatches and delay cuts.
6. Calculate using vector notation the value of $X(f-f_d)$.
7. Compute and store the vector resulting from the point by point multiplication $X^*(f)X(f-f_d)$.
8. Compute the inverse FFT of the product in step 7 for each delay value and store in a two-dimensional (2-D) array.
9. Plot the amplitude square of the resulting 2-D array to generate the ambiguity plot for the specific code under consideration.

An implementation of this algorithm using MATLAB was completed; this program is called *"ambiguity_code.m."* The listing of this program is as follows:

```
function [ambig] = ambiguity_code(uinput)
% Compute and plot the ambiguity function for any give code u
% Compute the ambiguity function by utilizing the FFT
% through combining multiple range cuts
N = size(uinput,2);
tau = N;
code = uinput;
```

```
samp_num = size(code,2) * 10;
n = ceil(log(samp_num) / log(2));
nfft = 2^n;
u(1:nfft) = 0;
j = 0;
for index = 1:10:samp_num
   index;
   j = j+1;
   u(index:index+10-1) = code(j);
end
% set-up the array v
v = u;
delay = linspace(0,5*tau,nfft);
freq_del = 12 / tau /100;
j = 0;
vfft = fft(v,nfft);
for freq = -6/tau:freq_del:6/tau;
   j = j+1;
   exf = exp(sqrt(-1) * 2. * pi * freq .* delay);
   u_times_exf = u .* exf;
   ufft = fft(u_times_exf,nfft);
   prod = ufft .* conj(vfft);
   ambig(j,:) = fftshift(abs(ifft(prod))');
end
freq = linspace(-6,6, size(ambig,1));
delay = linspace(-N,N,nfft);
figure(1)
mesh(delay,freq,(ambig ./ max(max(ambig))))
% colormap([.5 .5 .5])
% colormap(gray)
axis tight
ylabel('frequency')
xlabel('delay')
zlabel('ambiguity function a PRN code')
figure(2)
plot(delay,ambig(51,:)/(max(max(ambig))),'k')
xlabel('delay')
ylabel('normalized amibiguity cut for f=0')
grid
axis tight
figure(3)
contour(delay,freq,(ambig ./ max(max(ambig))))
axis tight
% colormap([.5 .5 .5])
% colormap(gray)
ylabel('frequency')
xlabel('delay')
grid
```

Problems

6.1. Define $\{x_I(n) = 1, -1, 1\}$ and $\{x_Q(n) = 1, 1, -1\}$. (a) Compute the discrete correlations: R_{x_I}, R_{x_Q}, $R_{x_I x_Q}$, and $R_{x_Q x_I}$. (b) A certain radar transmits the signal $s(t) = x_I(t)\cos 2\pi f_0 t - x_Q(t)\sin 2\pi f_0 t$. Assume that the autocorrelation $s(t)$ is equal to $y(t) = y_I(t)\cos 2\pi f_0 t - y_Q(t)\sin 2\pi f_0 t$. Compute and sketch $y_I(t)$ and $y_Q(t)$.

6.2. Consider the 7-bit Barker code, designated by the sequence $x(n)$. (a) Compute and plot the autocorrelation of this code. (b) A radar uses binary phase coded pulses of the form $s(t) = r(t)\cos(2\pi f_0 t)$, where $r(t) = x(0)$, *for* $0 < t < \Delta t$, $r(t) = x(n)$, *for* $n\Delta t < t < (n+1)\Delta t$, and $r(t) = 0$, *for* $t > 7\Delta t$. Assume $\Delta t = 0.5\,\mu s$. (a) Give an expression for the autocorrelation of the signal $s(t)$, and for the output of the matched filter when the input is $s(t - 10\Delta t)$; (b) compute the time bandwidth product, the increase in the peak SNR, and the compression ratio.

6.3. (a) Perform the discrete convolution between the sequence ϕ_{11} defined in Eq. (6.31), and the transversal filter impulse response; and (b) sketch the corresponding transversal filter output.

6.4. Repeat the previous problem for $N = 13$ and $k = 6$. Use Barker code of length 13.

6.5. Develop a Barker code of length 35. Consider both B_{75} and B_{57}.

6.6. The smallest positive primitive root of $q = 11$ is $\gamma = 2$; for $N = 10$ generate the corresponding Costas matrix.

6.7. Compute the discrete autocorrelation for an F_{16} Frank code.

6.8. Generate a Frank code of length 8, i.e., F_8.

6.9. Using the MATLAB program developed in this chapter, plot the matched filter output for a 3-, 4-, and 5-bits Barker code.

Chapter 7 — *Target Detection and Pulse Integration*

7.1. Target Detection in the Presence of Noise

A simplified block diagram of a radar receiver that employs an envelope detector followed by a threshold decision is shown in Fig. 7.1. The input signal to the receiver is composed of the radar echo signal $s(t)$ and additive zero mean white Gaussian noise random process $n(t)$, with variance σ^2. The input noise is assumed to be spatially incoherent and uncorrelated with the signal.

The output of the bandpass intermediate frequency (IF) filter is the signal $v(t)$, which can be written as a bandpass random process. That is,

$$v(t) = v_I(t)\cos\omega_0 t + v_Q(t)\sin\omega_0 t = r(t)\cos(\omega_0 t - \Phi(t))$$
$$v_I(t) = r(t)\cos\Phi(t) \tag{7.1a}$$
$$v_Q(t) = r(t)\sin\Phi(t)$$

$$r(t) = \sqrt{[v_I(t)]^2 + [v_Q(t)]^2}$$
$$\Phi(t) = \left[\tan\left(\frac{v_Q(t)}{v_I(t)}\right)\right]^{-1} \tag{7.1b}$$

where $\omega_0 = 2\pi f_0$ is the radar operating frequency, $r(t)$ is the envelope of $v(t)$, the phase is $\Phi(t) = \mathrm{atan}(v_Q/v_I)$, and the subscripts I, and Q, respectively, refer to the in-phase and quadrature components.

A target is detected when $r(t)$ exceeds the threshold value v_T, where the decision hypotheses are

$$s(t) + n(t) > v_T \Rightarrow Detection$$
$$n(t) > v_T \Rightarrow False\ alarm$$

259

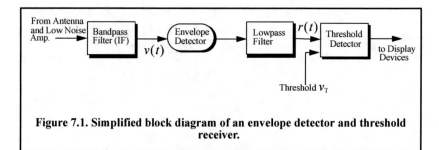

Figure 7.1. Simplified block diagram of an envelope detector and threshold receiver.

The case when the noise subtracts from the signal (while a target is present) to make $r(t)$ smaller than the threshold is called a miss. Radar designers seek to maximize the probability of detection for a given probability of false alarm.

The IF filter output is a complex random variable that is composed of either noise alone or noise plus target return signal (sine wave of amplitude A). The quadrature components corresponding to the case of noise alone are

$$v_I(t) = n_I(t)$$
$$v_Q(t) = n_Q(t) \tag{7.2}$$

and for the second case,

$$v_I(t) = A + n_I(t) = r(t)\cos\Phi(t) \Rightarrow n_I(t) = r(t)\cos\Phi(t) - A$$
$$v_Q(t) = n_Q(t) = r(t)\sin\Phi(t) \tag{7.3}$$

where the noise quadrature components $n_I(t)$ and $n_Q(t)$ are uncorrelated zero mean lowpass Gaussian noise with equal variances, σ^2. The joint Probability Density Function (*pdf*) of the two random variables $n_I; n_Q$ is

$$f_{n_I n_Q}(n_I, n_Q) = \frac{1}{2\pi\sigma^2}\exp\left(-\frac{n_I^2 + n_Q^2}{2\sigma^2}\right) \tag{7.4}$$

$$= \frac{1}{2\pi\sigma^2}\exp\left(-\frac{(r\cos\varphi - A)^2 + (r\sin\varphi)^2}{2\sigma^2}\right)$$

The *pdfs* of the random variables $r(t)$ and $\Phi(t)$, respectively, represent the modulus and phase of $v(t)$. The joint *pdf* for the two random variables $r(t); \Phi(t)$ are derived using a similar approach to that developed in Chapter 3. More precisely,

$$f_{R\Phi}(r, \varphi) = f_{n_I n_Q}(n_I, n_Q)|\mathbf{J} \tag{7.5}$$

where $\mathbf{J}$ is a matrix of derivatives defined by

$$
\boldsymbol{J} = \begin{bmatrix} \dfrac{\partial n_I}{\partial r} & \dfrac{\partial n_I}{\partial \varphi} \\ \dfrac{\partial n_Q}{\partial r} & \dfrac{\partial n_Q}{\partial \varphi} \end{bmatrix} = \begin{bmatrix} \cos\varphi & -r\sin\varphi \\ \sin\varphi & r\cos\varphi \end{bmatrix} \tag{7.6}
$$

The determinant of the matrix of derivatives is called the Jacobian, and in this case it is equal to

$$
|\boldsymbol{J}| = r(t) \tag{7.7}
$$

Substituting Eq. (7.4) and Eq. (7.7) into Eq. (7.5) and collecting terms yield

$$
f_{R\Phi}(r, \varphi) = \frac{r}{2\pi\sigma^2} \exp\left(-\frac{r^2 + A^2}{2\sigma^2}\right) \exp\left(\frac{rA\cos\varphi}{\sigma^2}\right) \tag{7.8}
$$

The *pdf* for $r(t)$ alone is obtained by integrating Eq. (7.8) over φ

$$
f_R(r) = \int_0^{2\pi} f_{R\Phi}(r, \varphi)d\varphi = \frac{r}{\sigma^2}\exp\left(-\frac{r^2 + A^2}{2\sigma^2}\right)\frac{1}{2\pi}\int_0^{2\pi}\exp\left(\frac{rA\cos\varphi}{\sigma^2}\right)d\varphi \tag{7.9}
$$

where the integral inside Eq. (7.9) is known as the modified Bessel function of zero order,

$$
I_0(\beta) = \frac{1}{2\pi}\int_0^{2\pi} e^{\beta\cos\theta}\, d\theta \tag{7.10}
$$

Thus,

$$
f_R(r) = \frac{r}{\sigma^2}I_0\left(\frac{rA}{\sigma^2}\right)\exp\left(-\frac{r^2 + A^2}{2\sigma^2}\right) \tag{7.11}
$$

which is the Rician probability density function. The case when $A/\sigma^2 = 0$ (noise alone) was analyzed in Chapter 3 and the resulting *pdf* is a Rayleigh probability density function

$$
f_R(r) = \frac{r}{\sigma^2}\exp\left(-\frac{r^2}{2\sigma^2}\right) \tag{7.12}
$$

When (A/σ^2) is very large, Eq. (7.11) becomes a Gaussian probability density function of mean A and variance σ^2:

$$f_R(r) \approx \frac{1}{\sqrt{2\pi\sigma^2}} \exp\left(-\frac{(r-A)^2}{2\sigma^2}\right) \tag{7.13}$$

Figure 7.2 shows plots for the Rayleigh and Gaussian densities.

The density function for the random variable Φ is obtained from

$$f_\Phi(\varphi) = \int_0^r f_{R\Phi}(r, \varphi) \, dr \tag{7.14}$$

While the detailed derivation is left as an exercise, the result of Eq. (7.14) is

$$f_\Phi(\varphi) = \frac{1}{2\pi} \exp\left(\frac{-A^2}{2\sigma^2}\right) + \frac{A\cos\varphi}{\sqrt{2\pi\sigma^2}} \exp\left(\frac{-(A\sin\varphi)^2}{2\sigma^2}\right) F\left(\frac{A\cos\varphi}{\sigma}\right) \tag{7.15}$$

where

$$F(x) = \int_{-\infty}^{x} \frac{1}{\sqrt{2\pi}} e^{-\zeta^2/2} \, d\xi \tag{7.16}$$

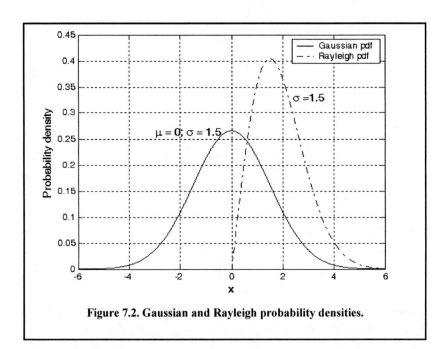

Figure 7.2. Gaussian and Rayleigh probability densities.

The function $F(x)$ can be found tabulated in most mathematical formula reference books. Note that for the case of noise alone ($A = 0$), Eq. (7.15) collapses to a uniform *pdf* over the interval $\{0, 2\pi\}$. One excellent approximation for the function $F(x)$ is

$$F(x) = 1 - \left(\frac{1}{0.661x + 0.339\sqrt{x^2 + 5.51}}\right) \frac{1}{\sqrt{2\pi}} e^{-x^2/2} \qquad x \geq 0 \qquad (7.17)$$

and for negative values of x

$$F(-x) = 1 - F(x) \qquad (7.18)$$

7.2. Probability of False Alarm

The probability of false alarm P_{fa} is defined as the probability that a sample r of the signal $r(t)$ will exceed the threshold voltage v_T when noise alone is present in the radar:

$$P_{fa} = \int_{v_T}^{\infty} \frac{r}{\sigma^2} \exp\left(-\frac{r^2}{2\sigma^2}\right) dr = \exp\left(\frac{-v_T^2}{2\sigma^2}\right) \qquad (7.19)$$

$$v_T = \sqrt{2\sigma^2 \ln\left(\frac{1}{P_{fa}}\right)} \qquad (7.20)$$

Figure 7.3 shows a plot of the normalized threshold versus the probability of false alarm. It is evident from this figure that P_{fa} is very sensitive to small changes in the threshold value. The false alarm time T_{fa} is related to the probability of false alarm by

$$T_{fa} = t_{int}/P_{fa} \qquad (7.21)$$

where t_{int} represents the radar integration time, or the average time that the output of the envelope detector will pass the threshold voltage. Since the radar operating bandwidth B is the inverse of t_{int}, by substituting Eq. (7.19) into Eq. (7.20), we can write T_{fa} as

$$T_{fa} = \frac{1}{B} \exp\left(\frac{v_T^2}{2\sigma^2}\right) \qquad (7.22)$$

Minimizing T_{fa} means increasing the threshold value, and as a result the radar maximum detection range is decreased. The choice of an acceptable value for T_{fa} becomes a compromise depending on the radar mode of operation.

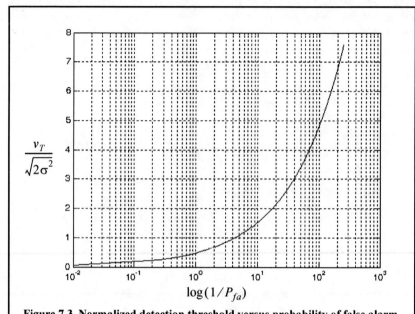

Figure 7.3. Normalized detection threshold versus probability of false alarm.

The false alarm number is defined as

$$n_{fa} = \frac{-\ln(2)}{\ln(1 - P_{fa})} \approx \frac{\ln(2)}{P_{fa}} \tag{7.23}$$

Other slightly different definitions for the false alarm number exist in the literature, causing a source of confusion for many non-expert readers. Other than the definition in Eq. (7.23), the most commonly used definition for the false alarm number is the one introduced by Marcum (1960). Marcum defines the false alarm number as the reciprocal of P_{fa}. In this text, the definition given in Eq. (7.23) is always assumed. Hence, a clear distinction is made between Marcum's definition of the false alarm number and the definition in Eq. (7.23).

7.3. Probability of Detection

The probability of detection P_D is the probability that a sample r of $r(t)$ will exceed the threshold voltage in the case of noise plus signal,

$$P_D = \int_{v_T}^{\infty} \frac{r}{\sigma^2} I_0\left(\frac{rA}{\sigma^2}\right) \exp\left(-\frac{r^2 + A^2}{2\sigma^2}\right) dr \tag{7.24}$$

If we assume that the radar signal is a sine waveform with amplitude A, then its power is $A^2/2$. Now, by using $SNR = A^2/2\sigma^2$ (single-pulse SNR) and $(v_T^2/2\sigma^2) = \ln(1/P_{fa})$, then Eq. (7.24) can be rewritten as

$$P_D = \int_{\sqrt{2\sigma^2\ln(1/p_{fa})}}^{\infty} \frac{r}{\sigma^2}I_0\left(\frac{rA}{\sigma^2}\right)\exp\left(-\frac{r^2+A^2}{2\sigma^2}\right)dr = Q\left[\sqrt{\frac{A^2}{\sigma^2}}, \sqrt{2\ln\left(\frac{1}{P_{fa}}\right)}\right] \quad (7.25)$$

$$Q[\alpha, \beta] = \int_{\beta}^{\infty} \zeta I_0(\alpha\zeta)e^{-(\zeta^2+\alpha^2)/2} \, d\zeta \quad (7.26)$$

Q is called Marcum's Q-function. When P_{fa} is small and P_D is relatively large so that the threshold is also large, Eq. (7.25) can be approximated by

$$P_D \approx F\left(\frac{A}{\psi} - \sqrt{2\ln\left(\frac{1}{P_{fa}}\right)}\right) \quad (7.27)$$

where $F(x)$ is given by Eq. (7.16). Many approximations for computing Eq. (7.25) can be found throughout the literature. One very accurate approximation presented by North (1963) is given by

$$P_D \approx 0.5 \times erfc(\sqrt{-\ln P_{fa}} - \sqrt{SNR + 0.5}) \quad (7.28)$$

where the complementary error function is

$$erfc(z) = 1 - \frac{2}{\sqrt{\pi}}\int_0^z e^{-v^2} dv \quad (7.29)$$

The integral given in Eq. (7.25) is complicated and can be computed using numerical integration techniques. Parl[1] developed an excellent algorithm to numerically compute this integral. It is summarized as follows:

$$Q[a, b] = \begin{cases} \dfrac{\alpha_n}{2\beta_n}\exp\left(\dfrac{(a-b)^2}{2}\right) & a < b \\[3mm] 1 - \left(\dfrac{\alpha_n}{2\beta_n}\exp\left(\dfrac{(a-b)^2}{2}\right)\right) & a \geq b \end{cases} \quad (7.30)$$

---

1. Parl, S., A New Method of Calculating the Generalized Q Function, *IEEE Trans. Information Theory*, Vol. IT-26, January 1980, pp. 121-124.

$$\alpha_n = d_n + \frac{2n}{ab}\alpha_{n-1} + \alpha_{n-2} \tag{7.31}$$

$$\beta_n = 1 + \frac{2n}{ab}\beta_{n-1} + \beta_{n-2} \tag{7.32}$$

$$d_{n+1} = d_n d_1 \tag{7.33}$$

$$\alpha_0 = \begin{Bmatrix} 1 & a < b \\ 0 & a \geq b \end{Bmatrix} \tag{7.34}$$

$$d_1 = \begin{Bmatrix} a/b & a < b \\ b/a & a \geq b \end{Bmatrix} \tag{7.35}$$

$\alpha_{-1} = 0.0$, $\beta_0 = 0.5$, and $\beta_{-1} = 0$. The recursive Eq. (7.30) through Eq. (7.33) are computed continuously until $\beta_n > 10^p$ for values of $p \geq 3$. The accuracy of the algorithm is enhanced as the value of p is increased. The MATLAB function *"marcumsq.m"* implements Parl's algorithm to calculate the probability of detection defined in Eq. (7.24). The syntax is as follows:

$$Pd = marcumsq(alpha, beta)$$

where *alpha* and *beta* are from Eq. (7.26). Figure 7.4 shows plots of the probability of detection, P_D, versus the single pulse SNR, with the P_{fa} as a parameter using this function. The following MATLAB program can be used to reproduce Fig. 7.4. It uses the function *"marcumsq.m."*

```
% This program is used to produce Fig. 7.4
close all; clear all;
for nfa = 2:2:12
  b = sqrt(-2.0 * log(10^(-nfa)));
  index = 0;
  hold on
  for snr = 0:.1:18
    index = index +1;
    a = sqrt(2.0 * 10^(.1*snr));
    pro(index) = marcumsq(a,b);
  end
  x = 0:.1:18;
  set(gca,'ytick',[.1 .2 .3 .4 .5 .6 .7 .75 .8 .85 .9 .95 .9999])
  set(gca,'xtick',[1 2 3 4 5 6 7 8 9 10 11 12 13 14 15 16 17 18])
  loglog(x, pro,'k');
end
hold off
xlabel ('Single pulse SNR in dB'); ylabel ('Probability of detection')
grid
```

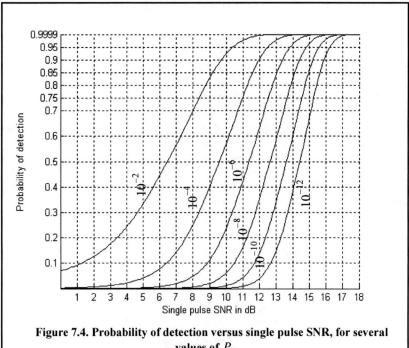

Figure 7.4. Probability of detection versus single pulse SNR, for several values of P_{fa}.

7.4. Pulse Integration

When a target is located within the radar beam during a single scan, it may reflect several pulses. By adding the returns from all pulses returned by a given target during a single scan, the radar sensitivity (SNR) can be increased. The number of returned pulses depends on the antenna scan rate and the radar PRF. More precisely, the number of pulses returned from a given target is given by

$$n_P = \frac{\theta_a T_{sc} f_r}{2\pi} \tag{7.36}$$

where θ_a is the azimuth antenna beamwidth, T_{sc} is the scan time, and f_r is the radar PRF. The number of reflected pulses may also be expressed as

$$n_P = \frac{\theta_a f_r}{\dot{\theta}_{scan}} \tag{7.37}$$

where $\dot{\theta}_{scan}$ is the antenna scan rate in degrees per second. Note that when using Eq. (7.36), θ_a is expressed in radians, while when using Eq. (7.37), it is expressed in degrees. As an example, consider a radar with an azimuth antenna beamwidth $\theta_a = 3°$, antenna scan rate $\dot{\theta}_{scan} = 45°/\sec$ (antenna scan time, $T_{sc} = 8\sec$), and a PRF $f_r = 300Hz$. Using either Eq. (7.36) or Eq. (7.37) yields $n_P = 20$ pulses.

The process of adding radar returns from many pulses is called radar pulse integration. Pulse integration can be performed on the quadrature components prior to the envelope detector. This is called coherent integration or predetection integration. Coherent integration preserves the phase relationship between the received pulses. Thus a buildup in the signal amplitude is achieved. Alternatively, pulse integration performed after the envelope detector (where the phase relation is destroyed) is called noncoherent or postdetection integration.

Radar designers should exercise caution when utilizing pulse integration for the following reasons. First, during a scan a given target will not always be located at the center of the radar beam (i.e., have maximum gain). In fact, during a scan a given target will first enter the antenna beam at the 3-dB point, reach maximum gain, and finally leave the beam at the 3-dB point again. Thus, the returns do not have the same amplitude even though the target RCS may be constant and all other factors that may introduce signal loss remain the same.

Other factors that may introduce further variation to the amplitude of the returned pulses include target RCS and propagation path fluctuations. Additionally, when the radar employs a very fast scan rate, an additional loss term is introduced due to the motion of the beam between transmission and reception. This is referred to as scan loss. A distinction should be made between scan loss due to a rotating antenna (which is described here) and the term scan loss that is normally associated with phased array antennas (which takes on a different meaning in that context).

Finally, since coherent integration utilizes the phase information from all integrated pulses, it is critical that any phase variation between all integrated pulses be known with a great level of confidence. Consequently, target dynamics (such as target range, range rate, tumble rate, RCS fluctuation) must be estimated or computed accurately so that coherent integration can be meaningful. In fact, if a radar coherently integrates pulses from targets without proper knowledge of the target dynamics, it suffers a loss in SNR rather than the expected SNR buildup. Knowledge of target dynamics is not as critical when employing noncoherent integration; nonetheless, target range rate must be estimated so that only the returns from a given target within a specific range bin are integrated. In other words, one must avoid range walk (i.e., having a target cross between adjacent range bins during a single scan).

A comprehensive analysis of pulse integration should take into account issues such as the probability of detection P_D, probability of false alarm P_{fa}, the target statistical fluctuation model, and the noise or interference of statistical models. This is the subject of the rest of this chapter.

7.4.1. Coherent Integration

In coherent integration, when a perfect integrator is used (100% efficiency), to integrate n_P pulses, the SNR is improved by the same factor. Otherwise, integration loss occurs, which is always the case for noncoherent integration. Coherent integration loss occurs when the integration process is not optimum. This could be due to target fluctuation, instability in the radar local oscillator, or propagation path changes.

Denote the single pulse SNR required to produce a given probability of detection as $(SNR)_1$. The SNR resulting from coherently integrating n_P pulses is then given by

$$(SNR)_{CI} = n_P (SNR)_1 \qquad (7.38)$$

Coherent integration cannot be applied over a large number of pulses, particularly if the target RCS is varying rapidly. If the target radial velocity is known and no acceleration is assumed, the maximum coherent integration time is limited to

$$t_{CI} = \sqrt{\lambda / 2 a_r} \qquad (7.39)$$

where λ is the radar wavelength and a_r is the target radial acceleration. Coherent integration time can be extended if the target radial acceleration can be compensated for by the radar.

In order to demonstrate the improvement in the SNR using coherent integration, consider the case where the radar return signal contains both signal plus additive noise. The mth pulse is

$$y_m(t) = s(t) + n_m(t) \qquad (7.40)$$

where $s(t)$ is the radar signal return of interest and $n_m(t)$ is white uncorrelated additive noise signal with variance σ^2. Coherent integration of n_P pulses yields

$$z(t) = \frac{1}{n_P} \sum_{m=1}^{n_P} y_m(t) = \sum_{m=1}^{n_P} \frac{1}{n_P} [s(t) + n_m(t)] = s(t) + \sum_{m=1}^{n_P} \frac{1}{n_P} n_m(t) \qquad (7.41)$$

The total noise power in $z(t)$ is equal to the variance. More precisely,

$$\sigma_{n_P}^2 = E\left[\left(\sum_{m=1}^{n_P} \frac{1}{n_P} n_m(t) \right) \left(\sum_{l=1}^{n_P} \frac{1}{n_P} n_l(t) \right)^* \right] \qquad (7.42)$$

where E is the expected value operator. It follows that

$$\sigma_{n_P}^2 = \frac{1}{n_P^2} \sum_{m,l=1}^{n_P} E[n_m(t)n_l*(t)] = \frac{1}{n_P^2} \sum_{m,l=1}^{n_P} \sigma_{ny}^2 \delta_{ml} = \frac{1}{n_P} \sigma_{ny}^2 \qquad (7.43)$$

where σ_{ny}^2 is the single pulse noise power and δ_{ml} is equal to zero for $m \neq l$ and unity for $m = l$. Observation of Eqs. (7.41) and (7.42) shows that the desired signal power after coherent integration is unchanged, while the noise power is reduced by the factor $1/n_P$. Thus, the SNR after coherent integration is improved by n_P.

7.4.2. Noncoherent Integration

When the phase of the integrated pulses is not known so that coherent integration is no longer possible, another form of pulse integration is done. In this case, pulse integration is performed by adding (integrating) the individual pulses' envelopes or the square of their envelopes. Thus, the term noncoherent integration is adopted. A block diagram of radar receiver utilizing noncoherent integration is illustrated in Fig. 7.5.

The performance difference (measured in SNR) between the linear envelope detector and the quadratic (square law) detector is practically negligible. Robertson (1967) showed that this difference is typically less than $0.2dB$; he showed that the performance difference is higher than $0.2dB$ only for cases where $n_P > 100$ and $P_D < 0.01$. Both of these conditions are of no practical significance in radar applications. It is much easier to analyze and implement the square law detector in real hardware than is the case for the envelope detector. Therefore, most authors make no distinction between the type of detector used when referring to noncoherent integration, and the square law detector is almost always assumed. The analysis presented in this book will always assume, unless indicated otherwise, noncoherent integration using the square law detector.

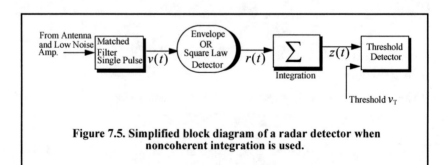

Figure 7.5. Simplified block diagram of a radar detector when noncoherent integration is used.

7.4.3. Improvement Factor and Integration Loss

Noncoherent integration is less efficient than coherent integration. Actually, the noncoherent integration gain is always smaller than the number of noncoherently integrated pulses. This loss in integration is referred to as postdetection or square-law detector loss.

Define $(SNR)_{NCI}$ as the SNR required to achieve a specific P_D given a particular P_{fa} when n_P pulses are integrated noncoherently. Also denote the single pulse SNR as $(SNR)_1$. It follows that

$$(SNR)_{NCI} = (SNR)_1 \times I(n_P) \tag{7.44}$$

where $I(n_P)$ is called the integration improvement factor. An empirically derived expression for the improvement factor that is accurate within $0.8dB$ is reported in Peebles (1998) as

$$[I(n_P)]_{dB} = 6.79(1 + 0.253P_D)\left(1 + \frac{\log(1/P_{fa})}{46.6}\right)\log(n_P) \tag{7.45}$$

$$(1 - 0.140\log(n_P) + 0.018310(\log n_P)^2)$$

The top part of Fig. 7.6 shows plots of the integration improvement factor as a function of the number of integrated pulses with P_D and P_{fa} as parameters using Eq. (7.45). The integration loss in dB is defined as

$$[L_{NCI}]_{dB} = 10\log n_P - [I(n_P)]_{dB} \tag{7.46}$$

The lower part of Fig. 7.6 shows plots of the corresponding integration loss versus n_P with P_D and P_{fa} as parameters. This figure can be reproduced using the following MATLAB code which uses MATLAB function *"improv_fac.m."*

```
% This program is used to produce Fig. 7.6
% It uses the function "improv_fac.m".
clear all;
close all;
Pfa = [1e-2, 1e-6, 1e-8, 1e-10];
Pd = [.5 .8 .95 .99];
np = linspace(1,1000,10000);
I(1,:) = improv_fac (np, Pfa(1), Pd(1));
I(2,:) = improv_fac (np, Pfa(2), Pd(2));
I(3,:) = improv_fac (np, Pfa(3), Pd(3));
I(4,:) = improv_fac (np, Pfa(4), Pd(4));
index = [1 2 3 4];
L(1,:) = 10.*log10(np) - I(1,:);
L(2,:) = 10.*log10(np) - I(2,:);
L(3,:) = 10.*log10(np) - I(3,:);
```

```
L(4,:) = 10.*log10(np) - I(4,:);
subplot(2,1,2);
semilogx (np, L(1,:), 'k:', np, L(2,:), 'k-.', ...
np, L(3,:), 'k-.', np, L(4,:), 'k')
xlabel ('Number of pulses');
ylabel ('Integration loss in dB')
axis tight; grid
subplot(2,1,1);
semilogx (np, I(1,:), 'k:', np, I(2,:), 'k-.', np, ...
 I(3,:), 'k--', np, I(4,:), 'k')
xlabel ('Number of pulses');
ylabel ('Improvement factor in dB')
legend ('pd=.5, Pfa=1e-2','pd=.8, Pfa=1e-6','pd=.95, ...
Pfa=1e-8','pd=.99, Pfa=1e-10');
grid;
axis tight
```

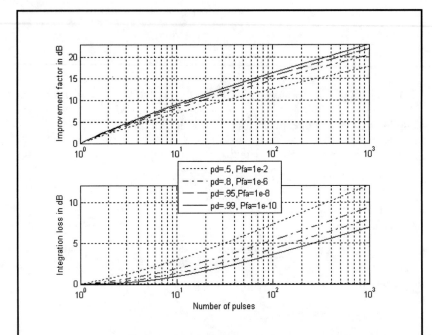

Figure 7.6. Typical plots for the improvement factor and integration loss versus number of noncoherently integrated pulses.

7.5. Target Fluctuation

Target detection utilizing the square law detector was first analyzed by Marcum[1], where he assumed a constant RCS (nonfluctuating target). This work was extended by Swerling[2] to four distinct cases of target RCS fluctuation. These cases have come to be known as Swerling models. They are Swerling I, Swerling II, Swerling III, and Swerling IV. The constant RCS case analyzed by Marcum is widely known as Swerling 0 or equivalently Swerling V. Target fluctuation introduces an additional loss factor in the SNR as compared to the case where fluctuation is not present given the same P_D and P_{fa}.

Swerling I targets have constant amplitude over one antenna scan or observation interval; however, a Swerling I target amplitude varies independently from scan to scan according to a chi-square probability density function with two degrees of freedom. The amplitude of Swerling II targets fluctuates independently from pulse to pulse according to a chi-square probability density function with two degrees of freedom. Target fluctuation associated with a Swerling III model is from scan to scan according to a chi-square probability density function with four degrees of freedom. Finally, the fluctuation of Swerling IV targets is from pulse to pulse according to a chi-square probability density function with four degrees of freedom.

Swerling showed that the statistics associated with Swerling I and II models apply to targets consisting of many small scatterers of comparable RCS values, while the statistics associated with Swerling III and IV models apply to targets consisting of one large RCS scatterer and many small equal RCS scatterers. Noncoherent integration can be applied to all four Swerling models; however, coherent integration cannot be used when the target fluctuation is either Swerling II or Swerling IV. This is because the target amplitude decorrelates from pulse to pulse (fast fluctuation) for Swerling II and IV models, and thus phase coherency cannot be maintained.

The chi-square *pdf* with $2N$ degrees of freedom can be written as

$$f_X(x) = \frac{N}{(N-1)! \sqrt{\sigma_x^2}} \left(\frac{Nx}{\sigma_x}\right)^{N-1} \exp\left(-\frac{Nx}{\sigma_x}\right) \qquad (7.47)$$

where σ_x is the standard deviation for the RCS value. Using this equation, the *pdf* associated with Swerling I and II targets can be obtained by letting $N = 1$, which yields a Rayleigh *pdf*. More precisely,

1. Marcum, J. I., A Statistical Theory of Target Detection by Pulsed Radar, *IRE Transactions on Information Theory,* Vol IT-6, pp. 59-267, April 1960.

2. Swerling, P., Probability of Detection for Fluctuating Targets, *IRE Transactions on Information Theory*, Vol IT-6, pp. 269-308, April 1960.

$$f_X(x) = \frac{1}{\sigma_x} \exp\left(-\frac{x}{\sigma_x}\right) \qquad x \ge 0 \qquad \text{(7.48)}$$

Letting $N = 2$ yields the *pdf* for Swerling III and IV type targets,

$$f_X(x) = \frac{4x}{\sigma_x^2} \exp\left(-\frac{2x}{\sigma_x}\right) \qquad x \ge 0 \qquad \text{(7.49)}$$

7.6. Probability of False Alarm Formulation for a Square Law Detector

Computation of the general formula for the probability of false alarm P_{fa} and subsequently the rest of square law detection theory requires knowledge and good understating of the incomplete Gamma function. Hence, those readers who are not familiar with this function are advised to read Appendix 7.A before proceeding with the rest of this chapter.

DiFranco and Rubin[1] derived a general form relating the threshold and P_{fa} for any number of pulses when noncoherent integration is used. The square law detector under consideration is shown in Fig. 7.7. There are $n_p \ge 2$ pulses integrated noncoherently and the noise power (variance) is σ^2.

The complex envelope in terms of the quadrature components is given by

$$\tilde{r}(t) = r_I(t) + jr_Q(t) \qquad \text{(7.50)}$$

thus, the square of the complex envelope is

$$|\tilde{r}(t)|^2 = r_I^2(t) + r_Q^2(t) \qquad \text{(7.51)}$$

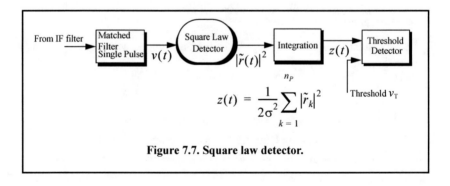

$$z(t) = \frac{1}{2\sigma^2} \sum_{k=1}^{n_P} |\tilde{r}_k|^2$$

Figure 7.7. Square law detector.

1. DiFranco, J. V. and Rubin, W. L., *Radar Detection*, Artech House, Norwood, MA 1980.

The samples $|\tilde{r}_k|^2$ are computed from the samples of $\tilde{r}(t)$ evaluated at $t = t_k$; $k = 1, 2, \ldots, n_P$. It follows that

$$Z = \frac{1}{2\sigma^2} \sum_{k=1}^{n_P} [r_I^2(t_k) + r_Q^2(t_k)] \tag{7.52}$$

The random variable Z is the sum of $2n_P$ squares of random variables, each of which is a Gaussian random variable with variance σ^2. Thus, using the analysis developed in Chapter 3, the *pdf* for the random variable Z is given by

$$f_Z(z) = \begin{cases} \dfrac{z^{n_P-1} e^{-z}}{\Gamma(n_P)} & z \geq 0 \\ 0 & z < 0 \end{cases} \tag{7.53}$$

Consequently, the probability of false alarm given a threshold value v_Y is

$$P_{fa} = Prob\{Z \geq v_T\} = \int_{v_T}^{\infty} \frac{z^{n_P-1} e^{-z}}{\Gamma(n_P)} \, dz \tag{7.54}$$

and using analysis provided in Appendix 7.A yields

$$P_{fa} = 1 - \Gamma_I\left(\frac{v_T}{\sqrt{n_P}}, n_P - 1\right) \tag{7.55}$$

Using the algebraic expression for the incomplete Gamma function, Eq. (7.55) can be written as

$$P_{fa} = e^{-v_T} \sum_{k=0}^{n_P-1} \frac{v_T^k}{k!} = 1 - e^{-v_T} \sum_{k=n_P}^{\infty} \frac{v_T^k}{k!} \tag{7.56}$$

The threshold value v_T can then be approximated by the recursive formula used in the Newton-Raphson method. More precisely,

$$v_{T,m} = v_{T,m-1} - \frac{G(v_{T,m-1})}{G'(v_{T,m-1})} \quad ; \quad m = 1, 2, 3, \ldots \tag{7.57}$$

The iteration is terminated when $|v_{T,m} - v_{T,m-1}| < v_{T,m-1}/10000.0$. The functions G and G' are

$$G(v_{T,m}) = (0.5)^{n_P/n_{fa}} - \Gamma_I(v_T, n_P) \tag{7.58}$$

$$G'(v_{T,\,m}) = -\,\frac{e^{-v_T}\,v_T^{\,n_P-1}}{(n_P-1)!}$$ (7.59)

The initial value for the recursion is

$$v_{T,\,0} = n_P - \sqrt{n_P} + 2.3\ \sqrt{-\log P_{fa}}\ (\sqrt{-\log P_{fa}} + \sqrt{n_P}\ -1)$$ (7.60)

Figure 7.8 shows plots of the threshold value versus the number of integrated pulses for several values of n_{fa}; remember that $P_{fa} \approx \ln(2)/n_{fa}$. This figure can be reproduced using the following MATLAB code which utilizes the MATLAB function *"threshold.m"*

```
% Use this program to reproduce Fig. 7.8 of text
clear all; close all;
for n= 1: 1:10000
  [pfa1 y1(n)] = threshold(1e4,n);
  [pfa2 y3(n)] = threshold(1e8,n);
  [pfa3 y4(n)] = threshold(1e12,n);
end
n =1:1:10000;
loglog(n,y1,'k',n,y3,'k--',n,y4,'k-.');
xlabel ('Number of pulses');
ylabel ('Threshold');
legend('nfa=1e4','nfa=1e8','nfa=1e12'); grid
```

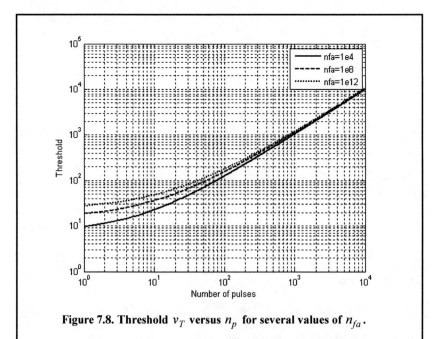

Figure 7.8. Threshold v_T versus n_p for several values of n_{fa}.

7.6.1. Square Law Detection

The *pdf* for the linear envelope $r(t)$ was derived earlier and it is given in Eq. (7.11). Define a new dimensionless variable y as

$$y_n = r_n / \sigma \tag{7.61}$$

and also define

$$\mathfrak{R}_p = A^2 / \sigma^2 = 2SNR \tag{7.62}$$

σ^2 is the noise variance. It follows that the *pdf* for the new variable is

$$f_{Y_n}(y_n) = f_{R_n}(r_n) \left| \frac{dr_n}{dy_n} \right| = y_n \ I_0(y_n \sqrt{\mathfrak{R}_p}) \ \exp\left(\frac{-(y_n^2 + \mathfrak{R}_p)}{2} \right) \tag{7.63}$$

The output of a square law detector for the *nth* pulse is proportional to the square of its input. Thus, it is convenient to define a new change variable,

$$z_n = \frac{1}{2} y_n^2 \tag{7.64}$$

The *pdf* for the variable at the output of the square law detector is given by

$$f_{Z_n}(x_n) = f(y_n) \left| \frac{dy_n}{dz_n} \right| = \exp\left(-\left(z_n + \frac{\mathfrak{R}_p}{2} \right) \right) I_0(\sqrt{2z_n \mathfrak{R}_p}) \tag{7.65}$$

Noncoherent integration of n_p pulses is implemented as

$$z = \sum_{n=1}^{n_P} \frac{1}{2} y_n^2 \tag{7.66}$$

Again, $n_P \geq 2$. Since the random variables y_n are independent, the *pdf* for the variable z is

$$f(z) = f((y_1) \otimes f(y_2) \otimes \ldots \otimes f(y_{n_p})) \tag{7.67}$$

The operator $\otimes$ symbolically indicates convolution. The characteristic functions for the individual *pdf*s can then be used to compute the joint *pdf* for Eq. (7.69). The result is

$$f_Z(z) = \left(\frac{2z}{n_P \mathfrak{R}_p} \right)^{(n_P - 1)/2} \exp\left(-z - \frac{1}{2} n_P \mathfrak{R}_p \right) I_{n_P - 1}(\sqrt{2n_P z \mathfrak{R}_p}) \tag{7.68}$$

$I_{n_p - 1}$ is the modified Bessel function of order $n_P - 1$. Substituting Eq. (7.62) into (7.68) yields

$$f_Z(z) = \left(\frac{z}{n_p SNR}\right)^{(n_p-1)/2} e^{(-z-n_p SNR)} I_{n_p-1}(2\sqrt{n_p z SNR}) \qquad (7.69)$$

When target fluctuation is not present (i.e., Swerling 0), the probability of detection is obtained by integrating $f_Z(z)$ from the threshold value to infinity. The probability of false alarm is obtained by letting $\Re_p$ be zero and integrating the *pdf* from the threshold value to infinity. More specifically,

$$P_D\big|_{SNR} = \int_{v_T}^{\infty} \left(\frac{z}{n_p SNR}\right)^{(n_p-1)/2} e^{(-z-n_p SNR)} I_{n_p-1}(2\sqrt{n_p z SNR}) dz \qquad (7.70)$$

Which can be rewritten as

$$P_D\big|_{SNR} = e^{-n_p SNR}\left(\sum_{k=0}^{\infty}\frac{(n_p SNR)^k}{k!}\right)\left(\sum_{j=0}^{n_p-1+k}\frac{e^{-v_T} v_T^j}{j!}\right) \qquad (7.71)$$

Alternatively, when target fluctuation is present, then the *pdf* is calculated using the conditional probability density function of Eq. (7.70) with respect to the SNR value of the target fluctuation type. In general, given a fluctuating target with SNR^F, where the superscript indicates fluctuation, the expression for the probability of detection is

$$P_D\big|_{SNR^F} = \int_0^{\infty} P_D\big|_{SNR} f_Z(z^F/SNR^F) dz = \qquad (7.72)$$

$$\int_0^{\infty} P_D\big|_{SNR}\left(\frac{z^F}{n_p SNR^F}\right)^{(n_p-1)/2} e^{(-z^F-n_p SNR^F)} I_{n_p-1}(2\sqrt{n_p z^F SNR^F}) dz$$

Remember that target fluctuation introduces an additional loss term in the SNR. It follows that for the same P_D given the same P_{fa} and the same n_p, $SNR^F > SNR$. One way to calculate this additional SNR is to first compute the required SNR given no fluctuation then add to it the amount of target fluctuation loss to get the required value for SNR^F. How to calculate this fluctuation loss will be addressed later on in this chapter. Meanwhile, hereon after, the superscript $\{^F\}$ will be dropped and it will always be assumed.

7.7. Probability of Detection Calculation

Marcum defined the probability of false alarm for the case when $n_p > 1$ as

$$P_{fa} \approx \ln(2)(n_p/n_{fa}) \qquad (7.73)$$

The single pulse probability of detection for nonfluctuating targets is given in Eq. (7.25). When $n_P > 1$, the probability of detection is computed using the Gram-Charlier series. In this case, the probability of detection is

$$P_D \cong \frac{erfc(V/\sqrt{2})}{2} - \frac{e^{-V^2/2}}{\sqrt{2\pi}}[C_3(V^2 - 1) + C_4 V(3 - V^2) \tag{7.74}$$

$$- C_6 V(V^4 - 10V^2 + 15)]$$

where the constants C_3, C_4, and C_6 are the Gram-Charlier series coefficients, and the variable V is

$$V = \frac{v_T - n_P(1 + SNR)}{\varpi} \tag{7.75}$$

In general, values for C_3, C_4, C_6, and ϖ vary depending on the target fluctuation type.

7.7.1. Swerling 0 Target Detection

For Swerling 0 (Swerling V) target fluctuations, the probability of detection is calculated using Eq. (7.74). In this case, the Gram-Charlier series coefficients are

$$C_3 = -\frac{SNR + 1/3}{\sqrt{n_p}(2SNR + 1)^{1.5}} \tag{7.76}$$

$$C_4 = \frac{SNR + 1/4}{n_p(2SNR + 1)^2} \tag{7.77}$$

$$C_6 = C_3^2/2 \tag{7.78}$$

$$\varpi = \sqrt{n_p(2SNR + 1)} \tag{7.79}$$

Figure 7.9 shows a plot for the probability of detection versus SNR for cases $n_p = 1, 10$. Note that it requires less SNR, with ten pulses integrated noncoherently, to achieve the same probability of detection as in the case of a single pulse. Hence, for any given P_D the SNR improvement can be read from the plot. Equivalently, using the function *"improv_fac.m"* leads to about the same result. For example, when $P_D = 0.8$, the function *"improv_fac.m"* gives an SNR improvement factor of $I(10) \approx 8.55dB$. Figure 7.9 shows that the ten pulse SNR is about $6.03dB$. Therefore, the single pulse SNR is about $14.5dB$, which can be read from the figure.

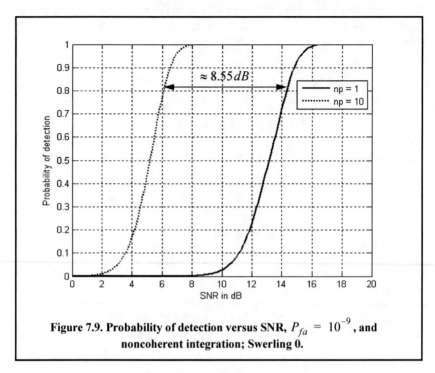

Figure 7.9. Probability of detection versus SNR, $P_{fa} = 10^{-9}$, and noncoherent integration; Swerling 0.

7.7.2. Detection of Swerling I Targets

The exact formula for the probability of detection for Swerling I type targets was derived by Swerling. It is

$$P_D = e^{-(v_T)/(1 + SNR)} \qquad ; \ n_p = 1 \qquad (7.80)$$

$$P_D = 1 - \Gamma_I(v_T, n_P - 1) + \left(1 + \frac{1}{n_P SNR}\right)^{n_p - 1} \Gamma_I\left(\frac{v_T}{1 + \frac{1}{n_P SNR}}, n_P - 1\right) \qquad (7.81)$$

$$\times \ e^{-v_T/(1 + n_P SNR)} \qquad ; \ n_P > 1$$

Figure 7.10 shows a plot of the probability of detection as a function of SNR for $n_p = 1$ and $P_{fa} = 10^{-9}$ for both Swerling I and V (Swerling 0) type fluctuations. Note that it requires more SNR, with fluctuation, to achieve the same P_D as in the case with no fluctuation. This figure can be reproduced using the following MATLAB code.

```
% Generate Figure 7.10
close all;
clear all;
pfa = 1e-9;
nfa = log(2) / pfa;
b = sqrt(-2.0 * log(pfa));
index = 0;
for snr = 0:.01:22
    index = index +1;
    a = sqrt(2.0 * 10^(.1*snr));
    swer0(index) = marcumsq(a,b);
    swer1(index) = pd_swerling1 (nfa, 1, snr);
end
x = 0:.01:22;
%figure(10)
plot(x, swer0,'k',x,swer1,'k:');
axis([2 22 0 1])
xlabel ('SNR in dB')
ylabel ('Probability of detection')
legend('Swerling 0','Swerling I')
grid
```

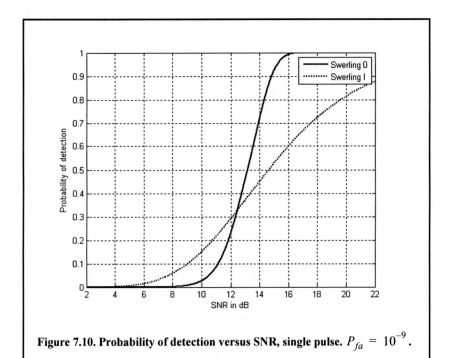

Figure 7.10. Probability of detection versus SNR, single pulse. $P_{fa} = 10^{-9}$.

Figure 7.11 is similar to Fig. 7.10 except in this case $P_{fa} = 10^{-6}$ and $n_P = 5$. This figure can be reproduced using the following MATLAB code

```
% Generate Figure 7.11
clear all
close all
pfa = 1e-6;
nfa = log(2) / pfa;
index = 0;
for snr = -10:.5:30
   index = index +1;
   prob1(index) = pd_swerling1 (nfa, 5, snr);
   prob0(index) = pd_swerling5 (nfa, 2, 5, snr);
   end
x = -10:.5:30;
plot(x, prob1, 'k',x,prob0, 'k:');
axis([-10 30 0 1])
xlabel ('SNR in dB')
ylabel ('Probability of detection')
legend('Swerling I', 'Swerling 0')
title('Pfa =1e-6; n=5')
grid
```

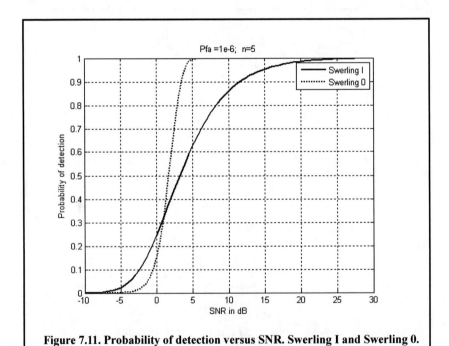

Figure 7.11. Probability of detection versus SNR. Swerling I and Swerling 0.

7.7.3. Detection of Swerling II Targets

In the case of Swerling II targets, the probability of detection is given by

$$P_D = 1 - \Gamma_I\left(\frac{v_T}{(1+SNR)}, n_p\right) \qquad ; \; n_P \leq 50 \qquad \textbf{(7.82)}$$

For the case when $n_P > 50$ the probability of detection is computed using the Gram-Charlier series. In this case,

$$C_3 = -\frac{1}{3\sqrt{n_p}} \qquad , \; C_6 = \frac{C_3^2}{2} \qquad \textbf{(7.83)}$$

$$C_4 = \frac{1}{4n_P} \qquad \textbf{(7.84)}$$

$$\varpi = \sqrt{n_P}\,(1+SNR) \qquad \textbf{(7.85)}$$

Figure 7.12a shows a plot of the probability of detection for Swerling 0, Swerling I, and Swerling II with $n_P = 5$, where $P_{fa} = 10^{-7}$. This figure can be reproduced using the following MATLAB code. Figure 7.12b is similar to Fig. 7.12a except in this case $n_P = 2$.

```
% Generate Figure 7.12
clc
clear all;
close all;
pfa = 1e-7;
nfa = log(2) / pfa;
index = 0;
for snr = -10:.5:30
    index = index +1;
    prob1(index) = pd_swerling1 (nfa, 5, snr); % Fig. 7.12a
    prob0(index) = pd_swerling5 (nfa, 2, 5, snr); % Fig. 7.12a
    prob2(index) = pd_swerling2 (nfa, 5, snr); % Fig. 7.12a
    % prob1(index) = pd_swerling1 (nfa, 2, snr); % Fig. 7.12b
    % prob0(index) = pd_swerling5 (nfa, 2, 2, snr); % Fig. 7.12b
    % prob2(index) = pd_swerling2 (nfa, 2, snr); % Fig. 7.12b
end
x = -10:.5:30;
plot(x, prob0,'k',x,prob1,'k:',x,prob2,'k--');
axis([-10 30 0 1])
xlabel ('SNR in dB')
ylabel ('Probability of detection')
legend('Swerling 0','Swerling I','Swerling II')
title('Pfa =1e-7;  n=5')
grid
```

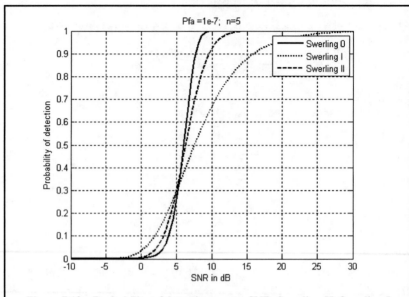

Figure 7.12a. Probability of detection versus SNR. Swerling II, Swerling I and Swerling 0.

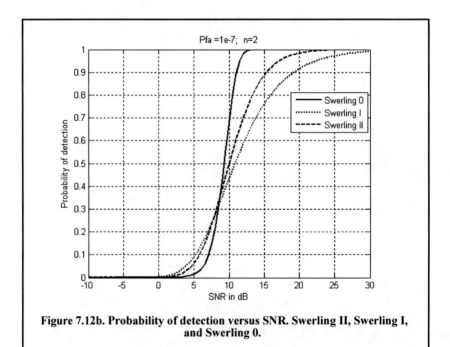

Figure 7.12b. Probability of detection versus SNR. Swerling II, Swerling I, and Swerling 0.

7.7.4. Detection of Swerling III Targets

The exact formulas, developed by Marcum, for the probability of detection
for Swerling III type targets when $n_P = 1, 2$

$$P_D = \exp\left(\frac{-v_T}{1 + n_P SNR/2}\right)\left(1 + \frac{2}{n_P SNR}\right)^{n_P - 2} \times K_0$$

$$K_0 = 1 + \frac{v_T}{1 + n_P SNR/2} - \frac{2}{n_P SNR}(n_P - 2)$$

(7.86)

For $n_P > 2$ the expression is

$$P_D = \frac{v_T^{n_P - 1} e^{-v_T}}{(1 + n_P SNR/2)(n_P - 2)!} + 1 - \Gamma_I(v_T, n_P - 1) + K_0$$

(7.87)

$$\times \ \Gamma_I\left(\frac{v_T}{1 + 2/n_P SNR}, n_P - 1\right)$$

Figure 7.13a shows a plot of the probability of detection as a function of
SNR for $n_P = 1, 10, 50, 100$, where $P_{fa} = 10^{-9}$. Figure 7.13b shows a plot
of the probability of detection for Swerling 0, Swerling I, Swerling II, and
Swerling III with $n_P = 5$ and $P_{fa} = 10^{-7}$. Figure 7.13a can be reproduced
using the following MATLAB code.

```
% Generate Figure 7.13a
close all;
clear all;
pfa = 1e-9;
nfa = log(2) / pfa;
index = 0;
for snr = -10:.5:30
  index = index +1;
  prob1(index) = pd_swerling3 (nfa, 1, snr);
  prob10(index) = pd_swerling3 (nfa, 10, snr);
  prob50(index) = pd_swerling3(nfa, 50, snr);
  prob100(index) = pd_swerling3 (nfa, 100, snr);
end
x = -10:.5:30;
plot(x, prob1,'k',x,prob10,'k:',x,prob50,'k--', x, prob100,'k-.');
axis([-10 30 0 1])
xlabel ('SNR in dB')
ylabel ('Probability of detection')
legend('np = 1','np = 10','np = 50','np = 100')
grid
```

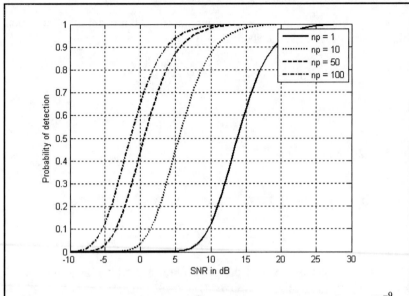

Figure 7.13a. Probability of detection versus SNR. Swerling III. $P_{fa} = 10^{-9}$.

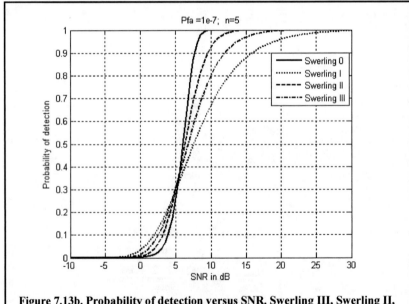

Figure 7.13b. Probability of detection versus SNR. Swerling III, Swerling II, Swerling I, and Swerling 0.

7.7.5. Detection of Swerling IV Targets

The expression for the probability of detection for Swerling IV targets for $n_P < 50$ is

$$P_D = 1 - \left[\gamma_0 + \left(\frac{SNR}{2}\right) n_P \gamma_1 + \left(\frac{SNR}{2}\right)^2 \frac{n_P(n_P - 1)}{2!} \gamma_2 + \dots + \right. \tag{7.88}$$

$$\left(\frac{SNR}{2}\right)^{n_P} \gamma_{n_P} \left] \left(1 + \frac{SNR}{2}\right)^{-n_P} \right.$$

$$\gamma_i = \Gamma_I \left(\frac{v_T}{1 + (SNR)/2}, n_P + i \right) \tag{7.89}$$

By using the recursive formula

$$\Gamma_I(x, i + 1) = \Gamma_I(x, i) - \frac{x^i}{i! \exp(x)} \tag{7.90}$$

then only γ_0 needs to be calculated using Eq. (7.89) and the rest of γ_i are calculated from the following recursion:

$$\gamma_i = \gamma_{i-1} - A_i \qquad ; \; i > 0 \tag{7.91}$$

$$A_i = \frac{v_T/(1 + (SNR)/2)}{n_P + i - 1} A_{i-1} \qquad ; \; i > 1 \tag{7.92}$$

$$A_1 = \frac{(v_T/(1 + (SNR)/2))^{n_P}}{n_P! \exp(v_T/(1 + (SNR)/2))} \tag{7.93}$$

$$\gamma_0 = \Gamma_I \left(\frac{v_T}{(1 + (SNR)/2)}, n_P \right) \tag{7.94}$$

For the case when $n_P \geq 50$, the Gram-Charlier series can be used to calculate the probability of detection. In this case,

$$C_3 = \frac{1}{3\sqrt{n_P}} \frac{2\beta^3 - 1}{(2\beta^2 - 1)^{1.5}} \qquad ; \; C_6 = \frac{C_3^2}{2} \tag{7.95}$$

$$C_4 = \frac{1}{4n_P} \frac{2\beta^4 - 1}{(2\beta^2 - 1)^2} \tag{7.96}$$

$$\omega = \sqrt{n_P(2\beta^2 - 1)} \tag{7.97}$$

$$\beta = 1 + (SNR)/2 \tag{7.98}$$

Figure 7.14 shows plots of the probability of detection as a function of SNR for $n_P = 1, 10, 25, 75$, where $P_{fa} = 10^{-6}$. This figure can be reproduced using the following MATLAB code.

```
clear all; close all;
pfa = 1e-6;
nfa = log(2) / pfa;
index = 0;
for snr = -7:.15:10
   index = index +1;
   prob1(index) = pd_swerling4 (nfa, 5, snr);
   prob10(index) = pd_swerling4 (nfa, 10, snr);
   prob25(index) = pd_swerling4(nfa, 25, snr);
   prob75(index) = pd_swerling4 (nfa, 75, snr);
end
x = -7:.15:10;
plot(x, prob1, 'k',x,prob10,'k.',x,prob25,'k:',x, prob75,'k-.','linewidth',1);
xlabel ('SNR - dB')
ylabel ('Probability of detection')
legend('np = 5','np = 10','np = 25','np = 75')
grid; axis tight
```

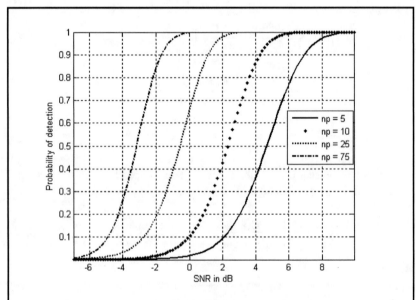

Figure 7.14. Probability of detection versus SNR. Swerling IV. $P_{fa} = 10^{-6}$.

7.8. Computation of the Fluctuation Loss

The fluctuation loss, L_f, can be viewed as the amount of additional SNR required to compensate for the SNR loss due to target fluctuation, given a specific P_D value. Kanter[1] developed an exact analysis for calculating the fluctuation loss. In this text the author will take advantage of the computational power of MATLAB and the MATLAB functions developed for this text to numerically calculate the amount of fluctuation loss. For this purpose consider the MATALB function "*fluct.m*", where its syntax is as follows:

$$[SNR] = fluct(pd, pfa, np, sw\_case)$$

where

Symbol	Description	Units	Status
pd	*desired probability of detection*	*none*	*input*
nfa	*desired number of false alarms*	*none*	*input*
np	*number of pulses*	*none*	*input*
sw_case	*0, 1, 2, 3, or 4 depending on the desired Swerling case*	*none*	*input*
SNR	*Resulting SNR*	*dB*	*output*

For example, using the syntax

$$[SNR0] = fluct(0.8, 1e6, 5, 0)$$

will calculate the *SNR0* corresponding to a Swerling 0. If one would use this *SNR* in the function "*pd_swerling5.m*" with following syntax

$$[pd] = pd\_swerling5 (1e6, 1, 5, SNR0)$$

the resulting P_D will be equal to 0.8 . Similarly, if the following syntax is used

$$[SNR1] = fluct(.8, 1\text{-}e\text{-}6, 5, 1)$$

then the value *SNR1* will be that of Swerling 1. Of course, if one would use this *SNR1* value in the function "*pd_swerling1.m*" with following syntax

$$[pd] = pd\_swerling1(1e6, 5, .8, SNR1)$$

the same P_D of 0.8 will be calculated. Therefore, the fluctuation loss for this case, is equal to *SNR0 - SNR1*.

---

1. Kanter, I., Exact Detection Probability for Partially Correlated Rayleigh Targets, *IEEE Trans*, AES-22, pp. 184-196, March 1986.

7.9. Cumulative Probability of Detection

Denote the range at which the single pulse SNR is unity (0 dB) as R_0, and refer to it as the reference range. Then, for a specific radar, the single pulse SNR at R_0 is defined by the radar equation and is given by

$$(SNR)_{R_0} = \frac{P_t G^2 \lambda^2 \sigma}{(4\pi)^3 k T_0 BFLR_0^4} = 1 \qquad (7.99)$$

The single pulse SNR at any range R is

$$SNR = \frac{P_t G^2 \lambda^2 \sigma}{(4\pi)^3 k T_0 BFLR^4} \qquad (7.100)$$

Dividing Eq. (7.100) by Eq. (7.99) yields

$$\frac{SNR}{(SNR)_{R_0}} = \left(\frac{R_0}{R}\right)^4 \qquad (7.101)$$

Therefore, if the range R_0 is known, then the SNR at any other range R is

$$(SNR)_{dB} = 40\log\left(\frac{R_0}{R}\right) \qquad (7.102)$$

Also, define the range R_{50} as the range at which $P_D = 0.5 = P_{50}$. Normally, the radar unambiguous range R_u is set equal to $2R_{50}$.

The cumulative probability of detection refers to detecting the target at least once by the time it is at range R. More precisely, consider a target closing on a scanning radar, where the target is illuminated only during a scan (frame). As the target gets closer to the radar, its probability of detection increases since the SNR is increased. Suppose that the probability of detection during the *nth* frame is P_{D_n}; then, the cumulative probability of detecting the target at least once during the *nth* frame (see Fig. 7.15) is given by

$$P_{C_n} = 1 - \prod_{i=1}^{n}(1 - P_{D_i}) \qquad (7.103)$$

P_{D_1} is usually selected to be very small. Clearly, the probability of not detecting the target during the *nth* frame is $1 - P_{C_n}$. The probability of detection for the *ith* frame, P_{D_i}, is computed as discussed in the previous section.

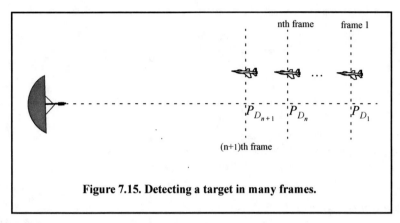

Figure 7.15. Detecting a target in many frames.

Example:

A radar detects a closing target at $R = 10Km$, with probability of detection P_D equal to 0.5. Assume $P_{fa} = 10^{-7}$. Compute and sketch the single look probability of detection as a function of normalized range (with respect to $R = 10Km$), over the interval $(2-20)Km$. If the range between two successive frames is $1Km$, what is the cumulative probability of detection at $R = 8Km$?

Solution:

From the function "marcumsq.m" the SNR corresponding to $P_D = 0.5$ and $P_{fa} = 10^{-7}$ is approximately 12dB. By using a similar analysis to that which led to Eq. (7.102), we can express the SNR at any range R as

$$(SNR)_R = (SNR)_{10} + 40 \ \log\frac{10}{R} = 52 - 40 \ \log R$$

By using the function "marcumsq.m" we can construct the following table:

R Km	(SNR) dB	P_D
2	39.09	0.999
4	27.9	0.999
6	20.9	0.999
8	15.9	0.999
9	13.8	0.9
10	12.0	0.5
11	10.3	0.25

R *Km*	(SNR) dB	P_D
12	8.8	0.07
14	6.1	0.01
16	3.8	ε
20	0.01	ε

where ε is very small. A sketch of P_D versus normalized range is shown in Fig. 7.16.

The cumulative probability of detection is given in Eq. (7.104), where the probability of detection of the first frame is selected to be very small. Thus, we can arbitrarily choose frame 1 to be at $R = 16Km$. Note that selecting a different starting point for frame 1 would have a negligible effect on the cumulative probability (we only need P_{D_1} to be very small). Below is a range listing for frames 1 through 9, where frame 9 corresponds to $R = 8Km$.

frame	1	2	3	4	5	6	7	8	9
range in Km	16	15	14	13	12	11	10	9	8

The cumulative probability of detection at 8 Km is then

$$P_{C_9} = 1 - (1-0.999)(1-0.9)(1-0.5)(1-0.25)(1-0.07)$$
$$(1-0.01)(1-\varepsilon)^2 \approx 0.9998$$

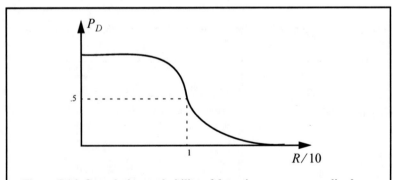

Figure 7.16. Cumulative probability of detection versus normalized range.

7.10. Constant False Alarm Rate (CFAR)

The detection threshold is computed so that the radar receiver maintains a constant predetermined probability of false alarm. Equation (7.20) gives the relationship between the threshold value V_T and the probability of false alarm P_{fa}, and for convenience is repeated here as Eq. (7.104):

$$v_T = \sqrt{2\sigma^2 \ln\left(\frac{1}{P_{fa}}\right)} \qquad (7.104)$$

If the noise power σ^2 is constant, then a fixed threshold can satisfy Eq. (7.104). However, due to many reasons this condition is rarely true. Thus, in order to maintain a constant probability of false alarm, the threshold value must be continuously updated based on the estimates of the noise variance. The process of continuously changing the threshold value to maintain a constant probability of false alarm is known as Constant False Alarm Rate (CFAR).

Three different types of CFAR processors are primarily used. They are adaptive threshold CFAR, nonparametric CFAR, and nonlinear receiver techniques. Adaptive CFAR assumes that the interference distribution is known and approximates the unknown parameters associated with these distributions. Nonparametric CFAR processors tend to accommodate unknown interference distributions. Nonlinear receiver techniques attempt to normalize the root-mean-square amplitude of the interference. In this book only analog Cell-Averaging CFAR (CA-CFAR) technique is examined. The analysis presented in this section closely follows Urkowitz[1].

7.10.1. Cell-Averaging CFAR (Single Pulse)

The CA-CFAR processor is shown in Fig. 7.17. Cell averaging is performed on a series of range and/or Doppler bins (cells). The echo return for each pulse is detected by a square-law detector. In analog implementation these cells are obtained from a tapped delay line. The Cell Under Test (CUT) is the central cell. The immediate neighbors of the CUT are excluded from the averaging process due to a possible spillover from the CUT. The output of M reference cells ($M/2$ on each side of the CUT) is averaged. The threshold value is obtained by multiplying the averaged estimate from all reference cells by a constant K_0 (used for scaling). A detection is declared in the CUT if

$$Y_1 \ge K_0 Z \qquad (7.105)$$

---

1. Urkowitz, H., Decision and Detection Theory, unpublished lecture notes. Lockheed Martin Co., Moorestown, NJ.

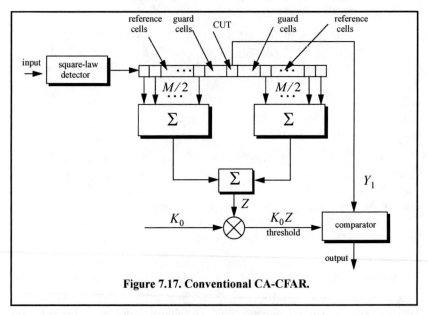

Figure 7.17. Conventional CA-CFAR.

CA-CFAR assumes that the target of interest is in the CUT and all reference cells contain zero-mean independent Gaussian noise of variance ψ^2. Therefore, the output of the reference cells, Z, represents a random variable with gamma probability density function (special case of the chi-square) with $2M$ degrees of freedom. In this case, the gamma *pdf* is

$$f(z) = \frac{z^{(M/2)-1} e^{(-z/2\psi^2)}}{2^{M/2}\, \sigma^M \Gamma(M/2)} \qquad ; \; z > 0 \tag{7.106}$$

The probability of false alarm corresponding to a fixed threshold was derived earlier. When CA-CFAR is implemented, then the probability of false alarm can be derived from the conditional false alarm probability, which is averaged over all possible values of the threshold in order to achieve an unconditional false alarm probability. The conditional probability of false alarm when $y = V_T$ can be written as

$$P_{fa}(v_T = y) = e^{-y/2\sigma^2} \tag{7.107}$$

It follows that the unconditional probability of false alarm is

$$P_{fa} = \int_0^\infty P_{fa}(v_T = y) f(y)\, dy \tag{7.108}$$

where $f(y)$ is the *pdf* of the threshold, which except for the constant K_0 is the same as that defined in Eq. (7.106). Therefore,

$$f(y) = \frac{y^{M-1} e^{(-y/2K_0\psi^2)}}{(2K_0\sigma^2)^M \Gamma(M)} \quad ; \ y \geq 0 \tag{7.109}$$

Performing the integration in Eq. (7.108) yields

$$P_{fa} = 1/(1+K_0)^M \tag{7.110}$$

Observation of Eq. (7.110) shows that the probability of false alarm is now independent of the noise power, which is the objective of CFAR processing.

7.10.2. Cell-Averaging CFAR with Noncoherent Integration

In practice, CFAR averaging is often implemented after noncoherent integration, as illustrated in Fig. 7.18. Now, the output of each reference cell is the sum of n_p squared envelopes. It follows that the total number of summed reference samples is Mn_p. The output Y_1 is also the sum of n_p squared envelopes. When noise alone is present in the CUT, Y_1 is a random variable whose *pdf* is a gamma distribution with $2n_p$ degrees of freedom. Additionally, the summed output of the reference cells is the sum of Mn_p squared envelopes. Thus, Z is also a random variable which has a gamma *pdf* with $2Mn_p$ degrees of freedom.

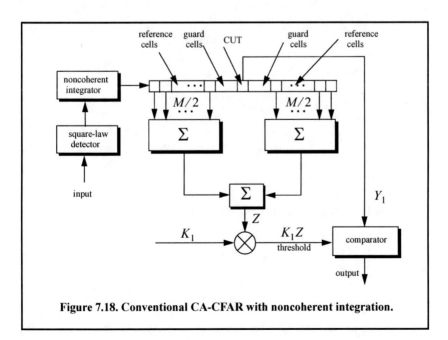

Figure 7.18. Conventional CA-CFAR with noncoherent integration.

The probability of false alarm is then equal to the probability that the ratio Y_1/Z exceeds the threshold. More precisely,

$$P_{fa} = Prob\{Y_1/Z > K_1\} \tag{7.111}$$

Equation (7.111) implies that one must first find the joint *pdf* for the ratio Y_1/Z. However, this can be avoided if P_{fa} is first computed for a fixed threshold value V_T, then averaged over all possible values of the threshold. Therefore, let the conditional probability of false alarm when $y = v_T$ be $P_{fa}(v_T = y)$. It follows that the unconditional false alarm probability is

$$P_{fa} = \int_0^\infty P_{fa}(v_T = y) f(y) dy \tag{7.112}$$

where $f(y)$ is the *pdf* of the threshold. In view of this, the probability density function describing the random variable $K_1 Z$ is given by

$$f(y) = \frac{(y/K_1)^{Mn_P - 1} e^{(-y/2K_0\sigma^2)}}{(2\sigma^2)^{Mn_P} K_1 \Gamma(Mn_P)} \quad ; \ y \geq 0 \tag{7.113}$$

It can be shown that in this case the probability of false alarm is independent of the noise power and is given by

$$P_{fa} = \frac{1}{(1 + K_1)^{Mn_P}} \sum_{k=0}^{n_P - 1} \frac{1}{k!} \frac{\Gamma(Mn_P + k)}{\Gamma(Mn_P)} \left(\frac{K_1}{1 + K_1}\right)^k \tag{7.114}$$

which is identical to Eq. (7.110) when $K_1 = K_0$ and $n_P = 1$.

7.11. MATLAB Programs and Routines

This section presents listings for all the MATLAB programs used to produce all of the MATLAB-generated figures in this chapter. Additionally, other specific MATLAB functions are also presented. They are listed in the same order they appear in the chapter.

7.11.1. MATLAB Function "que_func.m"

The function "que_func.m" computes $F(x)$ using Eqs. (7.17) and (7.18). The syntax is as follows:

$$fofx = que\_func\ (x)$$

MATLAB Function "que_func.m" Listing

```
function fofx = que_func(x)
% This function computes the value of the Q-function
% It uses the approximation in Eqs. (7.17) and (7.18)
if (x >= 0)
  denom = 0.661 * x + 0.339 * sqrt(x^2 + 5.51);
  expo = exp(-x^2 /2.0);
  fofx = 1.0 - (1.0 / sqrt(2.0 * pi)) * (1.0 / denom) * expo;
else
  denom = 0.661 * x + 0.339 * sqrt(x^2 + 5.51);
  expo = exp(-x^2 /2.0);
  value = 1.0 - (1.0 / sqrt(2.0 * pi)) * (1.0 / denom) * expo;
  fofx = 1.0 - value;
end
```

7.11.2. MATLAB Function "marcumsq.m"

This function utilizes Parl's method to compute P_D. The syntax is as follows:

$$Pd = marcumsq(a,b)$$

MATLAB Function *"marcumsq.m"* **Listing**

```
function Pd = marcumsq (a,b); % This function uses Parl's method to compute PD
max_test_value = 5000.;
if (a < b)
  alphan0 = 1.0;
  dn = a / b;
else
  alphan0 = 0.;
  dn = b / a;
end
alphan_1 = 0.;
betan0 = 0.5;
betan_1 = 0.;
D1 = dn;
n = 0;
ratio = 2.0 / (a * b);
r1 = 0.0;
betan = 0.0;
alphan = 0.0;
while betan < 1000.,
  n = n + 1;
  alphan = dn + ratio * n * alphan0 + alphan;
  betan = 1.0 + ratio * n * betan0 + betan;
  alphan_1 = alphan0;
  alphan0 = alphan;
  betan_1 = betan0;
```

```
betan0 = betan;
  dn = dn * D1;
end
PD = (alphan0 / (2.0 * betan0)) * exp( -(a-b)^2 / 2.0),
if ( a >= b)
  PD = 1.0 - PD;
end
return
```

7.11.3. MATLAB Function "improv_fac.m"

The function *"improv_fac.m"* calculates the improvement factor using Eq. (7.45). The syntax is as follows:

$$[impr\_of\_np] = improv\_fac\ (np,\ pfa,\ pd)$$

where

Symbol	Description	Units	Status
np	*number of integrated pulses*	*none*	*input*
pfa	*probability of false alarms*	*none*	*input*
pd	*probability of detection*	*none*	*input*
impr_of_np	*improvement factor*	*output*	*dB*

MATLAB Function *"improv_fac.m"* **Listing**

```
function impr_of_np = improv_fac (np, pfa, pd)
% This function computes the noncoherent integration improvement
% factor using the empirical formula defined in Eq. (7.54)
fact1 = 1.0 + log10( 1.0 / pfa) / 46.6;
fact2 = 6.79 * (1.0 + 0.235 * pd);
fact3 = 1.0 - 0.14 * log10(np) + 0.0183 * (log10(np))^2;
impr_of_np = fact1 * fact2 * fact3 * log10(np);
return
```

7.11.4. MATLAB Function "threshold.m"

The function *"threshold.m"* calculates the threshold value given the algorithm described in Section 7.6. The syntax is as follows:

$$[pfa,\ vt] = threshold\ (nfa,\ np)$$

where

Symbol	Description	Units	Status
nfa	*number of false alarm*	*none*	*input*
np	*number of pulses*	*none*	*input*
pfa	*probability of alarm*	*none*	*output*
vt	*threshold value*	*none*	*output*

MATLAB Function *"threshold.m"* **Listing**

```
function [pfa, vt] = threshold (nfa, np)
% This function calculates the threshold value from nfa and np.
% The Newton-Raphson recursive formula is used
% This function uses "gammainc.m".
delmax = .00001;
eps = 0.000000001;
delta =10000.;
pfa = np * log(2) / nfa;
sqrtpfa = sqrt(-log10(pfa));
sqrtnp = sqrt(np);
vt0 = np - sqrtnp + 2.3 * sqrtpfa * (sqrtpfa + sqrtnp - 1.0);
vt = vt0;
while (abs(delta) >= vt0)
  igf = gammainc(vt0,np);
  num = 0.5^(np/nfa) - igf;
  temp = (np-1) * log(vt0+eps) - vt0 - factor(np-1);
  deno = exp(temp);
  vt = vt0 + (num / (deno+eps));
  delta = abs(vt - vt0) * 10000.0;
  vt0 = vt;
```

7.11.5. MATLAB Function *"pd_swerling5.m"*

The function *"pd_swerling5.m"* calculates the probability of detection for Swerling 0 targets. The syntax is as follows:

$$[pd] = pd\_swerling5 \ (input1, indicator, np, snr)$$

where

Symbol	Description	Units	Status
input1	P_{fa} or n_{fa}	*none*	*input*
indicator	*1 when input1* $= P_{fa}$ *2 when input1* $= n_{fa}$	*none*	*input*

Symbol	Description	Units	Status
np	*number of integrated pulses*	*none*	*input*
snr	*SNR*	*dB*	*input*
pd	*probability of detection*	*none*	*output*

MATLAB Function "pd_swerling5.m" Listing

```
function pd = pd_swerling5 (input1, indicator, np, snrbar)
% This function is used to calculate the probability of detection
% for Swerling 5 or 0 targets for np>1.
if(np == 1)
  'Stop, np must be greater than 1'
  return
end
format long
snrbar - 10.0.^(snrbar./10.);
eps = 0.00000001;
delmax = .00001;
delta =10000.;
% Calculate the threshold Vt
if (indicator ~=1)
  nfa = input1;
  pfa =  np * log(2) / nfa;
else
  pfa = input1;
  nfa = np * log(2) / pfa;
end
sqrtpfa = sqrt(-log10(pfa));
sqrtnp = sqrt(np);
vt0 = np - sqrtnp + 2.3 * sqrtpfa * (sqrtpfa + sqrtnp - 1.0);
vt = vt0;
while (abs(delta) >= vt0)
  igf = incomplete_gamma(vt0,np);
  num = 0.5^(np/nfa) - igf;
  temp = (np-1) * log(vt0+eps) - vt0 - factor(np-1);
  deno = exp(temp);
  vt = vt0 + (num / (deno+eps));
  delta = abs(vt - vt0) * 10000.0;
  vt0 = vt;
end
% Calculate the Gram-Chrlier coefficients
temp1 = 2.0 .* snrbar + 1.0;
omegabar = sqrt(np .* temp1);
c3 = -(snrbar + 1.0 / 3.0) ./ (sqrt(np) .* temp1.^1.5);
c4 = (snrbar + 0.25) ./ (np .* temp1.^2.);
c6 = c3 .* c3 ./2.0;
```

```
V = (vt - np .* (1.0 + snrbar)) ./ omegabar;
Vsqr = V .*V;
val1 = exp(-Vsqr ./ 2.0) ./ sqrt( 2.0 * pi);
val2 = c3 .* (V.^2 -1.0) + c4 .* V .* (3.0 - V.^2) -...
   c6 .* V .* (V.^4 - 10. .* V.^2 + 15.0);
q = 0.5 .* erfc (V./sqrt(2.0));
pd = q - val1 .* val2;
return
```

7.11.6. MATLAB Function "pd_swerling1.m"

The function *"pd_swerling1.m"* calculates the probability of detection for Swerling I type targets. The syntax is as follows:

$$[pd] = pd\_swerling1 \ (nfa, \ np, \ snr)$$

where

Symbol	Description	Units	Status
nfa	*Marcum's false alarm number*	*none*	*input*
np	*number of integrated pulses*	*none*	*input*
snr	*SNR*	*dB*	*input*
pd	*probability of detection*	*none*	*output*

MATLAB Function "pd_swerling1.m" Listing

```
function [pd] = pd_swerling1 (nfa, np, snrbar)
% This function is used to calculate the probability of detection
% for Swerling 1 targets.
format long
snrbar = 10.0^(snrbar/10.);
eps = 0.00000001;
delmax = .00001;
delta =10000.;
% Calculate the threshold Vt
pfa = np * log(2) / nfa;
sqrtpfa = sqrt(-log10(pfa));
sqrtnp = sqrt(np);
vt0 = np - sqrtnp + 2.3 * sqrtpfa * (sqrtpfa + sqrtnp - 1.0);
vt = vt0;
while (delta < (vt0/10000));
   igf = gammainc(vt0,np);
   num = 0.5^(np/nfa) - igf;
   deno = -exp(-vt0) * vt0^(np-1) /factorial(np-1);
   vt = vt0 - (num / (deno+eps));
   delta = abs(vt - vt0);
   vt0 = vt;
```

```
end
if (np == 1)
  temp = -vt / (1.0 + snrbar);
  pd = exp(temp);
  return
end
  temp1 = 1.0 + np * snrbar;
  temp2 = 1.0 / (np *snrbar);
  temp = 1.0 + temp2;
  val1 = temp^(np-1.);
  igf1 = gammainc(vt,np-1);
  igf2 = gammainc(vt/temp,np-1);
  pd = 1.0 - igf1 + val1 * igf2 * exp(-vt/temp1);
  return
```

7.11.7. MATLAB Function "pd_swerling2.m"

The function *"pd_swerling2.m"* calculates P_D for Swerling II type targets. The syntax is as follows:

$$[pd] = pd\_swerling2 \ (nfa, \ np, \ snr)$$

where

Symbol	Description	Units	Status
nfa	Marcum's false alarm number	none	input
np	number of integrated pulses	none	input
snr	SNR	dB	input
pd	probability of detection	none	output

MATLAB Function *"pd_swerling2.m"* Listing

```
function [pd] = pd_swerling2 (nfa, np, snrbar)
% This function is used to calculate the probability of detection
% for Swerling 2 targets.
format long
snrbar = 10.0^(snrbar/10.);
eps = 0.00000001;
delmax = .00001;
delta =10000.;
% Calculate the threshold Vt
pfa =  np * log(2) / nfa;
sqrtpfa = sqrt(-log10(pfa));
sqrtnp = sqrt(np);
vt0 = np - sqrtnp + 2.3 * sqrtpfa * (sqrtpfa + sqrtnp - 1.0);
vt = vt0;
while (delta < (vt0/10000));
```

```
igf = gammainc(vt0,np);
num = 0.5^(np/nfa) - igf;
deno = -exp(-vt0) * vt0^(np-1) /factorial(np-1);
vt = vt0 - (num / (deno+eps));
delta = abs(vt - vt0);
vt0 = vt;
end
if (np <= 50)
  temp = vt / (1.0 + snrbar);
  pd = 1.0 - gammainc(temp,np);
  return
else
  temp1 = snrbar + 1.0;
  omegabar = sqrt(np) * temp1;
  c3 = -1.0 / sqrt(9.0 * np);
  c4 = 0.25 / np;
  c6 = c3 * c3 /2.0;
  V = (vt - np * temp1) / omegabar;
  Vsqr = V *V;
  val1 = exp(-Vsqr / 2.0) / sqrt( 2.0 * pi);
  val2 = c3 * (V^2 -1.0) + c4 * V * (3.0 - V^2) - ...
    c6 * V * (V^4 - 10. * V^2 + 15.0);
  q = 0.5 * erfc (V/sqrt(2.0));
  pd = q - val1 * val2;
end
return
```

7.11.8. MATLAB Function "pd_swerling3.m"

The function *"pd_swerling3.m"* calculates P_D for Swerling III type targets. The syntax is as follows:

$$[pd] = pd\_swerling3 \ (nfa, np, snr)$$

where

Symbol	Description	Units	Status
nfa	Marcum's false alarm number	none	input
np	number of integrated pulses	none	input
snr	SNR	dB	input
pd	probability of detection	none	output

MATLAB Function "pd_swerling3.m" Listing

```
function [pd] = pd_swerling3 (nfa, np, snrbar)
% This function is used to calculate the probability of detection
% for Swerling 3 targets.
```

```
format long
snrbar = 10.0^(snrbar/10.);
eps = 0.00000001;
delmax = .00001;
delta =10000.;
% Calculate the threshold Vt
pfa =  np * log(2) / nfa;
sqrtpfa = sqrt(-log10(pfa));
sqrtnp = sqrt(np);
vt0 = np - sqrtnp + 2.3 * sqrtpfa * (sqrtpfa + sqrtnp - 1.0);
vt = vt0;
while (delta < (vt0/10000));
  igf = gammainc(vt0,np);
  num = 0.5^(np/nfa) - igf;
  deno = -exp(-vt0) * vt0^(np-1) /factorial(np-1);
  vt = vt0 - (num / (deno+eps));
  delta = abs(vt - vt0);
  vt0 = vt;
end
temp1 = vt / (1.0 + 0.5 * np *snrbar);
temp2 = 1.0 + 2.0 / (np * snrbar);
temp3 = 2.0 * (np - 2.0) / (np * snrbar);
ko = exp(-temp1) * temp2^(np-2.) * (1.0 + temp1 - temp3);
if (np <= 2)
  pd = ko;
  return
else
  ko = exp(-temp1) * temp2^(np-2.) * (1.0 + temp1 - temp3);
  temp4 = vt^(np-1.) * exp(-vt) / (temp1 * (factorial(np-2.)));
  temp5 = vt / (1.0 + 2.0 / (np *snrbar));
  pd = temp4 + 1.0 - gammainc(vt,np-1.) + ko * gammainc(temp5,np-1.);
end
return
```

7.11.9. MATLAB Function "pd_swerling4.m"

The function *"pd_swerling4.m"* calculates P_D for Swerling IV type targets. The syntax is as follows:

$$[pd] = pd\_swerling4 (nfa, np, snr)$$

where

Symbol	Description	Units	Status
nfa	Marcum's false alarm number	none	input
np	number of integrated pulses	none	input

Symbol	Description	Units	Status
snr	*SNR*	*dB*	*input*
pd	*probability of detection*	*none*	*output*

MATLAB Function "pd_swerling4.m" Listing

```
function [pd] = pd_swerling4 (nfa, np, snrbar)
% This function is used to calculate the probability of detection
% for Swerling 4 targets.
format long
snrbar = 10.0^(snrbar/10.);
eps = 0.00000001;
delmax = .00001;
delta =10000.;
% Calculate the threshold Vt
pfa =  np * log(2) / nfa;
sqrtpfa = sqrt(-log10(pfa));
sqrtnp = sqrt(np);
vt0 = np - sqrtnp + 2.3 * sqrtpfa * (sqrtpfa + sqrtnp - 1.0);
vt = vt0;
while (delta < (vt0/10000));
   igf = gammainc(vt0,np);
   num = 0.5^(np/nfa) - igf;
   deno = -exp(-vt0) * vt0^(np-1) /factorial(np-1);
   vt = vt0 - (num / (deno+eps));
   delta = abs(vt - vt0);
   vt0 = vt;
end
h8 = snrbar /2.0;
beta = 1.0 + h8;
beta2 = 2.0 * beta^2 - 1.0;
beta3 = 2.0 * beta^3;
if (np >= 50)
   temp1 = 2.0 * beta -1;
   omegabar = sqrt(np * temp1);
   c3 = (beta3 - 1.) / 3.0 / beta2 / omegabar;
   c4 = (beta3 * beta3 - 1.0) / 4. / np /beta2 /beta2;
   c6 = c3 * c3 /2.0;
   V = (vt - np * (1.0 + snrbar)) / omegabar;
   Vsqr = V *V;
   val1 = exp(-Vsqr / 2.0) / sqrt( 2.0 * pi);
   val2 = c3 * (V^2 -1.0) + c4 * V * (3.0 - V^2) - c6 * V * (V^4 - 10. * V^2 + 15.0);
   q = 0.5 * erfc (V/sqrt(2.0));
   pd =  q - val1 * val2;
   return
else
```

```
gamma0 = gammainc(vt/beta,np);
a1 = (vt / beta)^np / (factorial(np) * exp(vt/beta));
sum = gamma0;
for i = 1:1:np
  temp1 = gamma0;
  if (i == 1)
    ai = a1;
  else
    ai = (vt / beta) * a1 / (np + i -1);
  end
  gammai = gamma0 - ai;
  gamma0 = gammai;
  a1 = ai;
  for ii = 1:1:i
    temp1 = temp1 * (np + 1 - ii);
  end
  term = (snrbar /2.0)^i * gammai * temp1 / (factorial(i));
  sum - sum | term;
end
pd = 1.0 - (sum / beta^np);
end
pd = max(pd,0.);
return
```

7.11.10. MATLAB Function "fluct_loss.m"

This functions has been described in Section 7.8.

MATLAB Function *"fluct_loss.m"*

```
function [SNR] = fluct(pd, nfa, np, sw_case)
% This function calculates the SNR fluctuation loss for Swerling models
% A negative Lf value indicates SNR gain instead of loss
format long
% ************** Swerling 5 case ****************
% check to make sure that np>1
pfa = np * log(2) / nfa;
if (sw_case == 0)
if (np ==1)
  nfa = 1/pfa;
  b = sqrt(-2.0 * log(pfa));
  Pd_Sw5 = 0.001;
  snr_inc = 0.1 - 0.005;
  while(Pd_Sw5 <= pd)
    snr_inc = snr_inc + 0.005;
    a = sqrt(2.0 * 10^(.1*snr_inc));
    Pd_Sw5 = marcumsq(a,b);
  end
```

```
  PD_SW5 = Pd_Sw5;
  SNR = snr_inc;
else
  % np > 1 use MATLAB function pd_swerling5.m
  snr_inc = 0.1 - 0.001;
  Pd_Sw5 = 0.001;
  while(Pd_Sw5 <= pd)
    snr_inc = snr_inc + 0.001;
    Pd_Sw5 = pd_swerling5(pfa, 1, np, snr_inc);
  end
  PD_SW5 = Pd_Sw5;
  SNR = snr_inc;
end
end
% ************** End Swerling 5 case ***********
% ************** Swerling 1 case ***************
% compute the false alarm number
if (sw_case==1)
  Pd_Sw1 = 0.001;
  snr_inc = 0.1 - 0.001;
  while(Pd_Sw1 <= pd)
    snr_inc = snr_inc + 0.001;
    Pd_Sw1 = pd_swerling1(nfa, np, snr_inc);
  end
  PD_SW1 = Pd_Sw1;
  SNR = snr_inc;
end
% ************** End Swerling 1 case ***********
% ************** Swerling 2 case ***************
if (sw_case == 2)
  Pd_Sw2 = 0.001;
  snr_inc = 0.1 - 0.001;
  while(Pd_Sw2 <= pd)
    snr_inc = snr_inc + 0.001;
    Pd_Sw2 = pd_swerling2(nfa, np, snr_inc);
  end
  PD_SW2 = Pd_Sw2;
  SNR = snr_inc;
end
% ************** End Swerling 2 case ***********
% ************** Swerling 3 case ***************
if (sw_case == 3)
  Pd_Sw3 = 0.001;
  snr_inc = 0.1 - 0.001;
  while(Pd_Sw3 <= pd)
    snr_inc = snr_inc + 0.001;
    Pd_Sw3 = pd_swerling3(nfa, np, snr_inc);
  end
```

```
  PD_SW3 = Pd_Sw3;
  SNR = snr_inc;
end
% ************* End Swerling 3 case ***********
% ************** Swerling 4 case ***************
if (sw_case == 4)
  Pd_Sw4 = 0.001;
  snr_inc = 0.1 - 0.001;
  while(Pd_Sw4 <= pd)
    snr_inc = snr_inc + 0.001;
    Pd_Sw4 = pd_swerling4(nfa, np, snr_inc);
  end
  PD_SW4 = Pd_Sw4;
  SNR = snr_inc;
end
```

Appendix 7.A The Incomplete Gamma Function

The Gamma Function

Define the Gamma function (not the incomplete Gamma function) of the variable z (generally complex) as

$$\Gamma(z) = \int_0^\infty x^{z-1} e^{-x} \, dx \qquad (7.115)$$

and when z is a positive integer, then

$$\Gamma(z) = (z-1)! \qquad (7.116)$$

One very useful and frequently used property is

$$\Gamma(z+1) = z\Gamma(z) \qquad (7.117)$$

The Incomplete Gamma Function

The incomplete gamma function. $\Gamma_I(u, q)$ used in this text is given by

$$\Gamma_I(u, q) = \int_0^{u\sqrt{q+1}} \frac{e^{-x} x^q}{q!} \, dx \qquad (7.118)$$

Another definition, which is often used in the literature, for the incomplete Gamma function is

$$\Gamma_l[z, q] = \int_q^\infty x^{z-1} e^{-x} \, dx \qquad (7.119)$$

It follows that

$$\Gamma(z) = \Gamma_l[z, 0] = \int_0^\infty x^{z-1} e^{-x} dx \qquad (7.120)$$

which is the same as Eq. (7.115). Furthermore, for a positive integer n, the incomplete Gamma function can be represented by

$$\Gamma_l[n, z] = (n-1)! e^{-z} \sum_{k=0}^{n-1} \frac{z^k}{k!} \qquad (7.121)$$

In order to relate $\Gamma_l[n, z]$ and $\Gamma_l(u, q)$ compute the following relation

$$\Gamma_l[a, 0] - \Gamma_l[a, z] = \int_0^\infty x^{a-1} e^{-x} dx - \int_z^\infty x^{a-1} e^{-x} dx = \int_0^z x^{a-1} e^{-x} dx \qquad (7.122)$$

Applying the change of variables $a = q + 1$ and $z = u\sqrt{q+1}$ yields

$$\Gamma_l[q+1, 0] - \Gamma_l[q+1, u\sqrt{q+1}] = \int_0^{u\sqrt{q+1}} x^q e^{-x} dx \qquad (7.123)$$

and if q is a positive integer then

$$\frac{\Gamma_l[q+1, 0] - \Gamma_l[q+1, u\sqrt{q+1}]}{q!} = \int_0^{u\sqrt{q+1}} \frac{x^q e^{-x}}{q!} dx = \Gamma_l(u, q) \qquad (7.124)$$

Using Eq. (7.116) and (7.121) in Eq. (7.124) yields

$$\Gamma_l(u, q) = 1 - \frac{(q+1-1)! e^{-u\sqrt{q+1}}}{q!} \sum_{k=0}^{q} \frac{(u\sqrt{q+1})^k}{k!} \qquad (7.125)$$

Finally, the incomplete Gamma function can be written as

$$\Gamma_I(u, q) = 1 - e^{-u\sqrt{q+1}} \sum_{k=0}^{q} \frac{(u\sqrt{q+1})^k}{k!} \tag{7.126}$$

The two limiting values for Eq. (7.126) are

$$\Gamma_I(0, q) = 0 \qquad \Gamma_I(\infty, q) = 1 \tag{7.127}$$

Figure 7A.1 shows the incomplete gamma function for $q = 1, 3, 5, 8$. This figure can be reproduced using the following MATLAB code which utilizes the built-in MATLAB function *"gammainc.m"*.

```
% This program can be used to reproduce Fig. 7A.1
close all; clear all
x=linspace(0,20,200);
y1 = gammainc(x,1);
y2 = gammainc(x,3);
y3 = gammainc(x,5);
y4 = gammainc(x,8);
plot(x,y1,'k',x,y2,'k:',x,y3,'k--',x,y4,'k-.')
legend('q = 1','q = 3','q = 5','q = 8')
xlabel('x'); ylabel('Incomplete Gamma function (x,q)')
grid
```

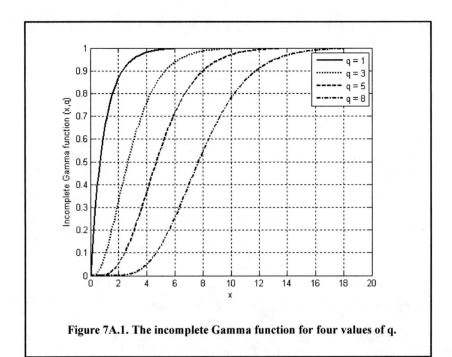

Figure 7A.1. The incomplete Gamma function for four values of q.

Problems

7.1. In the case of noise alone, the quadrature components of a radar return are independent Gaussian random variables with zero mean and variance σ^2. Assume that the radar processing consists of envelope detection followed by threshold decision. (a) Write an expression for the *pdf* of the envelope; (b) determine the threshold V_T as a function of σ that ensures a probability of false alarm $P_{fa} \leq 10^{-8}$.

7.2. (a) Derive Eq. (7.13); (b) derive Eq. (7.15).

7.3. A pulsed radar has the following specifications: time of false alarm $T_{fa} = 10$ *min*, probability of detection $P_D = 0.95$, operating bandwidth $B = 1MHz$. (a) What is the probability of false alarm P_{fa}? (b) What is the single pulse SNR? (c) Assuming noncoherent integration of 100 pulses, what is the SNR reduction so that P_D and P_{fa} remain unchanged?

7.4. An L-band radar has the following specifications: operating frequency $f_0 = 1.5GHz$, operating bandwidth $B = 2MHz$, noise figure $F = 8dB$, system losses $L = 4dB$, time of false alarm $T_{fa} = 12$ *minutes*, detection range $R = 12Km$, probability of detection $P_D = 0.5$, antenna gain $G = 5000$, and target RCS $\sigma = 1m^2$. (a) Determine the PRF f_r, the pulse width τ, the peak power P_t, the probability of false alarm P_{fa}, and the minimum detectable signal level S_{min}. (b) How can you reduce the transmitter power to achieve the same performance when 10 pulses are integrated noncoherently? (c) If the radar operates at a shorter range in the single pulse mode, find the new probability of detection when the range decreases to $9Km$.

7.5. (a) Show how you can use the radar equation to determine the PRF f_r, the pulse width τ, the peak power P_t, the probability of false alarm P_{fa}, and the minimum detectable signal level S_{min}. Assume the following specifications: operating frequency $f_0 = 1.5MHz$, operating bandwidth $B = 1MHz$, noise figure $F = 10dB$, system losses $L = 5dB$, time of false alarm $T_{fa} = 20$ *min*, detection range $R = 12Km$, probability of detection $P_D = 0.5$ (three pulses). (b) If post detection integration is assumed, determine the SNR.

7.6. Show that when computing the probability of detection at the output of an envelope detector, it is possible to use Gaussian probability approximation when the SNR is very large.

7.7. A radar system uses a threshold detection criterion. The probability of false alarm $P_{fa} = 10^{-10}$. (a) What must be the average SNR at the input of a linear detector so that the probability of miss is $P_m = 0.15$? Assume large SNR approximation. (b) Write an expression for the *pdf* at the output of the envelope detector.

7.8. An X-band radar has the following specifications: received peak power $10^{-10}W$, probability of detection $P_D = 0.95$, time of false alarm $T_{fa} = 8\ min$, pulse width $\tau = 2\mu s$, operating bandwidth $B = 2MHz$, operating frequency $f_0 = 10GHz$, and detection range $R = 100Km$. Assume single pulse processing. (a) Compute the probability of false alarm P_{fa}. (b) Determine the SNR at the output of the IF amplifier. (c) At what SNR would the probability of detection drop to 0.9 (P_{fa} does not change)? (d) What is the increase in range that corresponds to this drop in the probability of detection?

7.9. A certain radar utilizes 10 pulses for noncoherent integration. The single pulse SNR is $15dB$ and the probability of miss is $P_m = 0.15$. (a) Compute the probability of false alarm P_{fa}. (b) Find the threshold voltage V_T.

7.10. Consider a scanning low PRF radar. The antenna half-power beam width is $1.5°$, and the antenna scan rate is $35°$ per second. The pulse width is $\tau = 2\mu s$, and the PRF is $f_r = 400Hz$. (a) Compute the radar operating bandwidth. (b) Calculate the number of returned pulses from each target illumination. (c) Compute the SNR improvement due to post-detection integration (assume 100% efficiency). (d) Find the number of false alarms per minute for a probability of false alarm $P_{fa} = 10^{-6}$.

7.11. Using the equation

$$P_D = 1 - e^{-SNR}\int_{P_{fa}}^{1} I_0(\sqrt{-4SNR\ln u})du$$

calculate P_D when $SNR = 10dB$ and $P_{fa} = 0.01$. Perform the integration numerically.

7.12. Write a MATLAB program to compute the CA-CFAR threshold value. Use similar approach to that used in the case of a fixed threshold.

7.13. A certain radar has the following specifications: single pulse SNR corresponding to a reference range $R_0 = 200Km$ is $10dB$. The probability of detection at this range is $P_D = 0.95$. Assume a Swerling I type target. Use the radar equation to compute the required pulse widths at ranges $R = 220Km, 250Km, 175Km$, so that the probability of detection is maintained.

7.14. Repeat Problem 7.14 for Swerling IV type target.

7.15. Utilizing the MATLAB functions presented in this chapter, plot the actual value for the improvement factor versus the number of integrated pulses. Pick three different values for the probability of false alarm.

7.16. Develop a MATLAB program to calculate the cumulative probability of detection.

7.17. A certain radar has the following parameters: Peak power $P_t = 500KW$, total losses $L = 12dB$, operating frequency $f_o = 5.6GHZ$, PRF $f_r = 2KHz$, pulse width $\tau = 0.5\mu s$, antenna beamwidth $\theta_{az} = 2°$ and $\theta_{el} = 7°$, noise figure $F = 6dB$, scan time $T_{sc} = 2s$. The radar can experience one false alarm per scan. (a) What is the probability of false alarm? Assume that the radar searches a minimum range of 10 Km to its maximum unambiguous range. (b) Plot the detection range versus RCS in dBsm. The detection range is defined as the range at which the single scan probability of detection is equal to 0.94. Generate curves for Swerling I, II, III, and IV type targets. (c) Repeat part (b) above when noncoherent integration is used.

7.18. A certain circularly scanning radar with a fan beam has a rotation rate of 3 seconds per revolution. The azimuth beamwidth is 3 degrees and the radar uses a PRI of 600 microseconds. The radar pulse width is 2 microseconds and the radar searches a range window that extends from 15 Km to 100 Km. It is desired that the false alarm rate not be higher than two false alarms per revolution. What is the required probability of false alarm? What is the minimum SNR so that minimum probability of false alarm can be maintained?

7.19. Derive Eq(7.63).

Chapter 8   *Pulse Compression*

Range resolution for a given radar can be significantly improved by using very short pulses. Unfortunately, utilizing short pulses decreases the average transmitted power, hence reducing the SNR. Since the average transmitted power is directly linked to the receiver SNR, it is often desirable to increase the pulse width (i.e., the average transmitted power) while simultaneously maintaining adequate range resolution. This can be made possible by using pulse compression techniques and the matched filter receiver. Pulse compression allows us to achieve the average transmitted power of a relatively long pulse, while obtaining the range resolution corresponding to a short pulse. In this chapter, two pulse compression techniques are discussed. The first technique is known as correlation processing which is predominantly used for narrow band and some medium band radar operations. The second technique is called stretch processing and is normally used for extremely wide band radar operations.

8.1. Time-Bandwidth Product

Consider a radar system that employs a matched filter receiver. Let the matched filter receiver bandwidth be denoted as B. Then the noise power available within the matched filter bandwidth is given by

$$N_i = 2 \frac{\eta_0}{2} B \qquad (8.1)$$

where the factor of two is used to account for both negative and positive frequency bands, as illustrated in Fig. 8.1. The average input signal power over a pulse duration τ_0 is

$$S_i = \frac{E_x}{\tau_0} \qquad (8.2)$$

315

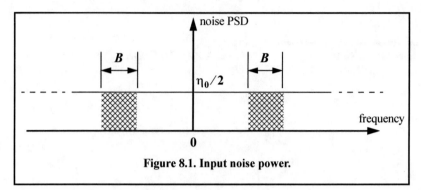

Figure 8.1. Input noise power.

E_x is the signal energy. Consequently, the matched filter input SNR is given by

$$(SNR)_i = \frac{S_i}{N_i} = \frac{E}{\eta_0 B \tau_0} \tag{8.3}$$

The output peak instantaneous SNR to the input SNR ratio is

$$\frac{SNR(t_0)}{(SNR)_i} = 2 B \tau_0 \tag{8.4}$$

The quantity $B\tau_0$ is referred to as the time-bandwidth product for a given waveform or its corresponding matched filter. The factor $B\tau_0$ by which the output SNR is increased over that at the input is called the matched filter gain, or simply the compression gain.

In general, the time-bandwidth product of an unmodulated pulse approaches unity. The time-bandwidth product of a pulse can be made much greater than unity by using frequency or phase modulation. If the radar receiver transfer function is perfectly matched to that of the input waveform, then the compression gain is equal to $B\tau_0$. Clearly, the compression gain becomes smaller than $B\tau_0$ as the spectrum of the matched filter deviates from that of the input signal.

8.2. Radar Equation with Pulse Compression

The radar equation for a pulsed radar can be written as

$$SNR = \frac{P_t \tau_0 G^2 \lambda^2 \sigma}{(4\pi)^3 R^4 k T_0 F L} \tag{8.5}$$

where P_t is peak power, τ_0 is pulse width, G is antenna gain, σ is target RCS, R is range, k is Boltzmann's constant, T_0 is 290 degrees Kelvin, F is noise figure, and L is total radar losses.

Pulse compression radars transmit relatively long pulses (with modulation) and process the radar echo into very short pulses (compressed). One can view the transmitted pulse as being composed of a series of very short subpulses (duty is 100%), where the width of each subpulse is equal to the desired compressed pulse width. Denote the compressed pulse width as τ_c. Thus, for an individual subpulse, Eq. (8.5) can be written as

$$(SNR)_{\tau_c} = \frac{P_t \tau_c G^2 \lambda^2 \sigma}{(4\pi)^3 R^4 k T_0 F L} \tag{8.6}$$

The SNR for the uncompressed pulse is then derived from Eq. (8.6) as

$$SNR = \frac{P_t(\tau_0 = n_P \tau_c) G^2 \lambda^2 \sigma}{(4\pi)^3 R^4 k T_0 F L} \tag{8.7}$$

where n_P is the number of subpulses. Equation (8.7) is denoted as the radar equation with pulse compression.

Observation of Eq. (8.5) and Eq.(8.7) indicates the following (note that both equations have the same form): For a given set of radar parameters, and as long as the transmitted pulse remains unchanged, the SNR is also unchanged regardless of the signal bandwidth. More precisely, when pulse compression is used, the detection range is maintained while the range resolution is drastically improved by keeping the pulse width unchanged and by increasing the bandwidth. Remember that range resolution is proportional to the inverse of the signal bandwidth:

$$\Delta R = \frac{c}{2B} \tag{8.8}$$

8.3. Basic Principal of Pulse Compression

For this purpose, consider a long pulse with LFM modulation and assume a matched filter receiver. The output of the matched filter (along the delay axis, i.e., range) is an order of magnitude narrower than that at its input. More precisely, the matched filter output is compressed by a factor $\xi = B\tau_0$, where τ_0 is the pulse width and B is the bandwidth. Thus, by using long pulses and wideband LFM modulation, large compression ratios can be achieved.

Figure 8.2 shows an ideal LFM pulse compression process. Part (a) shows the envelope of a pulse, part (b) shows the frequency modulation (in this case it is an upchirp LFM) with bandwidth $B = f_2 - f_1$. Part (c) shows the matched filter time-delay characteristic while part (d) shows the compressed pulse envelope. Finally part (e) shows the matched filter input/output waveforms.

Figure 8.3 illustrates the advantage of pulse compression using a more realistic LFM waveform. In this example, two targets with RCS, $\sigma_1 = 1m^2$ and $\sigma_2 = 0.5m^2$, are detected. The two targets are not separated enough in time to be resolved. Figure 8.3a shows the composite echo signal from those targets. Clearly, the target returns overlap, and thus, they are not resolved. However, after pulse compression the two pulses are completely separated and are resolved as two distinct targets. In fact, when using LFM, returns from neighboring targets are resolved as long as they are separated in time by τ_c, the compressed pulse width.

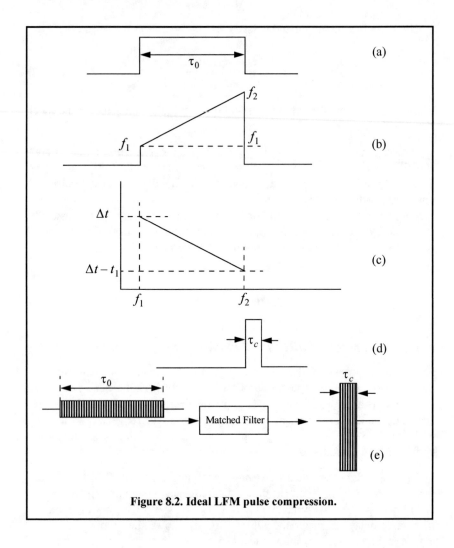

Figure 8.2. Ideal LFM pulse compression.

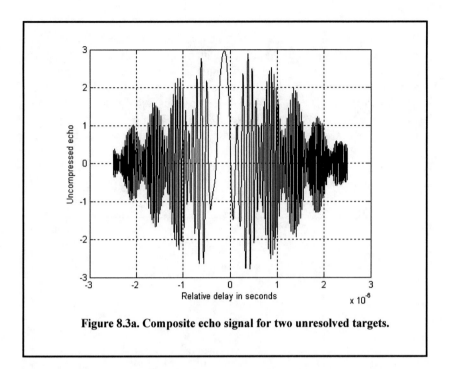

Figure 8.3a. Composite echo signal for two unresolved targets.

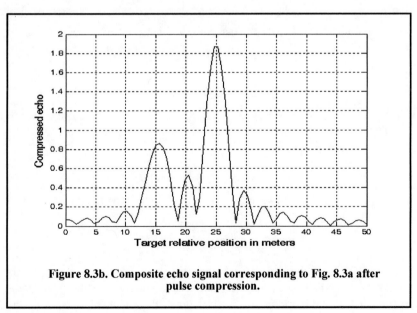

Figure 8.3b. Composite echo signal corresponding to Fig. 8.3a after pulse compression.

8.4. Correlation Processor

Radar operations (search, track, etc.) are usually carried out over a specified range window, referred to as the receive window and defined by the difference between the radar maximum and minimum range. Returns from all targets within the receive window are collected and passed through matched filter circuitry to perform pulse compression. One implementation of such analog processors is the Surface Acoustic Wave (SAW) devices. Because of the recent advances in digital computer development, the correlation processor is often performed digitally using the FFT. This digital implementation is called Fast Convolution Processing (FCP) and can be implemented at base band. The fast convolution process is illustrated in Fig. 8.4.

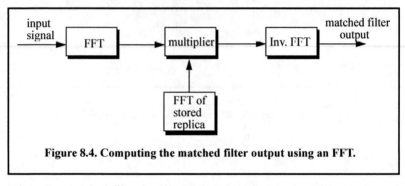

Figure 8.4. Computing the matched filter output using an FFT.

Since the matched filter is a linear time invariant system, its output can be described mathematically by the convolution between its input and its impulse response,

$$y(t) = s((t) \otimes h(t))$$ (8.9)

where $s(t)$ is the input signal, $h(t)$ is the matched filter impulse response (replica), and the ($\otimes$) operator symbolically represents convolution. From the Fourier transform properties,

$$FFT\{s((t) \otimes h(t))\} = S(f) \cdot H(f)$$ (8.10)

and when both signals are sampled properly, the compressed signal $y(t)$ can be computed from

$$y = FFT^{-1}\{S \cdot H\}$$ (8.11)

where FFT^{-1} is the inverse FFT. When using pulse compression, it is desirable to use modulation schemes that can accomplish a maximum pulse compression ratio and can significantly reduce the sidelobe levels of the compressed waveform. For the LFM case the first sidelobe is approximately

13.4dB below the main peak, and for most radar applications this may not be sufficient. In practice, high sidelobe levels are not preferable because noise and/or jammers located at the sidelobes may interfere with target returns in the main lobe.

Weighting functions (windows) can be used on the compressed pulse spectrum in order to reduce the sidelobe levels. The cost associated with such an approach is a loss in the main lobe resolution, and a reduction in the peak value (i.e., loss in the SNR). Weighting the time domain transmitted or received signal instead of the compressed pulse spectrum will theoretically achieve the same goal. However, this approach is rarely used, since amplitude modulating the transmitted waveform introduces extra burdens on the transmitter.

Consider a radar system that utilizes a correlation processor receiver (i.e., matched filter). The receive window in meters is defined by

$$R_{rec} = R_{max} - R_{min} \tag{8.12}$$

where R_{max} and R_{min}, respectively, define the maximum and minimum range over which the radar performs detection. Typically R_{rec} is limited to the extent of the target complex. The normalized complex transmitted signal has the form

$$s(t) = \exp\left(j2\pi\left(f_0 t + \frac{\mu}{2}t^2\right)\right) \qquad 0 \le t \le \tau_0 \tag{8.13}$$

τ_0 is the pulse width, $\mu = B/\tau_0$, and B is the bandwidth.

The radar echo signal is similar to the transmitted one with the exception of a time delay and an amplitude change that correspond to the target RCS. Consider a target at range R_1. The echo received by the radar from this target is

$$s_r(t) = a_1 \exp\left(j2\pi\left(f_0(t-t_1) + \frac{\mu}{2}(t-t_1)^2\right)\right) \tag{8.14}$$

where a_1 is proportional to target RCS, antenna gain, and range attenuation. The time delay t_1 is given by

$$t_1 = 2R_1/c \tag{8.15}$$

The first step of the processing consists of removing the frequency f_0. This is accomplished by mixing $s_r(t)$ with a reference signal whose phase is $2\pi f_0 t$. The phase of the resultant signal, after lowpass filtering, is then given by

$$\phi(t) = 2\pi\left(-f_0 t_1 + \frac{\mu}{2}(t-t_1)^2\right) \tag{8.16}$$

and the instantaneous frequency is

$$f_i(t) = \frac{1}{2\pi} \frac{d}{dt}\phi(t) = \mu(t - t_1) = \frac{B}{\tau_0}\left(t - \frac{2R_1}{c}\right) \tag{8.17}$$

The quadrature components are

$$\begin{pmatrix} x_I(t) \\ x_Q(t) \end{pmatrix} = \begin{pmatrix} \cos\phi(t) \\ \sin\phi(t) \end{pmatrix} \tag{8.18}$$

Sampling the quadrature components is performed next. The number of samples, N, must be chosen so that foldover (ambiguity) in the spectrum is avoided. For this purpose, the sampling frequency, f_s (based on the Nyquist sampling rate), must be

$$f_s \geq 2B \tag{8.19}$$

and the sampling interval is

$$\Delta t \leq 1/2B \tag{8.20}$$

Using Eq. (8.17) it can be shown that (the proof is left as an exercise) the frequency resolution of the FFT is

$$\Delta f = 1/\tau_0 \tag{8.21}$$

The minimum required number of samples is

$$N = \frac{1}{\Delta f \Delta t} = \frac{\tau_0}{\Delta t} \tag{8.22}$$

Equating Eqs. (8.20) and (8.22) yields

$$N \geq 2B\tau_0 \tag{8.23}$$

Consequently, a total of $2B\tau_0$ real samples, or $B\tau_0$ complex samples, is sufficient to completely describe an LFM waveform of duration τ_0 and bandwidth B. For example, an LFM signal of duration $\tau_0 = 20$ μs and bandwidth $B = 5$ MHz requires 200 real samples to determine the input signal (100 samples for the I-channel and 100 samples for the Q-channel).

For better implementation of the FFT N is extended to the next power of two, by zero padding. Thus, the total number of samples, for some positive integer m, is

$$N_{FFT} = 2^m \geq N \tag{8.24}$$

The final steps of the FCP processing include (1) taking the FFT of the sampled sequence, (2) multiplying the frequency domain sequence of the signal with the FFT of the matched filter impulse response, and (3) performing the inverse FFT of the composite frequency domain sequence in order to generate the time domain compressed pulse. Of course, weighting, antenna gain, and range attenuation compensation must also be performed.

Assume that I targets at ranges R_1, R_2, and so forth are within the receive window. From superposition, the phase of the down-converted signal is

$$\phi(t) = \sum_{i=1}^{I} 2\pi\left(-f_0 t_i + \frac{\mu}{2}(t - t_i)^2\right) \qquad (8.25)$$

The times $\{t_i = (2R_i/c); \; i = 1, 2, ..., I\}$ represent the two-way time delays, where t_1 coincides with the start of the receive window. As an example, consider the case where

# targets	R_{rec}	pulse width	Band-width	targets range	Target RCS	Window type
3	200m	0.005ms	100e6 Hz	[10 75 120] m	[1 2 1]m²	Hamming

Note that the compressed pulsed range resolution is $\Delta R = 1.5m$. Figure 8.5 and Fig. 8.6 shows the real part and the amplitude spectrum of the replica used for this example. Figure 8.7 shows the uncompressed echo, while Fig. 8.8 shows the compressed MF output. Note that the scatterer amplitude attenuation is a function of the inverse of the scatterer's range within the receive window. Figure 8.9 is similar to Fig. 8.8, except in this case the first and second scatterers are less than 1.5 meter apart (they are at 70 and 71 meters).

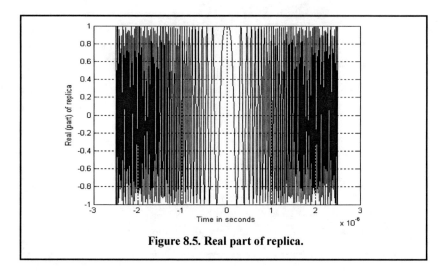

Figure 8.5. Real part of replica.

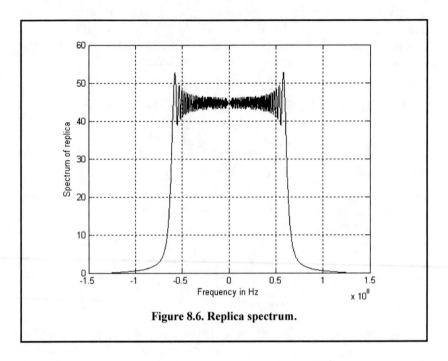

Figure 8.6. Replica spectrum.

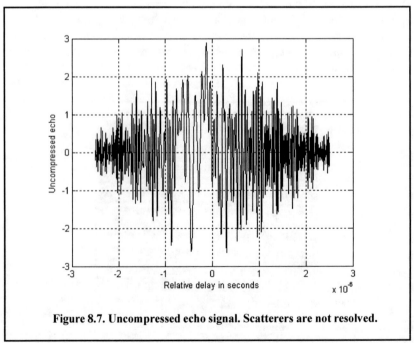

Figure 8.7. Uncompressed echo signal. Scatterers are not resolved.

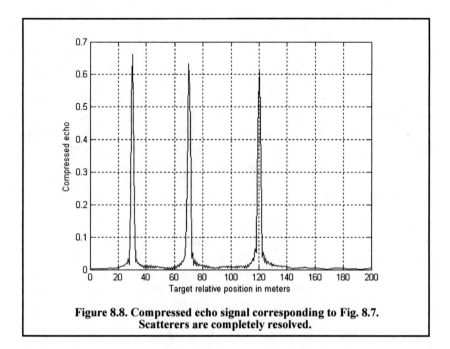

Figure 8.8. Compressed echo signal corresponding to Fig. 8.7. Scatterers are completely resolved.

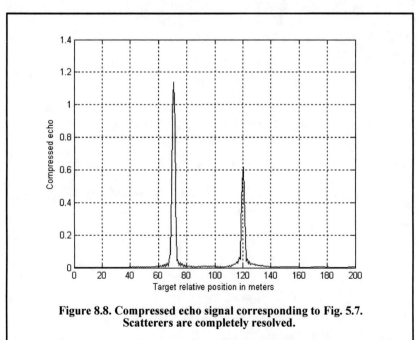

Figure 8.8. Compressed echo signal corresponding to Fig. 5.7. Scatterers are completely resolved.

8.5. Stretch Processor

Stretch processing, also known as *active correlation,* is normally used to process extremely high-bandwidth LFM waveforms. This processing technique consists of the following steps: First, the radar returns are mixed with a replica (reference signal) of the transmitted waveform. This is followed by Low Pass Filtering (LPF) and coherent detection. Next, Analog-to-Digital (A/D) conversion is performed; and finally, a bank of Narrow-Band Filters (NBFs) is used in order to extract the tones that are proportional to target range, since stretch processing effectively converts time delay into frequency. All returns from the same range bin produce the same constant frequency.

8.5.1. Single LFM Pulse

Figure 8.10 shows a block diagram for a stretch processing receiver. The reference signal is an LFM waveform that has the same LFM slope as the transmitted LFM signal. It exists over the duration of the radar "receive-window," which is computed from the difference between the radar maximum and minimum range. Denote the start frequency of the reference chirp as f_r. Consider the case when the radar receives returns from a few close (in time or range) targets, as illustrated in Fig. 8.10. Mixing with the reference signal and performing lowpass filtering are effectively equivalent to subtracting the return frequency chirp from the reference signal. Thus, the LPF output consists of constant tones corresponding to the targets' positions. The normalized transmitted signal can be expressed by

$$s_1(t) = \cos\left(2\pi\left(f_0 t + \frac{\mu}{2}t^2\right)\right) \qquad 0 \le t \le \tau_0 \qquad (8.26)$$

where $\mu = B/\tau_0$ is the LFM coefficient and f_0 is the chirp start frequency. Assume a point scatterer at range R. The signal received by the radar is

$$s_r(t) = a\cos\left[2\pi\left(f_0(t-t_0) + \frac{\mu}{2}(t-t_0)^2\right)\right] \qquad (8.27)$$

where a is proportional to target RCS, antenna gain, and range attenuation. The time delay t_0 is

$$t_0 = 2R/c \qquad (8.28)$$

The reference signal is

$$s_{ref}(t) = 2\cos\left(2\pi\left(f_r t + \frac{\mu}{2}t^2\right)\right) \qquad 0 \le t \le T_{rec} \qquad (8.29)$$

The receive window in seconds is

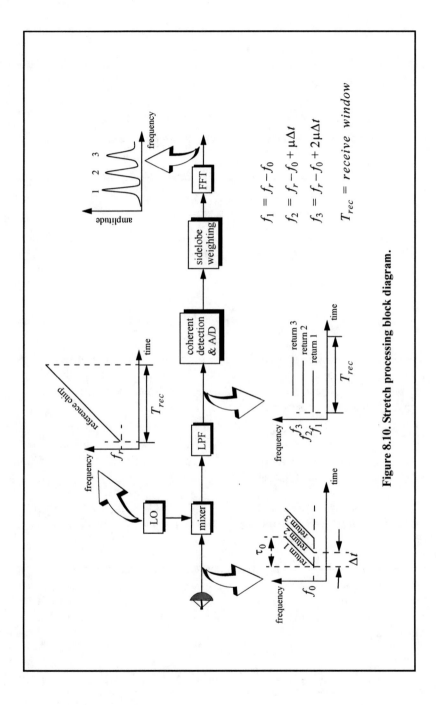

Figure 8.10. Stretch processing block diagram.

$$T_{rec} = \frac{2(R_{max} - R_{min})}{c} = \frac{2R_{rec}}{c} \tag{8.30}$$

It is customary to let $f_r = f_0$. The output of the mixer is the product of the received and reference signals. After lowpass filtering the signal is

$$s_0(t) = a\cos(2\pi f_0 t_0 + 2\pi \mu t_0 t - \pi \mu (t_0)^2) \tag{8.31}$$

Substituting Eq. (8.28) into Eq. (8.31) and collecting terms yield

$$s_0(t) = a \, \cos\left[\left(\frac{4\pi BR}{c\tau_0}\right)t + \frac{2R}{c}\left(2\pi f_0 - \frac{2\pi BR}{c\tau_0}\right)\right] \tag{8.32}$$

and since $\tau_0 \gg 2R/c$, Eq. (8.32) is approximated by

$$s_0(t) \approx a \, \cos\left[\left(\frac{4\pi BR}{c\tau_0}\right)t + \frac{4\pi R}{c}f_0\right] \tag{8.33}$$

The instantaneous frequency is

$$f_{inst} = \frac{1}{2\pi}\frac{d}{dt}\left(\frac{4\pi BR}{c\tau_0}t + \frac{4\pi R}{c}f_0\right)) = \frac{2BR}{c\tau_0} \tag{8.34}$$

which clearly indicates that target range is proportional to the instantaneous frequency. Therefore, proper sampling of the LPF output and taking the FFT of the sampled sequence lead to the following conclusion: a peak at some frequency f_1 indicates presence of a target at range

$$R_1 = f_1 c\tau_0/2B \tag{8.35}$$

Assume I close targets at ranges R_1, R_2, and so forth ($R_1 < R_2 < ... < R_I$). From superposition, the total signal is

$$s_r(t) = \sum_{i=1}^{I} a_i(t)\cos\left[2\pi\left(f_0(t - t_i) + \frac{\mu}{2}(t - t_i)^2\right)\right] \tag{8.36}$$

where $\{a_i(t); \; i = 1, 2, ..., I\}$ are proportional to the targets' cross sections, antenna gain, and range. The times $\{t_i = (2R_i/c); \; i = 1, 2, ..., I\}$ represent the two-way time delays, where t_1 coincides with the start of the receive window. Using Eq. (8.32) the overall signal at the output of the LPF can then be described by

$$s_0(t) = \sum_{i=1}^{I} a_i\cos\left[\left(\frac{4\pi BR_i}{c\tau_0}\right)t + \frac{2R_i}{c}\left(2\pi f_0 - \frac{2\pi BR_i}{c\tau_0}\right)\right] \tag{8.37}$$

Hence, target returns appear as constant frequency tones that can be resolved using the FFT. Consequently, determining the proper sampling rate and FFT size is very critical. The rest of this section presents a methodology for computing the proper FFT parameters required for stretch processing.

Assume a radar system using a stretch processor receiver. The pulse width is τ_0 and the chirp bandwidth is B. Since stretch processing is normally used in extreme bandwidth cases (i.e., very large B), the receive window over which radar returns will be processed is typically limited to from a few meters to possibly less than 100 meters. The compressed pulse range resolution is computed from Eq. (8.8). Declare the FFT size to be N and its frequency resolution to be Δf. The frequency resolution can be computed using the following procedure: Consider two adjacent point scatterers at ranges R_1 and R_2. The minimum frequency separation, Δf, between those scatterers so that they are resolved can be computed from Eq. (8.34). More precisely,

$$\Delta f = f_2 - f_1 = \frac{2B}{c\tau_0}(R_2 - R_1) = \frac{2B}{c\tau_0}\Delta R \qquad (8.38)$$

Substituting Eq. (8.8) into Eq. (8.38) yields

$$\Delta f = \frac{2B}{c\tau_0} \frac{c}{2B} = \frac{1}{\tau_0} \qquad (8.39)$$

The maximum frequency resolvable by the FFT is limited to the region $\pm N\Delta f/2$. Thus, the maximum resolvable frequency is

$$\frac{N\Delta f}{2} > \frac{2B(R_{max} - R_{min})}{c\tau_0} = \frac{2BR_{rec}}{c\tau_0} \qquad (8.40)$$

Using Eqs. (8.30) and (8.39) into Eq. (8.40) and collecting terms yield

$$N > 2BT_{rec} \qquad (8.41)$$

For better implementation of the FFT, choose an FFT of size

$$N_{FFT} \geq N = 2^m \qquad (8.42)$$

where m is a nonzero positive integer. The sampling interval is then given by

$$\Delta f = \frac{1}{T_s N_{FFT}} \Rightarrow T_s = \frac{1}{\Delta f N_{FFT}} \qquad (8.43)$$

As an example, consider the case where

# targets	3
pulsewidth	10 ms
center frequency	5.6 GHz
bandwidth	1 GHz
receive window	30 m
relative target's range	[2 5 10] m
target's RCS	[1, 1, 2] m²
window	2 (Kaiser)

Note that the compressed pulse range resolution, without using a window, is $\Delta R = 0.15m$. Figure 8.11 and Fig. 8.12, respectively, show the uncompressed and compressed echo signals corresponding to this example. Figures 8.13 a and b are similar to Fig. 8.11 and Fig. 8.12 except in this case two of the scatterers are less than 15 cm apart (i.e., unresolved targets at $R_{relative} = [3, 3.1]m$).

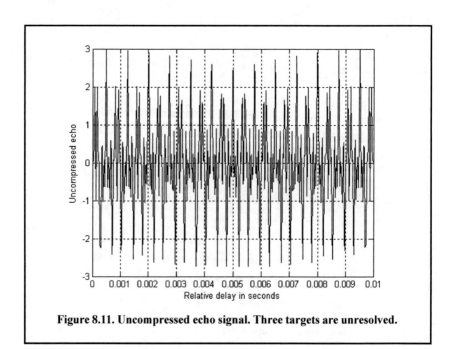

Figure 8.11. Uncompressed echo signal. Three targets are unresolved.

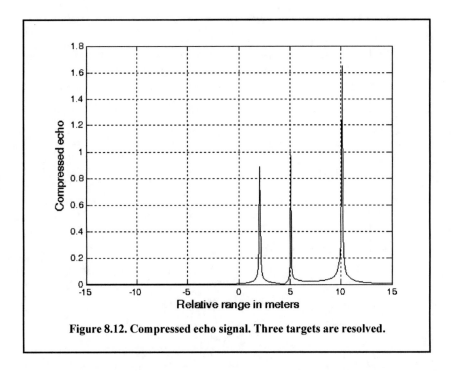

Figure 8.12. Compressed echo signal. Three targets are resolved.

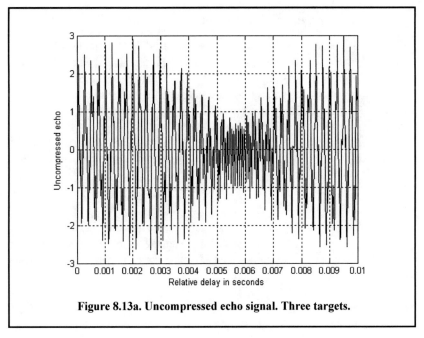

Figure 8.13a. Uncompressed echo signal. Three targets.

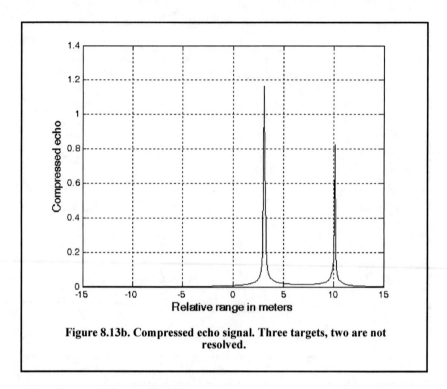

Figure 8.13b. Compressed echo signal. Three targets, two are not resolved.

8.5.2. Stepped Frequency Waveforms

Stepped Frequency Waveforms (SFW) are used in extremely wide band radar applications where very large time bandwidth product is required. Generation of SFW was discussed in Chapter 5. For this purpose, consider an LFM signal whose bandwidth is B_i and whose pulsewidth is T_i and refer to it as the primary LFM. Divide this long pulse into N subpulses each of width τ_0 to generate a sequence of pulses whose PRI is denoted by T. It follows that $T_i = (n-1)T$. Define the beginning frequency for each subpulse as that value measured from the primary LFM at the leading edge of each subpulse, as illustrated in Fig. 8.14. That is

$$f_i = f_0 + i\Delta f; \quad i = 0, N-1 \tag{8.44}$$

where Δf is the frequency step from one subpulse to another. The set of n subpulses is often referred to as a burst. Each subpulse can have its own LFM modulation. To this end, assume that each subpluse is of width τ_0 and bandwidth B, then the LFM slope of each pulse is

$$\mu = \frac{B}{\tau_0} \tag{8.45}$$

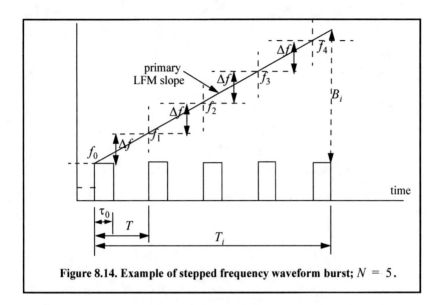

Figure 8.14. Example of stepped frequency waveform burst; $N = 5$.

The SFW operation and processing involve the following steps:

1. A series of N narrow-band LFM pulses is transmitted. The chirp beginning frequency from pulse to pulse is stepped by a fixed frequency step Δf, as defined in Eq. (8.44). Each group of N pulses is referred to as a burst.
2. The LFM slope (quadratic phase term) is first removed from the received signal, as described in Fig. 8.10. The reference slope must be equal to the combined primary LFM and single subpulse slopes. Thus, the received signal is reduced to a series of subpulses.
3. These subpulses are then sampled at a rate that coincides with the center of each pulse, sampling rate equivalent to ($1/T$).
4. The quadrature components for each burst are collected and stored.
5. Spectral weighting (to reduce the range sidelobe levels) is applied to the quadrature components. Corrections for target velocity, phase, and amplitude variations are applied.
6. The IDFT of the weighted quadrature components of each burst is calculated to synthesize a range profile for that burst. The process is repeated for M bursts to obtain consecutive high resolution range profiles.

Within a burst, the transmitted waveform for the ith step can be described as

$$x_i(t) = \left(\begin{array}{l} C_i \dfrac{1}{\sqrt{\tau_0}} Rect\left(\dfrac{t}{\tau_0}\right) e^{j2\pi\left(f_i t + \frac{\mu}{2} t^2\right)} \\[4mm] 0 \end{array} \right. \quad ; \quad \left. \begin{array}{l} iT \leq t \leq iT + \tau_0 \\[4mm] elsewhere \end{array} \right) \qquad \textbf{(3.46)}$$

where C_i are constants. The received signal from a target located at range R_0 is then given by

$$x_{ri}(t) = C_i' e^{j2\pi\left[f_i(t-\Delta(t)) - \frac{\mu}{2}(t-\Delta(t))^2\right]} \quad , \quad iT + \Delta(t) \leq t \leq iT + \tau_0 + \Delta(t) \quad \text{(8.47)}$$

where C_i' are constant and the round trip delay $\Delta(t)$ is given by

$$\Delta(t) = \frac{R_0 - vt}{c/2} \quad \text{(8.48)}$$

where c is the speed of light and v is the target radial velocity.

In order to remove the quadratic phase term, mixing is first performed with the reference signal given by

$$y_i(t) = e^{j2\pi\left(f_i t + \frac{\mu}{2}t^2\right)} \quad ; \quad iT \leq t \leq iT + \tau_0 \quad \text{(8.49)}$$

Next lowpass filtering is performed to extract the quadrature components. More precisely, the quadrature components are given by

$$\begin{pmatrix} x_I(t) \\ x_Q(t) \end{pmatrix} = \begin{pmatrix} A_i \cos\phi_i(t) \\ A_i \sin\phi_i(t) \end{pmatrix} \quad \text{(8.50)}$$

where A_i are constants, and

$$\phi_i(t) = -2\pi f_i\left(\frac{2R_0}{c} - \frac{2vt}{c}\right) \quad \text{(8.51)}$$

where now $f_i = \Delta f$. For each pulse, the quadrature components are then sampled at

$$t_i = iT + \frac{\tau_r}{2} + \frac{2R_0}{c} \quad \text{(8.52)}$$

τ_r is the time delay associated with the range that corresponds to the start of the range profile.

The quadrature components can then be expressed in complex form as

$$X_i = A_i e^{j\phi_i} \quad \text{(8.53)}$$

Equation (8.53) represents samples of the target reflectivity, due to a single burst, in the frequency domain. This information can then be transformed into a series of range delay reflectivity (i.e., range profile) values by using the IDFT. It follows that

$$H_l = \frac{1}{N}\sum_{i=0}^{N-1} X_i \ \exp\!\left(j\frac{2\pi l i}{N}\right) \qquad ; \ 0 \le l \le N-1 \tag{8.54}$$

Substituting Eq. (8.51) and Eq. (8.53) into (8.54) and collecting terms yield

$$H_l = \frac{1}{N}\sum_{i=0}^{N-1} A_i \ \exp\!\left\{j\!\left(\frac{2\pi l i}{N} - 2\pi f_i\!\left(\frac{2R_0}{c} - \frac{2vt_i}{c}\right)\right)\right\} \tag{8.55}$$

By normalizing with respect to N and by assuming that $A_i = 1$ and that the target is stationary (i.e., $v = 0$), then Eq. (8.55) can be written as

$$H_l = \sum_{i=0}^{N-1} \exp\!\left\{j\!\left(\frac{2\pi l i}{N} - 2\pi f_i\frac{2R_0}{c}\right)\right\} \tag{8.56}$$

Using $f_i = i\Delta f$ inside Eq. (8.56) yields

$$H_l = \sum_{i=0}^{N-1} \exp\!\left\{j\frac{2\pi i}{N}\!\left(-\frac{2NR_0\Delta f}{c} + l\right)\right\} \tag{8.57}$$

which can be simplified to

$$H_l = \frac{\sin\pi\zeta}{\sin\dfrac{\pi\zeta}{N}} \ \exp\!\left(j\frac{N-1}{2}\ \frac{2\pi\zeta}{N}\right) \tag{8.58}$$

where

$$\zeta = \frac{-2NR_0\Delta f}{c} + l \tag{8.59}$$

Finally, the synthesized range profile is

$$|H_l| = \left|\frac{\sin\pi\zeta}{\sin\dfrac{\pi\zeta}{N}}\right| \tag{8.60}$$

Range Resolution and Range Ambiguity in SFW

As usual, range resolution is determined from the overall system bandwidth. Assuming an SFW with N steps and step size Δf, then the corresponding range resolution is equal to

$$\Delta R = \frac{c}{2N\Delta f} \tag{8.61}$$

Range ambiguity associated with an SFW can be determined by examining the phase term that corresponds to a point scatterer located at range R_0. More precisely,

$$\phi_i(t) = 2\pi f_i \frac{2R_0}{c} \tag{8.62}$$

It follows that

$$\frac{\Delta \phi}{\Delta f} = \frac{4\pi(f_{i+1}-f_i)R_0}{(f_{i+1}-f_i)} \frac{1}{c} = \frac{4\pi R_0}{c} \tag{8.63}$$

or equivalently,

$$R_0 = \frac{\Delta \phi}{\Delta f} \frac{c}{4\pi} \tag{8.64}$$

It is clear from Eq. (8.64) that range ambiguity exists for $\Delta \phi = \Delta \phi + 2N\pi$. Therefore,

$$R_0 = \frac{\Delta \phi + 2N\pi}{\Delta f} \frac{c}{4\pi} = R_0 + N\left(\frac{c}{2\Delta f}\right) \tag{8.65}$$

and the unambiguous range window is

$$R_u = \frac{c}{2\Delta f} \tag{8.66}$$

A range profile synthesized using a particular SFW represents the relative range reflectivity for all scatterers within the unambiguous range window, with respect to the absolute range that corresponds to the burst time delay. Additionally, if a specific target extent is larger than R_u, then all scatterers falling outside the unambiguous range window will fold over and appear in the synthesized profile. This fold-over problem is identical to the spectral fold-over that occurs when using a Fast Fourier Transform (FFT) to resolve certain signal frequency contents. For example, consider an FFT with frequency resolution $\Delta f = 50 Hz$ and size $NFFT = 64$. In this case, this FFT can resolve frequency tones between $-1600 Hz$ and $1600 Hz$. When this FFT is used to resolve the frequency content of a sine-wave tone equal to $1800 Hz$, fold-over occurs and a spectral line at the fourth FFT bin (i.e., $200 Hz$) appears. Therefore, in order to avoid fold-over in the synthesized range profile, the frequency step Δf must be

$$\Delta f \leq c/2E \tag{8.67}$$

where E is the target extent in meters.

Additionally, the pulsewidth must also be large enough to contain the whole target extent. Thus,

$$\Delta f \le 1/\tau_0 \qquad\qquad (8.68)$$

and in practice,

$$\Delta f \le 1/2\tau_0 \qquad\qquad (8.69)$$

This is necessary in order to reduce the amount of contamination of the synthesized range profile caused by the clutter surrounding the target under consideration.

For example, assume that the range profile starts at $R_0 = 900m$ and that

# targets	pulsewidth	N	Δf	1/T	v
3	$100\mu\,sec$	64	$10MHz$	$100KHz$	0.0

In this case,

$$\Delta R = \frac{3 \times 10^8}{2 \times 64 \times 10 \times 10^6} = 0.235m \text{, and } R_u = \frac{3 \times 10^8}{2 \times 10 \times 10^6} = 15m$$

Thus, scatterers that are more than 0.235 meters apart will appear as distinct peaks in the synthesized range profile. Assume two cases; in the first case, [scat_range] = [908, 910, 912] meters, and in the second case, [scat_range] = [908, 910, 910.2] meters. In both cases, let [scat_rcs] = [100, 10, 1] meters squared. Figure 8.15 shows the synthesized range profiles generated using the function "SWF.m" and the first case when the Hamming window is not used. Figure 8.16 is similar to Fig. 8.15, except in this case the Hamming window is used. Figure 8.17 shows the synthesized range profile that corresponds to the second case (Hamming window is used). Note that all three scatterers were resolved in Fig. 8.15 and Fig. 8.16; however, the last two scatterers are not resolved in Fig. 8.17, because they are separated by less than ΔR.

Next, consider another case where *[scat_range] = [908, 912, 916] meters*. Figure 8.18 shows the corresponding range profile. In this case, foldover occurs, and the last scatterer appears at the lower portion of the synthesized range profile. Also, consider the case where [scat_range] = [908, 910, 923] meters. Figure 8.19 shows the corresponding range profile. In this case, ambiguity is associated with the first and third scatterers since they are separated by $15m$. Both appear at the same range bin.

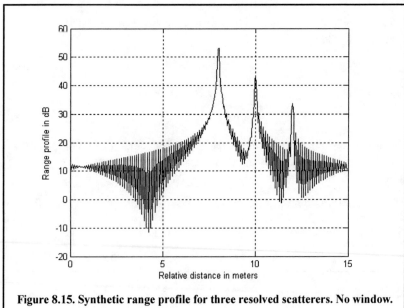

Figure 8.15. Synthetic range profile for three resolved scatterers. No window.

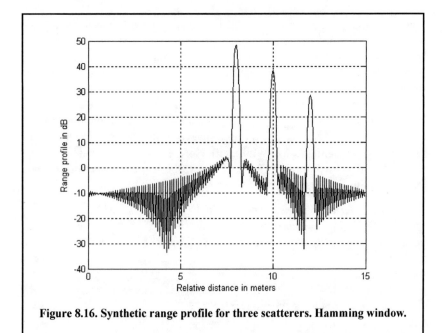

Figure 8.16. Synthetic range profile for three scatterers. Hamming window.

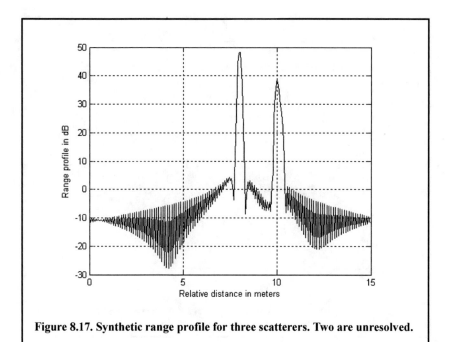

Figure 8.17. Synthetic range profile for three scatterers. Two are unresolved.

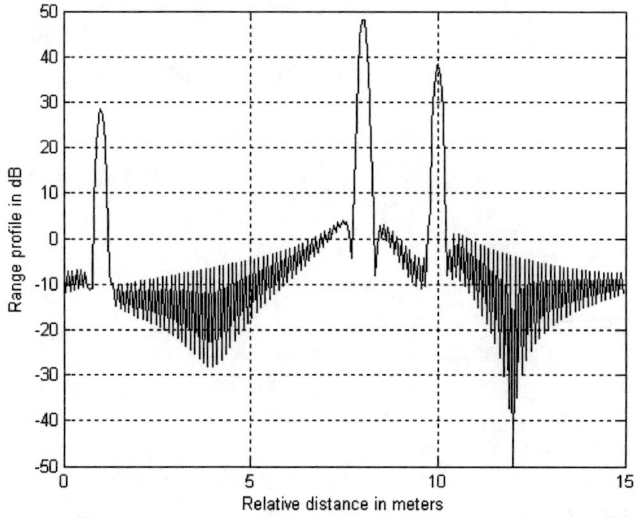

Figure 8.18. Synthetic range profile for three scatterers. Third scatterer folds over.

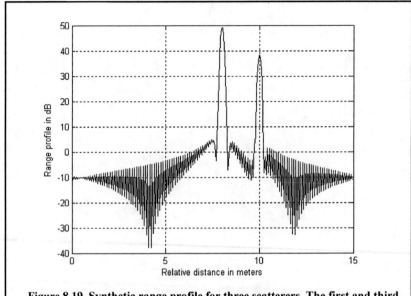

Figure 8.19. Synthetic range profile for three scatterers. The first and third scatterers appear in the same FFT bin.

8.5.2.1. Effect of Target Velocity

The range profile defined in Eq. (8.60) is obtained by assuming that the target under examination is stationary. The effect of target velocity on the synthesized range profile can be determined by starting with Eq. (8.55) and assuming that $v \neq 0$. Performing similar analysis as that of the stationary target case yields a range profile given by

$$H_l = \sum_{i=0}^{N-1} A_i \exp\left\{ j\frac{2\pi l i}{N} - j2\pi f_i \left[\frac{2R}{c} - \frac{2v}{c}\left(iT + \frac{\tau_r}{2} + \frac{2R}{c} \right) \right] \right\} \qquad \text{(8.70)}$$

The additional phase term present in Eq. (8.70) distorts the synthesized range profile. In order to illustrate this distortion, consider the SFW described in the previous section, and assume the three scatterers of the first case. Also, assume that $v = 200m/s$. Figure 8.20 shows the synthesized range profile for this case. Comparisons of Figs. 8.16 and 8.20 clearly show the distortion effects caused by the uncompensated target velocity. Figure 8.21 is similar to Fig. 8.20 except in this case, $v = -200m/s$. Note in either case, the targets have moved from their expected positions (to the left or right) by $Disp = 2 \times n \times v / PRF$ *(1.28 m)*.

This distortion can be eliminated by multiplying the complex received data at each pulse by the phase term

$$\Phi = \exp\left(-j2\pi f_i\left[\frac{2\hat{v}}{c}\left(iT + \frac{\tau_r}{2} + \frac{2\hat{R}}{c}\right)\right]\right) \tag{3.71}$$

$\hat{v}$ and R are, respectively, estimates of the target velocity and range. This process of modifying the phase of the quadrature components is often referred to as "phase rotation." In practice, when good estimates of $\hat{v}$ and $\hat{R}$ are not available, then the effects of target velocity are reduced by using frequency hopping between the consecutive pulses within the SFW. In this case, the frequency of each individual pulse is chosen according to a predetermined code. Waveforms of this type are often called Frequency Coded Waveforms (FCW). Costas waveforms or signals are a good example of this type of waveform.

Figure 8.22 shows a synthesized range profile for a moving target whose RCS is $\sigma = 10m^2$ and $v = 10m/s$. The initial target range is at $R = 912m$. All other parameters are as before. This figure can be reproduced using the following MATLAB code.

```
clear all;
close all;
nscat = 1;
scat_range = 912;
scat_rcs = 10;
n =64;
deltaf = 10e6;
prf = 10e3;
v = 10;
rnote = 900,
winid = 1;
count = 0;
for time = 0:.05:3
    count = count +1;
    hl = SFW (nscat, scat_range, scat_rcs, n, deltaf, prf, v, rnote, winid);
    array(count,:) = transpose(hl);
    hl(1:end) = 0;
    scat_range = scat_range - 2 * n * v / prf;
end
figure (1)
numb = 2*256;% this number matches that used in hrr_profile.
delx_meter = 15 / numb;
xmeter = 0:delx_meter:15-delx_meter;
imagesc(xmeter, 0:0.05:4,array)
colormap(gray)
ylabel ('Time in seconds')
xlabel('Relative distance in meters')
```

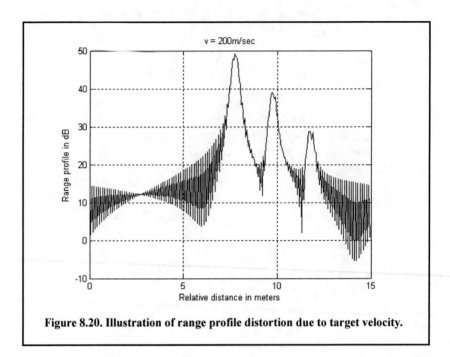

Figure 8.20. Illustration of range profile distortion due to target velocity.

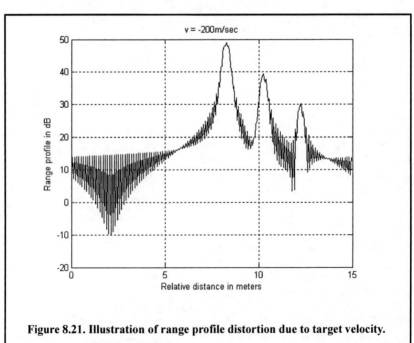

Figure 8.21. Illustration of range profile distortion due to target velocity.

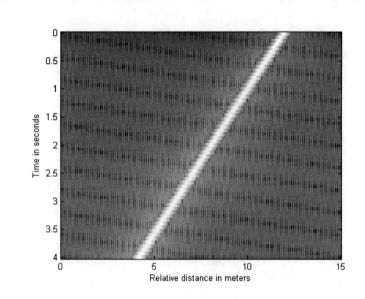

Figure 8.22. Synthesized range profile for a moving target (4 seconds long).

8.6. MATLAB Program Listings

This section presents listings for all the MATLAB programs used to produce all of the MATLAB-generated figures in this chapter.

8.6.1. MATLAB Function "matched_filter.m"

The function *"matched_filter.m"* performs fast convolution processing. The user can access this function either by a MATLAB function call or by executing the MATLAB program *"matched_filter_gui.m,"* which utilizes a MATLAB-based GUI. The work space associated with this program is shown in Fig. 8.23. The outputs for this function include plots of the compressed and uncompressed signals as well as the replica used in the pulse compression process. This function utilizes the function *"power_integer_2.m."*

The function *"matched_filter.m"* syntax is as follows:

[y] = matched_filter(nscat, rrec, taup, b, scat_range, scat_rcs, win)

where

Symbol	Description	Units	Status
nscat	number of point scatterers within the received window	none	input
rrec	receive window size	m	input
taup	uncompressed pulse width	seconds	input
b	chirp bandwidth	Hz	input
scat_range	vector of scatterers' relative range (within the receive window)	m	input
scat_rcs	vector of scatterers' RCS	m²	input
win	0 = no window 1 = Hamming 2 = Kaiser with parameter pi 3 = Chebychev side-lobes at -60dB	none	input
y	normalized compressed output	volts	output

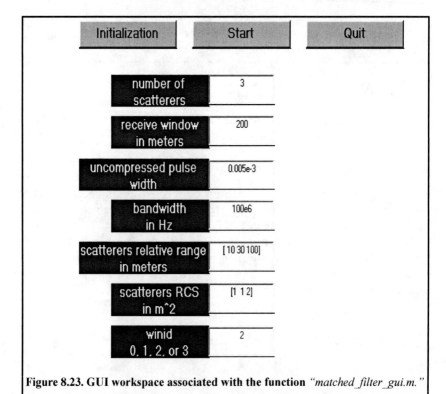

Figure 8.23. GUI workspace associated with the function *"matched_filter_gui.m."*

MATLAB Function *"matched_filter.m"* Listing

```
function [y] = matched_filter(nscat,taup,b,rrec,scat_range,scat_rcs,winid)
eps = 1.0e-16;
% time bandwidth product
time_B_product = b * taup;
if(time_B_product < 5 )
    fprintf('*********** Time Bandwidth product is TOO SMALL **************')
    fprintf('\n Change b and or taup')
  return
end
% speed of light
c = 3.e8;
% number of samples
n = fix(5 * taup * b);
% initialize input, output, and replica vectors
x(nscat,1:n) = 0.;
y(1:n) = 0.;
replica(1:n) = 0.;
% determine proper window
if( winid == 0.)
  win(1:n) = 1.;
end
if(winid == 1.);
  win = hamming(n)';
end
if( winid == 2.)
  win = kaiser(n,pi)';
end
if(winid == 3.)
  win = chebwin(n,60)';
end
% check to ensure that scatterers are within recieve window
index = find(scat_range > rrec);
if (index ~= 0)
   'Error. Receive window is too large; or scatterers fall outside window'
  return
end
% calculate sampling interval
t = linspace(-taup/2,taup/2,n);
replica = exp(i * pi * (b/taup) .* t.^2);
figure(1)
subplot(2,1,1)
plot(t,real(replica))
ylabel('Real (part) of replica')
xlabel('Time in seconds')
grid
subplot(2,1,2)
```

```
sampling_interval = taup / n;
freqlimit = 0.5/ sampling_interval;
freq = linspace(-freqlimit,freqlimit,n);
plot(freq,fftshift(abs(fft(replica))));
ylabel('Spectrum of replica')
xlabel('Frequency in Hz')
grid
 for j = 1:1:nscat
   range = scat_range(j) ;
   x(j,:) = scat_rcs(j) .* exp(i * pi * (b/taup) .* (t +(2*range/c)).^2) ;
   y = x(j,:)  + y;
end
figure(2)
 y = y .* win;
plot(t,real(y),'k')
xlabel ('Relative delay in seconds')
ylabel ('Uncompressed echo')
grid
out =xcorr(replica, y);
out = out ./ n;
s = taup * c /2;
Npoints = ceil(rrec * n /s);
dist =linspace(0, rrec, Npoints);
delr = c/2/b;
figure(3)
plot(dist,abs(out(n:n+Npoints-1)),'k')
xlabel ('Target relative position in meters')
ylabel ('Compressed echo')
grid
return
```

MATLAB Function *"power_integer_2.m"* Listing

```
function n = power_integer_2 (x)
m = 0.;
for j = 1:30
  m = m + 1.;
  delta = x - 2.^m;
  if(delta < 0.)
    n = m;
    return
  else
  end
end
return
```

8.6.2. MATLAB Function "stretch.m"

The function *"stretch.m"* presents a digital implementation of stretch processing. The syntax is as follows:

[y] = stretch (nscat, taup, f0, b, scat_range, rrec, scat_rcs, win)

where

Symbol	Description	Units	Status
nscat	number of point scatterers within the receive window	none	input
taup	uncompressed pulse width	seconds	input
f0	chirp start frequency	Hz	input
b	chirp bandwidth	Hz	input
scat_range	vector of scatterers' range	m	input
rrec	range receive window	m	input
scat_rcs	vector of scatterers' RCS	m^2	input
win	0 = no window 1 = Hamming 2 = Kaiser with parameter pi 3 = Chebychev side-lobes at -60dB	none	input
y	compressed output	volts	output

The user can access this function either by a MATLAB function call or by executing the MATLAB program *"stretch_gui.m,"* which utilizes MATLAB-based GUI and is shown in Fig. 8.24. The outputs of this function are the complex array *y* and plots of the uncompressed and compressed echo signal versus time.

MATLAB Function "stretch.m" Listing

```
function [y] = stretch(nscat, taup, f0, b, scat_range, rrec, scat_rcs, winid)
eps = 1.0e-16;
htau = taup / 2.;
c = 3.e8;
trec = 2. * rrec / c;
n = fix(2. * trec * b);
m = power_integer_2(n);
nfft = 2.^m;
x(nscat,1:n) = 0.;
y(1:n) = 0.;
if( winid == 0.)
   win(1:n) = 1.;
```

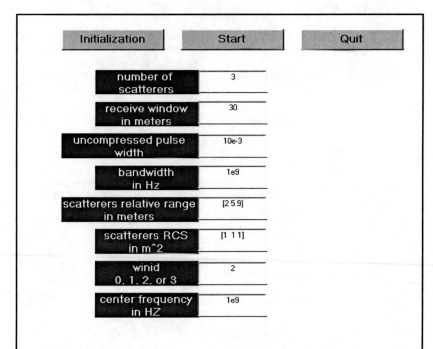

Figure 8.24. GUI workspace associated with the function *"stretch_gui.m."*

```
win =win';
else
  if(winid == 1.)
    win = hamming(n);
  else
    if( winid == 2.)
      win = kaiser(n,pi);
    else
      if(winid == 3.)
        win = chebwin(n,60);
      end
    end
  end
end
deltar = c / 2. / b;
max_rrec = deltar * nfft / 2.;
maxr = max(scat_range);
if(rrec > max_rrec | maxr >= rrec )
  'Error. Receive window is too large; or scatterers fall outside window'
  return
end
```

```
t = linspace(0,taup,n);
for j = 1:1:nscat
   range = scat_range(j);% + rmin;
   psi1 = 4. * pi * range * f0 / c - ...
      4. * pi * b * range * range / c / c/ taup;
   psi2 = (2*4. * pi * b * range / c / taup) .* t;
   x(j,:) = scat_rcs(j) .* exp(i * psi1 + i .* psi2);
   y = y + x(j,:);
end
figure(1)
plot(t,real(y),'k')
xlabel ('Relative delay in seconds')
ylabel ('Uncompressed echo')
grid
ywin = y .* win';
yfft = fft(y,n) ./ n;
out= fftshift(abs(yfft));
figure(2)
delinc = rrec/ n;
%dist = linspace(-delinc-rrec/2,rrec/2,n);
dist = linspace((-rrec/2), rrec/2,n);
plot(dist,out,'k')
xlabel ('Relative range in meters')
ylabel ('Compressed echo')
axis auto
grid
```

8.6.3. MATLAB Function "SFW.m"

The function *"SFW.m"* computes and plots the range profile for a specific SFW. This function utilizes an Inverse Fast Fourier Transform (IFFT) of a size equal to twice the number of steps. Hamming window of the same size is also assumed. The syntax is as follows:

[h1] = SFW (nscat, scat_range, scat_rcs, n, deltaf, prf, v, r0, winid)

where

Symbol	Description	Units	Status
nscat	number of scatterers that make up the target	none	input
scat_range	vector containing range to individual scatterers	meters	input
scat_rcs	vector containing RCS of individual scatterers	meter square	input
n	number of steps	none	input

Symbol	Description	Units	Status
deltaf	frequency step	Hz	input
prf	PRF of SFW	Hz	input
v	target velocity	meter/second	input
r0	profile starting range	meters	input
winid	number>0 for Hamming window number < 0 for no window	none	input
hl	range profile	dB	output

MATLAB Function "SFW.m" Listing

```
function [hl] = SFW (nscat, scat_range, scat_rcs, n, deltaf, prf, v, rnote,winid)
% Range or Time domain Profile
% Range_Profile returns the Range or Time domain plot of a simulated
% HRR SFWF returning from a predetermined number of targets with a predetermined
% RCS for each target.
c=3.0e8;  % speed of light (m/s)
num_pulses  = n;
SNR_dB = 40;
nfft = 256;
% carrier_freq = 9.5e9; %Hz (10GHz)
freq_step   = deltaf; %Hz (10MHz)
V = v; % radial velocity (m/s)  -- (+)=towards radar (-)=away
PRI = 1. / prf; % (s)
if (nfft > 2*num_pulses)
   num_pulses = nfft/2;
else
end
Inphase = zeros((2*num_pulses),1);
Quadrature = zeros((2*num_pulses),1);
Inphase_tgt    = zeros(num_pulses,1);
Quadrature_tgt = zeros(num_pulses,1);
IQ_freq_domain = zeros((2*num_pulses),1);
Weighted_I_freq_domain = zeros((num_pulses),1);
Weighted_Q_freq_domain = zeros((num_pulses),1);
Weighted_IQ_time_domain = zeros((2*num_pulses),1);
Weighted_IQ_freq_domain = zeros((2*num_pulses),1);
abs_Weighted_IQ_time_domain = zeros((2*num_pulses),1);
dB_abs_Weighted_IQ_time_domain = zeros((2*num_pulses),1);
taur = 2. * rnote / c;
for jscat = 1:nscat
  ii = 0;
  for i = 1:num_pulses
    ii = ii+1;
```

```
rec_freq = ((i-1)*freq_step);
Inphase_tgt(ii) = Inphase_tgt(ii) + sqrt(scat_rcs(jscat)) * cos(-2*pi*rec_freq*...
   (2.*scat_range(jscat)/c - 2*(V/c)*((i-1)*PRI + taur/2 + 2*scat_range(jscat)/c)));
        Quadrature_tgt(ii)  =  Quadrature_tgt(ii)  +  sqrt(scat_rcs(jscat))*sin(-
2*pi*rec_freq*...
   (2*scat_range(jscat)/c - 2*(V/c)*((i-1)*PRI + taur/2 + 2*scat_range(jscat)/c)));
  end
end
if(winid >= 0)
  window(1:num_pulses) = hamming(num_pulses);
else
  window(1:num_pulses) = 1;
end
Inphase = Inphase_tgt;
Quadrature = Quadrature_tgt;
Weighted_I_freq_domain(1:num_pulses) = Inphase(1:num_pulses).* window';
Weighted_Q_freq_domain(1:num_pulses) = Quadrature(1:num_pulses).* window';
Weighted_IQ_freq_domain(1:num_pulses)= Weighted_I_freq_domain + ...
  Weighted_Q_freq_domain*j;
Weighted_IQ_freq_domain(num_pulses:2*num_pulses)=0.+0.i;
Weighted_IQ_time_domain = (ifft(Weighted_IQ_freq_domain));
abs_Weighted_IQ_time_domain = (abs(Weighted_IQ_time_domain));
dB_abs_Weighted_IQ_time_domain =
20.0*log10(abs_Weighted_IQ_time_domain)+SNR_dB;
% calculate the unambiguous range window size
Ru = c /2/deltaf;
hl = dB_abs_Weighted_IQ_time_domain;
 numb = 2*num_pulses;
delx_meter = Ru / numb;
xmeter = 0:delx_meter:Ru-delx_meter;
plot(xmeter, dB_abs_Weighted_IQ_time_domain,'k')
xlabel ('Relative distance in meters')
ylabel ('Range profile in dB')
grid
```

Chapter 9   *Radar Clutter*

Clutter is a term used to describe any object that may generate unwanted radar returns that may interfere with normal radar operations. Parasitic returns that enter the radar through the antenna's mainlobe are called main-lobe clutter; otherwise they are called sidelobe clutter. Clutter can be classified into two main categories: surface clutter and airborne or volume clutter. Surface clutter includes trees, vegetation, ground terrain, man-made structures, and sea surface (sea clutter). Volume clutter normally has a large extent (size) and includes chaff, rain, birds, and insects. Surface clutter changes from one area to another, while volume clutter may be more predictable.

Clutter echoes are random and have thermal noise-like characteristics because the individual clutter components (scatterers) have random phases and amplitudes. In many cases, the clutter signal level is much higher than the receiver noise level. Thus, the radar's ability to detect targets embedded in high clutter background depends on the Signal-to-Clutter Ratio (SCR) rather than the SNR.

9.1. Clutter Cross Section Density

Since clutter returns are target-like echoes, the only way a radar can distinguish target returns from clutter echoes is based on the target RCS σ_t and the anticipated clutter RCS σ_c. Clutter RCS can be defined as the equivalent radar cross section attributed to reflections from a clutter area, A_c. The average clutter RCS is given by

$$\sigma_c = \sigma^0 A_c \qquad (9.1)$$

where σ^0 is the clutter scattering coefficient, a dimensionless quantity that is often expressed in dB. The equivalent of Eq. (9.1) for volume clutter is

$$\sigma_c = \eta^0 V_w \qquad (9.2)$$

where V_w is the clutter volume and η^0 is the volume clutter scattering coefficient. Note that η^0 units are m^{-1}, and because of this, it is typically expressed in dB/meter units.

9.2. Surface Clutter

Surface clutter includes both land and sea clutter, and is often called area clutter. Area clutter manifests itself in airborne radars in the look-down mode. It is also a major concern for ground-based radars when searching for targets at low grazing angles. The grazing angle ψ_g is the angle from the surface of the earth to the main axis of the illuminating beam, as illustrated in Fig. 9.1.

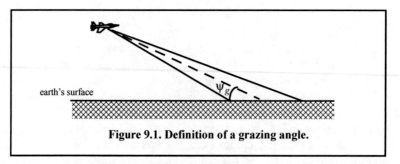

Figure 9.1. Definition of a grazing angle.

Factors that affect the radar performance due to the presence of clutter include clutter reflectivity which is function of radar wavelength, polarization, and of course shape and size of the clutter itself. The amount of clutter RCS in the radar beam depends heavily on the grazing angle, surface roughness, and spatial characteristics of clutter and its time fluctuation characteristics. Typically, the clutter scattering coefficient σ^0 is larger for smaller wavelengths. Figure 9.2 shows a sketch describing the dependency of σ^0 on the grazing angle. Three regions are identified; they are the low grazing angle region, the flat or plateau region, and the high grazing angle region.

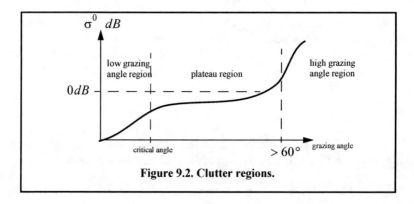

Figure 9.2. Clutter regions.

The low grazing angle region extends from zero to about the critical angle. The critical angle is defined by Rayleigh as the angle below which a surface is considered to be smooth and above which a surface is considered to be rough; Denote the root mean square (rms) of a surface height irregularity as h_{rms}; then according to the Rayleigh criteria, the surface is considered to be smooth if

$$\frac{4\pi h_{rms}}{\lambda} \sin \psi_g < \frac{\pi}{2} \tag{9.3}$$

Consider a wave incident on a rough surface, as shown in Fig. 9.3. Due to surface height irregularity (surface roughness), the rough path is longer than the smooth path by a distance $2h_{rms}\sin\psi_g$. This path difference translates into a phase differential $\Delta\psi$:

$$\Delta\psi = \frac{2\pi}{\lambda} 2h_{rms}\sin\psi_g \tag{9.4}$$

The critical angle ψ_{gc} is then computed when $\Delta\psi = \pi$ (first null); thus,

$$\frac{4\pi h_{rms}}{\lambda} \sin\psi_{gc} = \pi \tag{9.5}$$

or equivalently,

$$\psi_{gc} = a\sin\frac{\lambda}{4h_{rms}} \tag{9.6}$$

In the case of sea clutter, for example, the rms surface height irregularity is

$$h_{rms} \approx 0.025 + 0.046 \ S_{state}^{1.72} \tag{9.7}$$

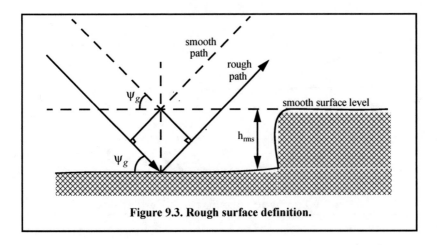

Figure 9.3. Rough surface definition.

where S_{state} is the sea state, which is tabulated in several cited references. The sea state is characterized by the wave height, period, length, particle velocity, and wind velocity. For example, $S_{state} = 3$ refers to a moderate sea state, in which the wave height is approximately 0.9144 *to* 1.2192 *m*, the wave period 6.5 *to* 4.5 seconds, wave length 1.9812 *to* 33.528 *m*, wave velocity 20.372 *to* 25.928 *Km/hr*, and wind velocity 22.224 *to* 29.632 *Km/hr*.

Clutter at low grazing angles is often referred to as diffuse clutter, where there are a large number of clutter returns in the radar beam (noncoherent reflections). In the flat region the dependency of σ^0 on the grazing angle is minimal. Clutter in the high grazing angle region is more specular (coherent reflections) and the diffuse clutter components disappear. In this region the smooth surfaces have larger σ^0 than rough surfaces, the opposite of the low grazing angle region.

9.2.1. Radar Equation for Surface Clutter

Consider an airborne radar in the look-down mode shown in Fig. 9.4. The intersection of the antenna beam with the ground defines an elliptically shaped footprint. The size of the footprint is a function of the grazing angle and the antenna $3dB$ beamwidth θ_{3dB}, as illustrated in Fig. 9.5. The footprint is divided into many ground range bins each of size $(c\tau/2)\sec\psi_g$, where τ is the pulse width. From Fig. 9.5, the clutter area A_c is

$$A_c \approx R\theta_{3dB} \frac{c\tau}{2} \sec\psi_g \qquad (9.8)$$

The power received by the radar from a scatterer within A_c is given by the radar equation as

$$S_t = \frac{P_t G^2 \lambda^2 \sigma_t}{(4\pi)^3 R^4} \qquad (9.9)$$

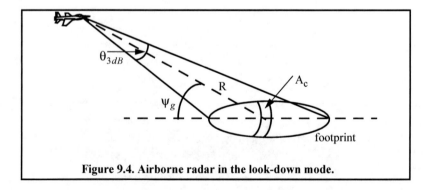

Figure 9.4. Airborne radar in the look-down mode.

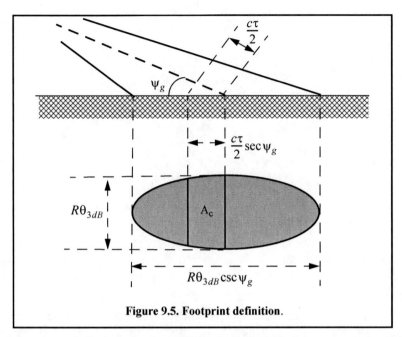

Figure 9.5. Footprint definition.

where, as usual, P_t is the peak transmitted power, G is the antenna gain, λ is the wavelength, and σ_t is the target RCS. Similarly, the received power from clutter is

$$S_C = \frac{P_t G^2 \lambda^2 \sigma_c}{(4\pi)^3 R^4} \qquad (9.10)$$

where the subscript C is used for area clutter. Substituting Eq. (9.1) for σ_c into Eq. (9.10), we can then obtain the SCR for area clutter by dividing Eq. (9.9) by Eq. (9.10). More precisely,

$$(SCR)_C = \frac{2\sigma_t \cos\psi_g}{\sigma^0 \theta_{3dB} R c \tau} \qquad (9.11)$$

Example:

Consider an airborne radar shown in Fig. 9.4. Let the antenna 3dB beam-width be $\theta_{3dB} = 0.02 rad$, the pulse width $\tau = 2\mu s$, range $R = 20 Km$, and grazing angle $\psi_g = 20°$. The target RCS is $\sigma_t = 1 m^2$. Assume that the clutter reflection coefficient is $\sigma^0 = 0.0136$. Compute the SCR.

Solution:

The SCR is given by Eq. (9.11) as

$$(SCR)_C = \frac{2\sigma_t \cos\psi_g}{\sigma^0 \theta_{3dB} R c \tau} \Rightarrow$$

$$(SCR)_C = \frac{(2)(1)(\cos 20°)}{(0.0136)(0.02)(20000)(3 \times 10^8)(2 \times 10^{-6})} = 5.76 \times 10^{-4}$$

It follows that

$$(SCR)_C = -32.4dB$$

Thus, for reliable detection the radar must somehow increase its SCR by at least $(32 + X)dB$, where X is on the order of 13 to 15dB or better.

9.3. Volume Clutter

Volume clutter has large extents and includes rain (weather), chaff, birds, and insects. The volume clutter coefficient is normally expressed in square meters (RCS per resolution volume). Birds, insects, and other flying particles are often referred to as angle clutter or biological clutter.

Weather or rain clutter can be suppressed by treating the rain droplets as perfect small spheres. We can use the Rayleigh approximation of a perfect sphere to estimate the rain droplets' RCS. The Rayleigh approximation, without regard to the propagation medium index of refraction is

$$\sigma = 9\pi r^2 (kr)^4 \qquad r \ll \lambda \qquad (9.12)$$

where $k = 2\pi/\lambda$, and r is radius of a rain droplet.

Electromagnetic waves when reflected from a perfect sphere become strongly co-polarized (have the same polarization as the incident waves). Consequently, if the radar transmits, for example, a right-hand-circular (RHC) polarized wave, then the received waves are left-hand-circular (LHC) polarized because they are propagating in the opposite direction. Therefore, the back-scattered energy from rain droplets retains the same wave rotation (polarization) as the incident wave, but has a reversed direction of propagation. It follows that radars can suppress rain clutter by co-polarizing the radar transmit and receive antennas.

Denote η as RCS per unit resolution volume V_w. It is computed as the sum of all individual scatterers RCS within the volume

$$\sigma_w = \sum_{i=1}^{N} \sigma_i \qquad (9.13)$$

where N is the total number of scatterers within the resolution volume. Thus, the total RCS of a single resolution volume is

$$\sigma_W = \sum_{i=1}^{N} \sigma_i V_W \qquad (9.14)$$

A resolution volume is shown in Fig. 9.6 and is approximated by

$$V_W \approx \frac{\pi}{8} \theta_a \theta_e R^2 c\tau \qquad (9.15)$$

where θ_a and θ_e are, respectively, the antenna azimuth and elevation beamwidths in radians, τ is the pulse width in seconds, c is the speed of light, and R is range.

Consider a propagation medium with an index of refraction m. The *ith* rain droplet RCS approximation in this medium is

$$\sigma_i \approx \frac{\pi^5}{\lambda^4} K^2 D_i^6 \qquad (9.16)$$

where

$$K^2 = \left| \frac{m^2 - 1}{m^2 + 2} \right|^2 \qquad (9.17)$$

and D_i is the *ith* droplet diameter. For example, temperatures between $32°F$ and $68°F$ yield

$$\sigma_i \approx 0.93 \frac{\pi^5}{\lambda^4} D_i^6 \qquad (9.18)$$

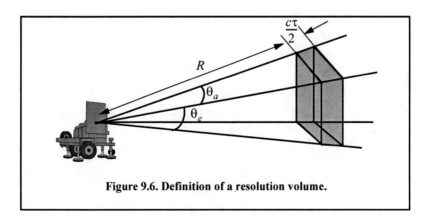

Figure 9.6. Definition of a resolution volume.

and for ice Eq. (9.18) can be approximated by

$$\sigma_i \approx 0.2 \frac{\pi^5}{\lambda^4} D_i^6 \tag{9.19}$$

Substituting Eq. (9.19) into Eq. (9.14) yields

$$\sigma_w = \frac{\pi^5}{\lambda^4} K^2 Z \tag{9.20}$$

where the weather clutter coefficient Z is defined as

$$Z = \sum_{i=1}^{N} D_i^6 \tag{9.21}$$

In general, a rain droplet diameter is given in millimeters and the radar resolution volume is expressed in cubic meters; thus the units of Z are often expressed in $millimeter^6 / m^3$.

9.3.1. Radar Equation for Volume Clutter

The radar equation gives the total power received by the radar from a σ_t target at range R as

$$S_t = \frac{P_t G^2 \lambda^2 \sigma_t}{(4\pi)^3 R^4} \tag{9.22}$$

where all parameters in Eq. (9.22) have been defined earlier. The weather clutter power received by the radar is

$$S_w = \frac{P_t G^2 \lambda^2 \sigma_w}{(4\pi)^3 R^4} \tag{9.23}$$

It follows that

$$S_w = \frac{P_t G^2 \lambda^2}{(4\pi)^3 R^4} \frac{\pi}{8} R^2 \theta_a \theta_e c\tau \sum_{i=1}^{N} \sigma_i \tag{9.24}$$

The SCR for weather clutter is then computed by dividing Eq. (9.22) by Eq. (9.24). More precisely,

$$(SCR)_V = \frac{S_t}{S_w} = (8\sigma_t) \Big/ \left(\pi \theta_a \theta_e c\tau R^2 \sum_{i=1}^{N} \sigma_i \right) \tag{9.25}$$

where the subscript V is used to denote volume clutter.

Example:

A certain radar has target RCS $\sigma_t = 0.1 m^2$, pulse width $\tau = 0.2 \mu s$, antenna beamwidth $\theta_a = \theta_e = 0.02 radians$. Assume the detection range to be $R = 50 Km$, and compute the SCR if $\sum \sigma_i = 1.6 \times 10^{-8} (m^2/m^3)$.

Solution:

From Eq. (9.25) we have

$$(SCR)_V = \frac{8\sigma_t}{\pi\theta_a\theta_e c\tau R^2 \sum_N \sigma_i}$$

Substituting the proper values we get

$$(SCR)_V = \frac{(8)(0.1)}{\pi(0.02)^2(3 \times 10^8)(0.2 \times 10^{-6})(50 \times 10^3)^2(1.6 \times 10^{-8})} = 0.265$$

$$(SCR)_V = -5.76 dB.$$

9.4. Clutter RCS

9.4.1. Single Pulse - Low PRF Case

Again the received power from clutter is also calculated using Eq. (9.9). However, in this case the clutter RCS σ_c is computed differently. It is

$$\sigma_c = \sigma_{MBc} + \sigma_{SLc} \tag{9.26}$$

where σ_{MBc} is the main-beam clutter RCS and σ_{SLc} is the sidelobe clutter RCS, as illustrated in Fig. 9.7.

In order to calculate the total clutter RCS given in Eq. (9.11), one must first compute the corresponding clutter areas for both the main beam and the sidelobes. For this purpose, consider the geometry shown in Fig. 9.8. The angles θ_A and θ_E represent the antenna 3-dB azimuth and elevation beamwidths, respectively. The radar height (from the ground to the phase center of the antenna) is denoted by h_r, while the target height is denoted by h_t. The radar slant range is R, and its ground projection is R_g. The range resolution is ΔR and its ground projection is ΔR_g. The main beam clutter area is denoted by A_{MBc} and the sidelobe clutter area is denoted by A_{SLc}.

From Fig. 9.8, the following relations can be derived

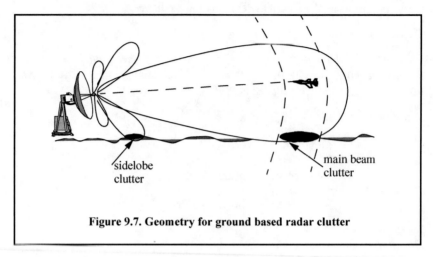

Figure 9.7. Geometry for ground based radar clutter

$$\theta_r = \operatorname{asin}(h_r / R) \tag{9.27}$$

$$\theta_e = \operatorname{asin}((h_t - h_r)/R) \tag{9.28}$$

$$\Delta R_g = \Delta R \cos\theta_r \tag{9.29}$$

where ΔR is the radar range resolution. The slant range ground projection is

$$R_g = R\cos\theta_r \tag{9.30}$$

It follows that the main beam and the sidelobe clutter areas are

$$A_{MBc} = \Delta R_g \ R_g \ \theta_A \tag{9.31}$$

$$A_{SLc} = \Delta R_g \ \pi R_g \tag{9.32}$$

Assume a radar antenna beam $G(\theta)$ of the form

$$G(\theta) = \exp\left(-\frac{2.776\theta^2}{\theta_E^2}\right) \Rightarrow Gaussian \tag{9.33}$$

$$G(\theta) = \begin{cases} \left(\dfrac{\sin\left(\dfrac{\theta}{\theta_E}\right)}{\left(\dfrac{\theta}{\theta_E}\right)}\right)^2 & ;|\theta| \le \dfrac{\pi\theta_E}{2.78} \\ \\ 0 & ;elsewhere \end{cases} \Rightarrow \left(\frac{\sin(x)}{x}\right)^2 \tag{9.34}$$

Then the main-beam clutter RCS is

$$\sigma_{MBc} = \sigma^0 A_{MBc} G^2(\theta_e + \theta_r) = \sigma^0 \Delta R_g \; R_g \; \theta_A G^2(\theta_e + \theta_r) \qquad \text{(9.35)}$$

and the sidelobe clutter RCS is

$$\sigma_{SLc} = \sigma^0 A_{SLc}(SL_{rms})^2 = \sigma^0 \Delta R_g \; \pi R_g (SL_{rms})^2 \qquad \text{(9.36)}$$

where the quantity SL_{rms} is the rms for the antenna sidelobe level.

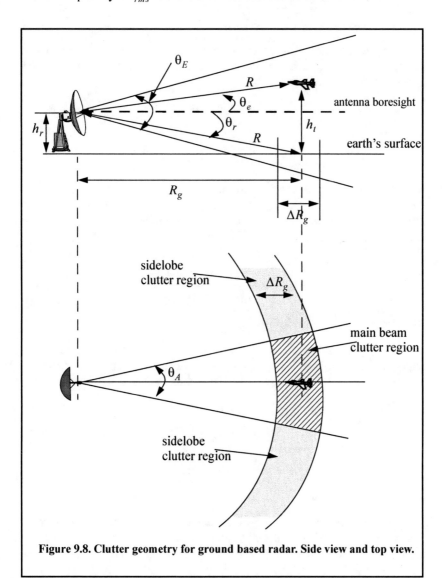

Figure 9.8. Clutter geometry for ground based radar. Side view and top view.

Finally, in order to account for the variation of the clutter RCS versus range, one can calculate the total clutter RCS as a function of range. It is given by

$$\sigma_c(R) = \frac{\sigma_{MBc} + \sigma_{SLc}}{(1 + (R/R_h)^4)} \qquad (9.37)$$

where R_h is the radar range to the horizon calculated as

$$R_h = \sqrt{8h_r r_e / 3} \qquad (9.38)$$

where r_e is the Earth's radius equal to $6371 Km$. The denominator in Eq. (9.37) is put in that format in order to account for refraction and for round (spherical) Earth effects.

The radar SNR due to a target at range R is

$$SNR = \frac{P_t G^2 \lambda^2 \sigma_t}{(4\pi)^3 R^4 k T_o BFL} \qquad (9.39)$$

where, as usual, P_t is the peak transmitted power, G is the antenna gain, λ is the wavelength, σ_t is the target RCS, k is Boltzmann's constant, T_0 is the effective noise temperature, B is the radar operating bandwidth, F is the receiver noise figure, and L is the total radar losses. Similarly, the Clutter-to-Noise Ratio (CNR) at the radar is

$$CNR = \frac{P_t G^2 \lambda^2 \sigma_c}{(4\pi)^3 R^4 k T_o BFL} \qquad (9.40)$$

where the σ_c is calculated using Eq. (9.37).

When the clutter statistic is Gaussian, the clutter signal return and the noise return can be combined, and a new value for determining the radar measurement accuracy is derived from the Signal-to-Clutter+Noise Ratio, denoted by SIR. It is given by

$$SIR = \frac{SNR}{1 + CNR} \qquad (9.41)$$

Note that the CNR is computed from Eq. (9.40).

9.4.2. High PRF Case

High PRFs are typically used by pulsed Doppler radars. Pulsed Doppler radars use very short unmodulated train of pulses, and hence, range resolution is limited by the pulsewidth, which forces the radar to use extremely short duration pulses. High PRF radars make up for the loss of average transmitted power due to using short pulses by coherently processing a train of these pulses within one coherent processing interval (integration time or dwell interval). Although high PRF radars although are ambiguous in range, they provide excellent capability to measuring Doppler frequency. Range ambiguity can be dealt with by using multiple PRF (PRF staggering) which will be addressed later section. One major drawback of using high PRFs (or pulsed Doppler radars) is the fact that pulsed Doppler radars have to contend with much more clutter than do low PRF radars.

Consider the illustrations shown in Fig. 9.9. The low PRF case is shown in Fig. 9.9a. In this case, the target is at maximum detection range which corresponds to an unambiguous range

$$R_u = \frac{cT}{2} = \frac{c}{2f_r} \tag{9.42}$$

where T is the pulse repetition interval and f_r is the radar PRF. The amount of clutter entering the radar through its main-beam corresponds only to the clutter patch located at the target's range. Alternatively, in Fig. 9.9b the high PRF case is depicted. In this case, the radar is range ambiguous and the amount of main-beam clutter entering the radar corresponds to many more clutter patches as shown in Fig. 9.9b. Consequently, the amount of clutter competing with target detection in an order of magnitude larger than the case of low PRF. This is typically referred to as clutter folding.

Denote the clutter power entering the radar due to a single pulse for the target at range R_0 as P_{C_1}, then because of the high PRF operation, the total clutter power entering the radar is

$$P_{C_{folded}} = \sum_{n=0}^{N-1} P_{C_1} Rect\left(\frac{t-nT}{\tau_0}\right) \tag{9.43}$$

where N is the number of pulses in one coherent processing interval (dwell), T is the PRI, and τ_0 is the pulsewidth. Note that since the radar receiver is shut off during transmission of a given pulse, Eq. (9.43) is computed only at delays (range) that correspond to

$$\{(nT+2\tau_0) < t < (n+1)T - \tau_0; \ 0 \le n \le N-1\} \tag{9.44}$$

where in this case, the transmitter is assumed to be shut off not only during the transmission of each pulse but also for one pulsewidth before and after each transmission. Thus, one would expect the folded clutter RCS to not be continuous versus the range, but rather to exist over intervals of length T seconds with gaps that correspond to three times the pulsewidth. This is illustrated in the following few examples for both low and high PRF cases.

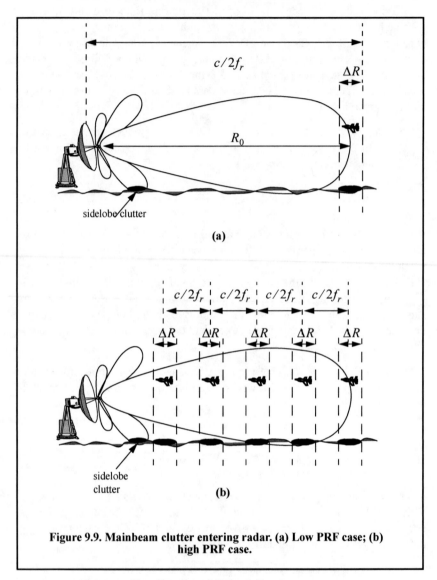

Figure 9.9. Mainbeam clutter entering radar. (a) Low PRF case; (b) high PRF case.

As an example consider the case with the following parameters

clutter back scatterer coefficient	*-20 dB*
antenna 3dB elevation beamwidth	*1.5 degrees*
antenna 3dB azimuth beamwidth	*2 degrees*
antenna sidelobe level	*-25 dB*
radar height	*3 meters*

target height	150 meters
radar peak power	45 KW
radar operating frequency	50 KHz
pulsewidth	1 micro sec
effective noise temperature	290 Kelvins
noise figure	6 dB
radar losses	10 dB
target RCS	-10 dBsm
radar center frequency	5 GHz

Figure 9.10 is concerned with a low PRF case (i.e, single pulse, no clutter folding). Figure 9.10a shows the clutter RCS versus range when a sin(x)/x antenna pattern is used, and Fig. 9.10b shows the resulting SNR, CNR, and SCR. Figure 9.11 is similar to Fig. 9.10 except in this case the antenna has a Gaussian shape. These plots can be reproduced using the following MATLAB code which uses the function *"clutter_rcs.m."*

```
%Use this code to generate Fig. 9.10 and 9.11
clear all;
close all;
k = 1.38e-23; % Boltzman's constant
pt = 45e3;
theta_AZ = 1.5;
theta_EL = 2;
F = 6;
L = 10;
tau = 1e-6;
B = 1/tau;
sigmmat = -10;
sigmma0 = -20;
SL = -25;
hr = 3;
ht = 150;
f0 = 5e9;
lambda = 3e8/f0;
range = linspace(2,50, 120);
[sigmmaC] = clutter_rcs(sigmma0, theta_EL, theta_AZ, SL, range, hr, ht, B,1);
sigmmaC = 10.^(sigmmaC./10);
range_m = 1000 .* range;
F = 10.^(F/10); % noise figure is 6 dB
T0 = 290; % noise temperature 290K
g = 26000 /theta_AZ /theta_EL; % antenna gain
Lt = 10.^(L/10); % total radar losses 13 dB
sigmmat = 10^(sigmmat/10)
```

CNR = pt*g*g*lambda^2 .* sigmmaC ./ ((4*pi)^3 .* (range_m).^4 .* k*T0*F*Lt*B); % CNR
SNR = pt*g*g*lambda^2 .* sigmmat ./ ((4*pi)^3 .* (range_m).^4 .* k*T0*F*L*B); % SNR
SCR = SNR ./ CNR; % Signal to clutter ratio
SIR = SNR ./ (1+CNR); % Signal to interference ratio
%%%%%%%%%%%%%%%%%%%%%%%%%%%%%%%%%%%%%%
figure(2)
subplot(3,1,1)
plot(range,10*log10(SNR));
ylabel('SNR in dB');
grid on;
axis tight
subplot(3,1,2)
plot(range,10*log10(CNR));
ylabel('CNR in dB');
grid on;
axis tight
subplot(3,1,3)
plot(range,10*log10(SCR));
ylabel('SCR in dB') ;
grid on;
axis tight
xlabel('Range in Km')

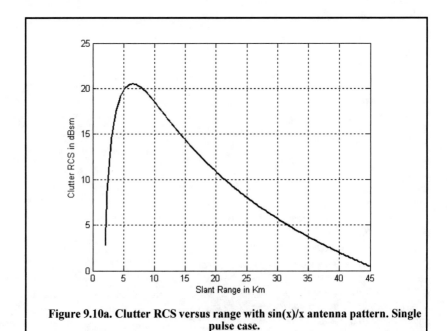

Figure 9.10a. Clutter RCS versus range with sin(x)/x antenna pattern. Single pulse case.

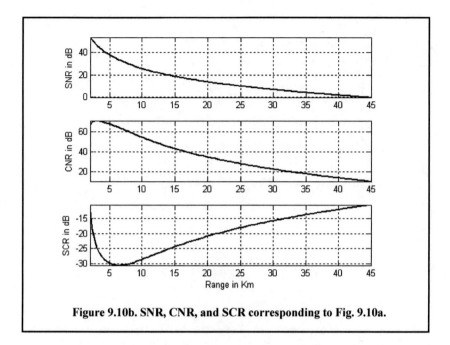

Figure 9.10b. SNR, CNR, and SCR corresponding to Fig. 9.10a.

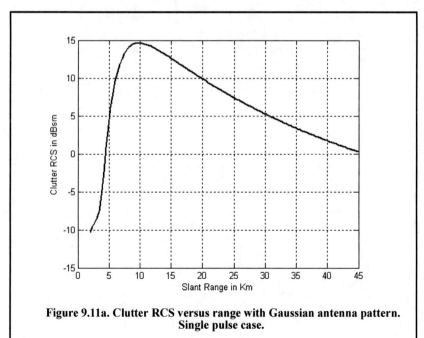

**Figure 9.11a. Clutter RCS versus range with Gaussian antenna pattern.
Single pulse case.**

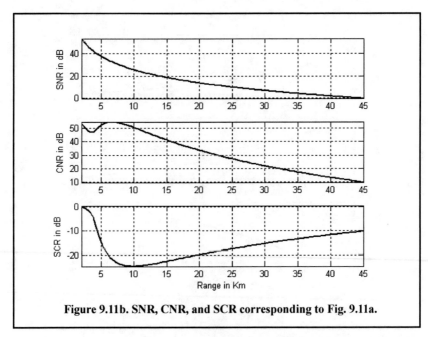

Figure 9.11b. SNR, CNR, and SCR corresponding to Fig. 9.11a.

Figure 9.12 shows the SNR, CNR, and SCR for the high PRF case (i.e, pulse Doppler radar, clutter folding). In this figure the antenna pattern has a sin(x)/x shape. Figure 9.13 is similar to Fig. 9.12 except in this case the antenna pattern is Gaussian. These plots can be reproduced using the following MATLAB code.

```
% Use this code to generate Fig. 9.12 or 9. 13 of text
clear all
close all
k = 1.38e-23; % Boltzmann's constant
T0 = 290; % degrees Kelvin
ant_id = 1; % use 1 for sin(x)/x antenna pattern and use 2 for Gaussian pattern
theta_ref = 0.75; % reference angle of radar antenna in degrees
re = 6371000 * 4 /3; % 4/3rd earth radius in Km
c = 3e8; % speed of light
theta_EL = 1.5; % Antenna elevation beamwidth in degrees
theta_AZ = 2.; % Antenna azimuth beamwidth in degrees
SL_dB = -25; % Antenna RMS sidelobe level
hr = 3; % Radar antenna height in meters
ht = 150; % Target height in meters
Sigmmat = -10; % Target RCS in dB
Sigmma0 = -20; % Clutter backscatter coefficient
P = 45e3; % Radar peak power in Watts
tau = 1e-6; % Pulse width (unmodulated)
```

```
fr = 50e3; % PRF in Hz
f0 = 5e9; % Radar center frequency
F = 6; % Noise figure in dB
L = 10; % Radar losses in dB
lambda = c /f0;
SL = 10^(SL_dB/10);
sigmma0 = 10^(Sigmma0/10);
F = 10^(F/10);
L = L^(L/10);
sigmmat = 10^(Sigmmat/10);
T = 1/fr; % PRI
B = 1/tau; % Bandwidth
delr = c * tau /2; % Range resolution;
Rh = sqrt(2*re*hr); % Range to Horizon
R1 = [2*delr:delr:c/2*(T-tau)];
Rclut = sqrt(R1.^2 + hr^2); % Range to clutter patches
G = 26000 /theta_EL /theta_AZ; % Antenna gain
for j = 0:40
    Rtgt = [c/2*(j*T+2*tau):delr:c/2*((j+1)*T-tau)];
    thetaR = asin(hr./Rclut); % Ele angle from radar to clutter patch target is present
    thetae = theta_ref*pi/180;
    d = Rclut .* cos(thetaR); % Ground range to center of clutter at range Rclut
    del_d = delr .* cos(thetaR);
    % claculte clutter RCS
    theta_sum = thetaR+thetae;
    if(ant_id ==1) % use sinc^2 antenna pattern
        ant_arg = ( theta_sum ) ./ (pi*theta_EL/180);
        gain = (sinc(ant_arg)).^2;
    else
        gain = exp(-2.776 .*(theta_sum./(pi*theta_EL/180)).^2);
    end
    % clutter RCS
    sigmmac = (pi*SL^2+(theta_AZ*pi/180).*gain.*sigmma0.*d.*del_d) ./ (1+(Rclut/
Rh).^4);
    CNR = P*G*G*lambda^2 .* sigmmac ./ ((4*pi)^3 .* Rclut.^4 .* k*T0*F*L*B); %
CNR
    SNR = P*G*G*lambda^2 .* sigmmat./ ((4*pi)^3 .* Rtgt.^4 .* k*T0*F*L*B); % SNR
    SCR = SNR ./ CNR; % Signal to clutter ratio
    SIR = SNR ./ (1+CNR); % Signal to infernce ratio
    figure(2)
    subplot(4,1,1),
    hold on
    plot(Rtgt/1000,10*log10(SNR));
    ylabel('SNR - dB');
    grid on
    subplot(4,1,2),
    hold on
    plot(Rtgt/1000,10*log10(CNR));
```

```
ylabel('CNR - dB');
grid on
subplot(4,1,3),
hold on
plot(Rtgt/1000,10*log10(SCR));
ylabel('SCR - dB') ;
grid on
subplot(4,1,4),
hold on
plot(Rtgt/1000,10*log10(SIR));
xlabel('Range - Km')
ylabel('SIR - dB');
grid on
end
subplot(4,1,1)
axis([0 50 -10 100])
subplot(4,1,2)
axis([0 50 60 90]);
subplot(4,1,3)
axis([0 50 -100 0])
subplot(4,1,4)
axis([0 50 -100 0])
```

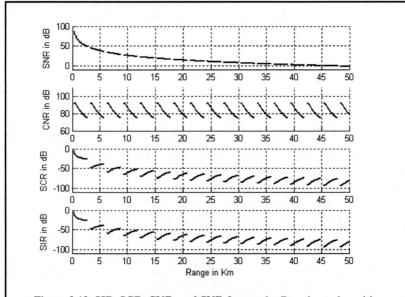

Figure 9.12. SIR, SCR, CNR, and SNR for a pulse Doppler radar with sin(x)/x antenna pattern.

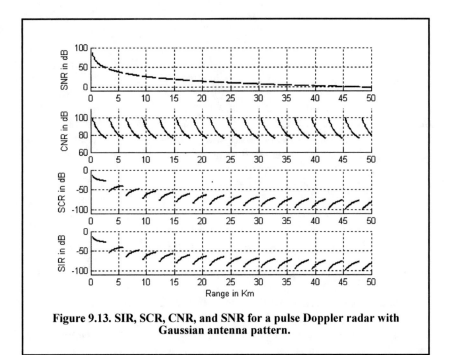

Figure 9.13. SIR, SCR, CNR, and SNR for a pulse Doppler radar with Gaussian antenna pattern.

9.5. Clutter Spectrum

9.5.1. Clutter Statistical Models

Since clutter within a resolution cell or volume is composed of a large number of scatterers with random phases and amplitudes, it is statistically described by a probability distribution function. The type of distribution depends on the nature of clutter itself (sea, land, volume), the radar operating frequency, and the grazing angle.

If sea or land clutter is composed of many small scatterers when the probability of receiving an echo from one scatterer is statistically independent of the echo received from another scatterer, then the clutter may be modeled using a Rayleigh distribution,

$$f(x) = \frac{2x}{x_0} \exp\left(\frac{-x^2}{x_0}\right) \; ; \; x \geq 0 \tag{9.45}$$

where x_0 is the mean-squared value of x.

The log-normal distribution best describes land clutter at low grazing angles. It also fits sea clutter in the plateau region. It is given by

$$f(x) = \frac{1}{\sigma\sqrt{2\pi}\,x}\,\exp\left(-\frac{(\ln x - \ln x_m)^2}{2\sigma^2}\right) \; ; \; x > 0 \qquad (9.46)$$

where x_m is the median of the random variable x, and σ is the standard deviation of the random variable $\ln(x)$.

The Weibull distribution is used to model clutter at low grazing angles (less than five degrees) for frequencies between 1 and $10 GHz$. The Weibull probability density function is determined by the Weibull slope parameter a (often tabulated) and a median scatter coefficient $\overline{\sigma_0}$, and is given by

$$f(x) = \frac{bx^{b-1}}{\overline{\sigma_0}}\,\exp\left(-\frac{x^b}{\overline{\sigma_0}}\right) \; ; \; x \geq 0 \qquad (9.47)$$

where $b = 1/a$ is known as the shape parameter. Note that when $b = 2$ the Weibull distribution becomes a Rayleigh distribution.

9.5.2. Clutter Components

It was established earlier that the complex envelope of the signal received by the radar comprise the target returns and additive bandlimited white noise. In the presence of clutter, the complex envelope is now composed of target, noise, and clutter returns. That is,

$$\tilde{x}(t) = \tilde{s}(t) + \tilde{n}(t) + \tilde{w}(t) \qquad (9.48)$$

where $\tilde{s}(t)$, $\tilde{n}(t)$, and $\tilde{w}(t)$ are, respectively, the target, noise, and clutter complex envelope echoes. Noise is typically modeled (as discussed in earlier chapters) as a bandlimited white Gaussian random process. Furthermore, noise samples are consider statistically independent of each other and of clutter measurements.

Clutter arises from reflections of unwanted objects within the radar beam. Since many objects comprose the clutter returns, clutter may also be molded as a Gaussian random process. In other words, clutter samples from one radar measurement to another constitute a joint set of Gaussian random variables. However, because of the clutter fluctuation and due to antenna mechanical scanning, wind speed, and radar platform motion (if applicable), these random variables are not statistically independent.

More precisely, because of the antenna mechanical scanning, clutter returns in the radar mainbeam do not have the same amplitude from pulse to pulse. This will effectively add amplitude modulation to the clutter returns. This additional modulation is governed by the shape of the antenna pattern, the rate of mechanical scanning, and the radar PRF. Denote the antenna two-way azimuth $3dB$ beamwidth as θ_a and the antenna scan rate as $\dot{\theta}_{scan}$. It follows that the contribution of antenna scanning to the standard deviation of the clutter fluctuation is

$$\sigma_s = 0.399 \frac{\dot{\theta}_{scan}}{\theta_a} \tag{9.49}$$

Another contributor to the clutter spectral spreading is caused by motion of the clutter itself, due to wind. Trees, vegetation, and sea waves are the main contributors to this effect. This relative motion, although relatively small, introduces additional Doppler shift in the clutter returns. Earlier, it was established that Doppler frequency due to a relative velocity v is given by

$$f_d = 2v/\lambda \tag{9.50}$$

where λ is the radar operating wavelength. It follows that if the apparent rms velocity due to wind is v_{rms}, then the standard deviation is

$$\sigma_w = 2v_{rms}/\lambda \tag{9.51}$$

Finally, if the radar platform is in motion, then the relative motion between the platform and the stationary clutter will cause a Doppler shift given by

$$f_c = (2v_{radar}\cos\theta)/\lambda \tag{9.52}$$

where $v_{radar}\cos\theta$ is the radial velocity component of the platform in the direction of clutter. Since the radar beam has a finite width, not all clutter components have the same radial velocity at all times. More specifically, if the angles θ_1 and θ_2 represent the edges of the radar beam, then Eq. (9.52) ca be written as

$$f_c = \frac{2v_{radar}}{\lambda}(\cos\theta_2 - \cos\theta_1) \approx \frac{2v_{radar}}{\lambda}\theta_a\sin\theta \tag{9.53}$$

and the standard deviation due to platform motion is given by

$$\sigma_v = \frac{v_{radar}}{\lambda}\sin\theta \tag{9.54}$$

Finally, the overall clutter spreading is denoted by σ_f, where

$$\sigma_f^2 = \sigma_v^2 + \sigma_s^2 + \sigma_w^2 \tag{9.55}$$

The overall value of the clutter spreading defined in Eq. (9.55) is relatively small.

9.5.3. Clutter Power Spectrum Density

Clutter primarily comprises stationary ground unwanted reflections with limited relative motion with respect to the radar. Therefore, its power spectrum density will be concentrated around $f = 0$. However, because σ_f (see Eq. (9.55)) is not always zero, clutter actually exhibits some Doppler frequency spread. The clutter power spectrum can be written as the sum of fixed (stationary) and random (due to frequency spreading) components, as

$$S_c(f) = \frac{P_c}{T\sigma_f\sqrt{2\pi}} \sum_{k=-\infty}^{\infty} \exp\left(-\frac{(f-k/T)^2}{2\sigma_f^2}\right) \tag{9.56}$$

where T is the PRI (i.e., $1/f_r$, f_r is the PRF), P_c is the clutter power or clutter mean square value, and σ_f is the clutter spectral spreading parameter as defined in Eq. (9.55). As clearly indicated by Eq. (9.56), the clutter PSD is periodic with period equal to f_r. Furthermore, the clutter PSD extends about each multiple integer of the PRF in accordance with Eq. (9.55). It must be noted that this spread is relatively small and thus the relation $\sigma_f \ll f_r$ is always true. This is illustrated in Fig. 9.14. The mean square value can be calculated from

$$P_c = T \int_{-f_r/2}^{f_r/2} S_c(f)df \tag{9.57}$$

Let $S_{c0}(f)$ denote the central portion of Eq. (9.56); then P_c is be expressed by

$$P_c = T \int_{-\infty}^{\infty} S_{c0}(f)df \tag{9.58}$$

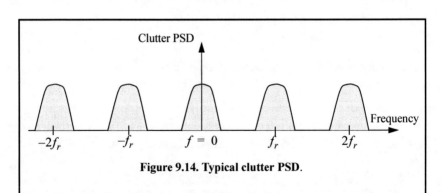

Figure 9.14. Typical clutter PSD.

where $S_{c0}(f)$ is a Gaussian shape function given by

$$S_{c0}(f) = \frac{k}{\sigma_f \sqrt{2\pi}} \exp\left(-\frac{f^2}{2\sigma_f^2}\right) \tag{9.59}$$

and $k = P_c / T$.

9.6. Moving Target Indicator (MTI)

The clutter spectrum is concentrated around DC ($f = 0$) and multiple integers of the radar PRF f_r, as was illustrated in Fig. 9.14. In CW radars, clutter is avoided or suppressed by ignoring the receiver output around DC, since most of the clutter power is concentrated about the zero frequency band. Pulsed radar systems may utilize special filters that can distinguish between slow-moving or stationary targets and fast-moving ones. This class of filter is known as the Moving Target Indicator (MTI). In simple words, the purpose of an MTI filter is to suppress target-like returns produced by clutter and allow returns from moving targets to pass through with little or no degradation. In order to effectively suppress clutter returns, an MTI filter needs to have a deep stop-band at DC and at integer multiples of the PRF. Figure 9.15b shows a typical sketch of an MTI filter response, while Fig. 9.15c shows its output when the PSD shown in Fig. 9.15a is the input.

MTI filters can be implemented using delay line cancelers. As we will show later in this chapter, the frequency response of this class of MTI filter is periodic, with nulls at integer multiples of the PRF. Thus, targets with Doppler frequencies equal to nf_r are severely attenuated. Since Doppler is proportional to target velocity ($f_d = 2v/\lambda$), target speeds that produce Doppler frequencies equal to integer multiples of f_r are known as blind speeds. More precisely,

$$v_{blind} = (n\lambda f_r)/2; \ \ n \geq 0 \tag{9.60}$$

Radar systems can minimize the occurrence of blind speeds either by employing multiple PRF schemes (PRF staggering) or by using high PRFs in which the radar may become range ambiguous. The main difference between PRF staggering and PRF agility is that the pulse repetition interval (within an integration interval) can be changed between consecutive pulses for the case of PRF staggering.

9.6.1. Single Delay Line Canceler

A single delay line canceler can be implemented as shown in Fig. 9.16. The canceler's impulse response is denoted as $h(t)$. The output $y(t)$ is equal to the convolution between the impulse response $h(t)$ and the input $x(t)$. The single delay canceler is often called a two-pulse canceler since it requires two distinct input pulses before an output can be read.

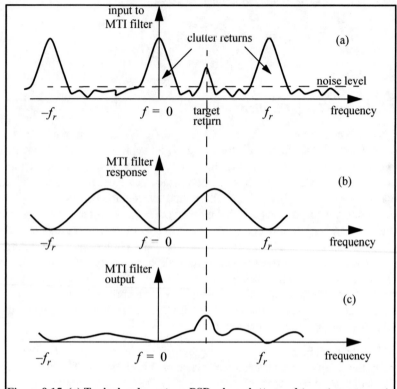

Figure 9.15. (a) Typical radar return PSD when clutter and target are present. (b) MTI filter frequency response. (c) Output from an MTI filter.

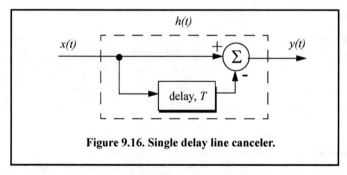

Figure 9.16. Single delay line canceler.

The delay T is equal to the radar PRI ($1/f_r$). The output signal $y(t)$ is

$$y(t) = x(t) - x(t - T)$$ (9.61)

The impulse response of the canceler is given by

$$h(t) = \delta(t) - \delta(t - T) \tag{9.62}$$

where $\delta(\ )$ is the delta function. It follows that the Fourier transform (FT) of $h(t)$ is

$$H(\omega) = 1 - e^{-j\omega T} \tag{9.63}$$

where $\omega = 2\pi f$. In the z-domain, the single delay line canceler response is

$$H(z) = 1 - z^{-1} \tag{9.64}$$

The power gain for the single delay line canceler is given by

$$|H(\omega)|^2 = H(\omega)H^*(\omega) = (1 - e^{-j\omega T})(1 - e^{j\omega T}) \tag{9.65}$$

It follows that

$$|H(\omega)|^2 = 1 + 1 - (e^{j\omega T} + e^{-j\omega T}) = 2(1 - \cos\omega T) \tag{9.66}$$

and using the trigonometric identity $(2 - 2\cos 2\vartheta) = 4(\sin\vartheta)^2$ yields

$$|H(\omega)|^2 = 4(\sin(\omega T/2))^2 \tag{9.67}$$

The amplitude frequency response for a single delay line canceller is shown in Fig. 9.17. Clearly, the frequency response of a single canceler is periodic with a period equal to f_r. The peaks occur at $f = (2n + 1)/(2f_r)$, and the nulls are at $f = nf_r$, where $n \geq 0$. In most radar applications the response of a single canceler is not acceptable since it does not have a wide notch in the stopband. A double delay line canceler has better response in both the stop- and pass-bands, and thus it is more frequently used than a single canceler. In this book, we will use the names *single delay line canceler* and *single canceler* interchangeably.

9.6.2. Double Delay Line Canceler

Two basic configurations of a double delay line canceler are shown in Fig. 9.18. Double cancelers are often called three-pulse cancelers since they require three distinct input pulses before an output can be read. The double line canceler impulse response is given by

$$h(t) = \delta(t) - 2\delta(t - T) + \delta(t - 2T) \tag{9.68}$$

Again, the names *double delay line canceler* and *double canceler* will be used interchangeably. The power gain for the double delay line canceler is

$$|H(\omega)|^2 = |H_1(\omega)|^2 |H_1(\omega)|^2 \tag{9.69}$$

where $|H_1(\omega)|^2$ is the single line canceler power gain given in Eq. (9.55). It follows that

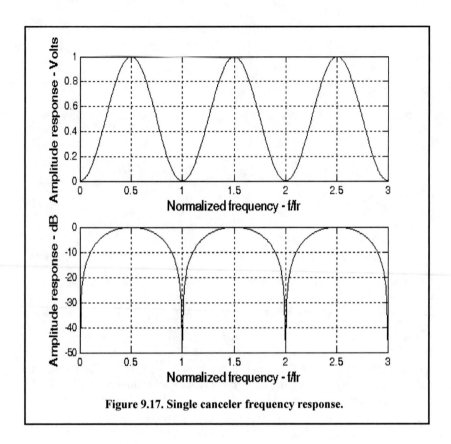

Figure 9.17. Single canceler frequency response.

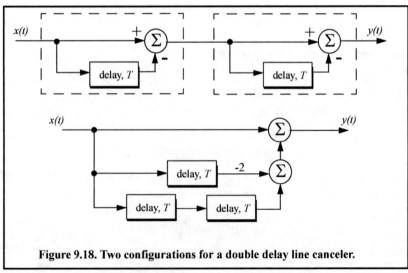

Figure 9.18. Two configurations for a double delay line canceler.

$$|H(\omega)|^2 = 16\left(\sin\left(\omega\frac{T}{2}\right)\right)^4 \qquad \text{(9.70)}$$

And in the z-domain, we have

$$H(z) = (1 - z^{-1})^2 = 1 - 2z^{-1} + z^{-2} \qquad \text{(9.71)}$$

Figure 9.19 shows typical output from this function. Note that the double canceler has a better response than the single canceler (deeper notch and flatter pass-band response).

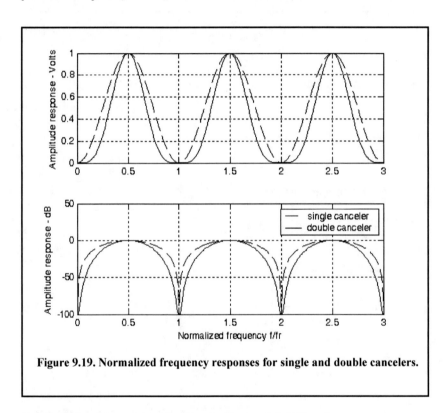

Figure 9.19. Normalized frequency responses for single and double cancelers.

9.6.3. Delay Lines with Feedback (Recursive Filters)

Delay line cancelers with feedback loops are known as recursive filters. The advantage of a recursive filter is that through a feedback loop, we will be able to shape the frequency response of the filter. As an example, consider the single canceler shown in Fig. 9.20. From the figure we can write

$$y(t) = x(t) - (1 - K)w(t) \qquad \text{(9.72)}$$

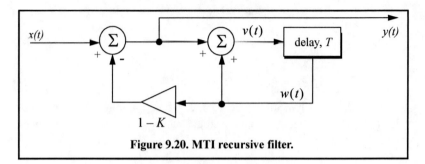

Figure 9.20. MTI recursive filter.

$$v(t) = y(t) + w(t) \tag{9.73}$$

$$w(t) = v(t - T) \tag{9.74}$$

Applying the z-transform to the above three equations yields

$$Y(z) = X(z) - (1 - K)W(z) \tag{9.75}$$

$$V(z) = Y(z) + W(z) \tag{9.76}$$

$$W(z) = z^{-1}V(z) \tag{9.77}$$

Solving for the transfer function $H(z) = Y(z)/X(z)$ yields

$$H(z) = \frac{1 - z^{-1}}{1 - Kz^{-1}} \tag{9.78}$$

The modulus square of $H(z)$ is then equal to

$$|H(z)|^2 = \frac{(1 - z^{-1})(1 - z)}{(1 - Kz^{-1})(1 - Kz)} = \frac{2 - (z + z^{-1})}{(1 + K^2) - K(z + z^{-1})} \tag{9.79}$$

Using the transformation $z = e^{j\omega T}$ yields

$$z + z^{-1} = 2\cos\omega T \tag{9.80}$$

Thus, Eq. (9.79) can now be rewritten as

$$|H(e^{j\omega T})|^2 = \frac{2(1 - \cos\omega T)}{(1 + K^2) - 2K\cos(\omega T)} \tag{9.81}$$

Note that when $K = 0$, Eq. (9.81) collapses to Eq. (9.67) (single line canceler). Figure 9.21 shows a plot of Eq. (9.81) for $K = 0.25, 0.7, 0.9$. Clearly, by changing the gain factor K one can control the filter response. This plot can be reproduced using the following MATLAB code.

```
clear all;
fofr = 0:0.001:1;
arg = 2.*pi.*fofr;
nume = 2.*(1.-cos(arg));
den11 = (1. + 0.25 * 0.25);
den12 = (2. * 0.25) .* cos(arg);
den1 = den11 - den12;
den21 = 1.0 + 0.7 * 0.7;
den22 = (2. * 0.7) .* cos(arg);
den2 = den21 - den22;
den31 = (1.0 + 0.9 * 0.9);
den32 = ((2. * 0.9) .* cos(arg));
den3 = den31 - den32;
resp1 = nume ./ den1;
resp2 = nume ./ den2;
resp3 = nume ./ den3;
plot(fofr,resp1,'k',fofr,resp2,'k-.',fofr,resp3,'k--');
xlabel('Normalized frequency')
ylabel('Amplitude response')
legend('K=0.25','K=0.7','K=0.9')
grid
axis tight
```

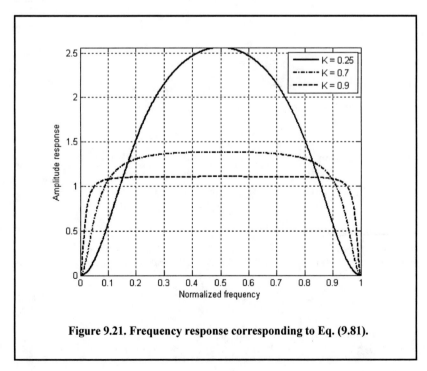

Figure 9.21. Frequency response corresponding to Eq. (9.81).

In order to avoid oscillation due to the positive feedback, the value of K should be less than unity. The value $(1 - K)^{-1}$ is normally equal to the number of pulses received from the target. For example, $K = 0.9$ corresponds to ten pulses, while $K = 0.98$ corresponds to about fifty pulses.

9.7. PRF Staggering

Target velocities that correspond to multiple integers of the PRF are referred to as blind speeds. This terminology is used since an MTI filter response is equal to zero at these values. Blind speeds can pose serious limitations on the performance of MTI radars and their ability to perform adequate target detection. Using PRF agility by changing the pulse repetition interval between consecutive pulses can extend the first blind speed to more tolerable values. In order to show how PRF staggering can alleviate the problem of blind speeds, let us first assume that two radars with distinct PRFs are utilized for detection. Since blind speeds are proportional to the PRF, the blind speeds of the two radars would be different. However, using two radars to alleviate the problem of blind speeds is a very costly option. A more practical solution is to use a single radar with two or more different PRFs.

For example, consider a radar system with two interpulse periods T_1 and T_2, such that

$$\frac{T_1}{T_2} = \frac{n_1}{n_2} \tag{9.82}$$

where n_1 and n_2 are integers. The first true blind speed occurs when

$$\frac{n_1}{T_1} = \frac{n_2}{T_2} \tag{9.83}$$

This is illustrated in Fig. 9.22 for $n_1 = 4$ and $n_2 = 5$. The ratio

$$k_s = \frac{n_1}{n_2} \tag{9.84}$$

is known as the stagger ratio. Using staggering ratios closer to unity pushes the first true blind speed farther out. However, the dip in the vicinity of $1/T_1$ becomes deeper. In general, if there are N PRFs related by

$$\frac{n_1}{T_1} = \frac{n_2}{T_2} = \ldots = \frac{n_N}{T_N} \tag{9.85}$$

and if the first blind speed to occur for any of the individual PRFs is v_{blind1}, then the first true blind speed for the staggered waveform is

$$v_{blind} = \frac{n_1 + n_2 + \dots + n_N}{N} \, v_{blind1} \qquad (9.86)$$

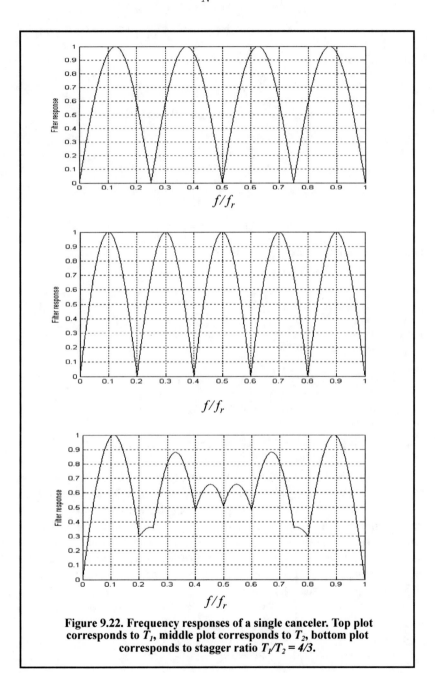

Figure 9.22. Frequency responses of a single canceler. Top plot corresponds to T_1, middle plot corresponds to T_2, bottom plot corresponds to stagger ratio $T_1/T_2 = 4/3$.

To better determine the frequency response of an MTI filter with staggered PRFs consider a three-pulse canceler with two PRFs, or equivalently two PRIs, T_1 and T_2. In this case, the impulse response will be given by

$$h(t) = [\delta(t) - \delta(t - T_1)] - [\delta(t - T_1) - \delta(t - T_1 - T_2)] \tag{9.87}$$

which can be written as

$$h(t) = \delta(t) - 2\delta(t - T_1) + \delta(t - T_1 - T_2) \tag{9.88}$$

Note that PRF staggering requires a minimum of two PRFs.

Make the change of variables $u = t - T_1$ in Eq. (9.88), and it follows

$$h(u + T_1) = \delta(u + T_1) - 2\delta(u) + \delta(u - T_2) \tag{9.89}$$

The Z-transform of the impulse response in Eq. (9.89) is then given by

$$H(z)z^{T_1} = z^{T_1} - 2 + z^{-T_2} \tag{9.90}$$

and the amplitude frequency response for the staggered double delay line canceller is then given by

$$|H(z)|^2 \Big|_{z = e^{j\omega T}} = (z^{T_1} - 2 + z^{-T_2})(z^{-T_1} - 2 + z^{T_2}) \tag{9.91}$$

Performing the algebraic manipulation in Eq. (9.91) and using the t trigonometric identity $(e^{j\omega T} + e^{-j\omega T}) = 2\cos\omega T$ yields

$$|H(\omega)|^2 = 6 - 4\cos(2\pi f T_1) - 4\cos(2\pi f T_2) + 2\cos(2\pi f(T_1 + T_2)) \tag{9.92}$$

It is customary to normalize the amplitude frequency response, thus

$$|H(\omega)|^2 = 1 - \frac{2}{3}\cos(2\pi f T_1) - \frac{2}{3}\cos(2\pi f T_2) + \frac{1}{3}\cos(2\pi f(T_1 + T_2)) \tag{9.93}$$

To determine the characteristics of higher stagger ratio MTI filters, adopt the notion of having several MTI filters, one for each combination of two staggered PRFs. Then the overall filter response is computed as the average of all individual filters. For example, consider the case where a PRF stagger is required with PRIs T_1, T_2, T_3, and T_4. First, compute the filter response using T_1 T_2 and denote by H_1. Then compute H_2 using T_2 and T_3, the filter H_3 is computed using T_3 T_4 and the filter H_4 is computed using T_4 and T_1. Finally compute the overall response as

$$H(f) = \frac{1}{4}[H_1(f) + H_2(f) + H_3(f) + H_4(f)] \tag{9.94}$$

Figure 9.23 shows the MTI filter response for a 4 stagger ratio defined. The overall response is computed as the average of 4 individual filters each corresponding to one combination of the stagger ratio. In the top portion of the figure the individual filters used were 2-pulse MTIs, while the bottom portion used 4-pulse individual MTI filters. This plot can be reproduced using the following MATLAB code.

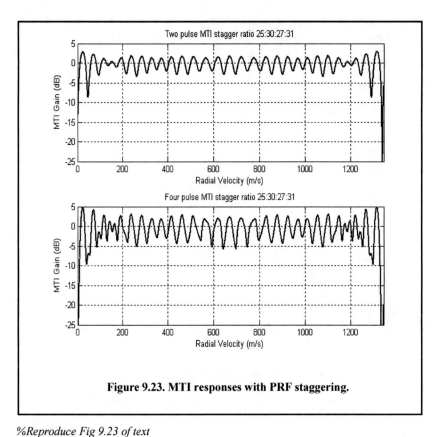

Figure 9.23. MTI responses with PRF staggering.

```
%Reproduce Fig 9.23 of text
k = .00035/25; a = 25*k; b = 30*k; c = 27*k; d = 31*k;
v2 = linspace(0,1345,10000);
f2 = (2.*v2)/.0375;
% H1(f)
T1 = exp(-j*2*pi.*f2*a); X1 = 1/2.*(1 - T1).*conj(1 - T1); H1 = 10*log10(abs(X1));
% H2(f)
T2 = exp(-j*2*pi.*f2*b); X2 = 1/2.*(1 - T2).*conj(1 - T2); H2 = 10*log10(abs(X2));
% H3(f)
T3 = exp(-j*2*pi.*f2*c); X3 = 1/2.*(1 - T3).*conj(1 - T3); H3 = 10*log10(abs(X3));
% H4(f)
T4 = exp(-j*2*pi.*f2*d); X4 = 1/2.*(1 - T4).*conj(1 - T4); H4 = 10*log10(abs(X4));
```

```
% Plot of the four components of H(f)
figure(1)
subplot(2,1,1)
% H(f) Average
ave2 = abs((X1 + X2 + X3 + X4)./4);
Have2 = 10*log10(abs((X1 + X2 + X3 + X4)./4));
plot(v2,Have2);
axis([0 1345 -25 5]);
 title('Two pulse MTI stagger ratio 25:30:27:31');
xlabel('Radial Velocity (m/s)');
 ylabel('MTI Gain (dB)'); grid on
% %Mean value of H(f)
v4 = v2; f4 = (2.*v4)/.0375;
% H1(f)
T1 = exp(-j*2*pi.*f4*a);
 T2 = exp(-j*2*pi.*f4*(a + b));
T3 = exp(-j*2*pi.*f4*(a + b + c));
X1 = 1/20.*(1 - 3.*T1 + 3.*T2 - T3).*conj(1 - 3.*T1 + 3.*T2 - T3);
H1 = 10*log10(abs(X1));
% H2(f)
T3 = exp(-j*2*pi.*f4*b);
T4 = exp(-j*2*pi.*f4*(b + c));
T5 = exp(-j*2*pi.*f4*(b + c + d));
X2 = 1/20.*(1 - 3.*T3 + 3.*T4 - T5).*conj(1 - 3.*T3 + 3.*T4 - T5);
H2 = 10*log10(abs(X2));
% H3(f)
T6 = exp(-j*2*pi.*f4*c);
T7 = exp(-j*2*pi.*f4*(c + d));
T8 = exp(-j*2*pi.*f4*(c + d + a));
X3 = 1/20.*(1 - 3.*T6 + 3.*T7 - T8).*conj(1 - 3.*T6 + 3.*T7 - T8);
H3 = 10*log10(abs(X3));
% H4(f)
T9 = exp(-j*2*pi.*f4*d); T10 = exp(-j*2*pi.*f4*(d + a));
T11 = exp(-j*2*pi.*f4*(d + a + b));
X4 = 1/20.*(1 - 3.*T9 + 3.*T10 - T11).*conj(1 - 3.*T9 + 3.*T10 - T11);
H4 = 10*log10(abs(X4));
% H(f) Average
ave4 = abs((X1 + X2 + X3 + X4)./4);
Have4 = 10*log10(abs((X1 + X2 + X3 + X4)./4));
% Plot of H(f) Average
subplot(2,1,2)
plot(v4,Have4);
axis([0 1345 -25 5]);
title('Four pulse MTI stagger ratio 25:30:27:31');
xlabel('Radial Velocity (m/s)');
ylabel('MTI Gain (dB)');
grid on
```

9.8. MTI Improvement Factor

In this section two quantities that are normally used to define the performance of MTI systems are introduced. They are Clutter Attenuation (CA) and the Improvement Factor. The MTI CA is defined as the ratio between the MTI filter input clutter power C_i to the output clutter power C_o,

$$CA = C_i/C_o \tag{9.95}$$

The MTI improvement factor is defined as the ratio of the SCR at the output to the SCR at the input,

$$I = \left(\frac{S_o}{C_o}\right) \Big/ \left(\frac{S_i}{C_i}\right) \tag{9.96}$$

which can be rewritten as

$$I = \frac{S_o}{S_i}CA \tag{9.97}$$

The ratio S_o/S_i is the average power gain of the MTI filter, and it is equal to $|H(\omega)|^2$. In this section, a closed form expression for the improvement factor using a Gaussian-shaped power spectrum (see Eq. (9.59)) is developed. A Gaussian-shaped clutter power spectrum is given by

$$S(f) = \frac{P_c}{\sqrt{2\pi}\ \sigma_f}\exp(-f^2/2\sigma_f^2) \tag{9.98}$$

where P_c is the clutter power (constant), and σ_f is the clutter rms frequency (which describes the clutter spectrum spread in the frequency domain, see Eq. (9.55)).

The clutter power at the input of an MTI filter is

$$C_i = \int_{-\infty}^{\infty}\frac{P_c}{\sqrt{2\pi}\ \sigma_f}\exp\left(-\frac{f^2}{2\sigma_f^2}\right)df \tag{9.99}$$

Factoring out the constant P_c yields

$$C_i = P_c\int_{-\infty}^{\infty}\frac{1}{\sqrt{2\pi}\sigma_f}\exp\left(-\frac{f^2}{2\sigma_f^2}\right)df \tag{9.100}$$

It follows that

$$C_i = P_c \tag{9.101}$$

The clutter power at the output of an MTI is

$$C_o = \int_{-\infty}^{\infty} S(f) |H(f)|^2 \, df \qquad (9.102)$$

9.8.1. Two-Pulse MTI Case

In this section we will continue the analysis using a single delay line canceler. The frequency response for a single delay line canceler is

$$|H(f)|^2 = 4\left(\sin\left(\frac{\pi f}{f_r}\right) \right)^2 \qquad (9.103)$$

It follows that

$$C_o = \int_{-\infty}^{\infty} \frac{P_c}{\sqrt{2\pi}\,\sigma_f} \exp\left(-\frac{f^2}{2\sigma_f^2} \right) 4\left(\sin\left(\frac{\pi f}{f_r}\right) \right)^2 \, df \qquad (9.104)$$

Now, since clutter power will only be significant for small f, the ratio f/f_r is very small (i.e., $\sigma_f \ll f_r$). Consequently, by using the small angle approximation, Eq. (9.104) is approximated by

$$C_o \approx \int_{-\infty}^{\infty} \frac{P_c}{\sqrt{2\pi}\,\sigma_f} \exp\left(-\frac{f^2}{2\sigma_f^2} \right) 4\left(\frac{\pi f}{f_r}\right)^2 \, df \qquad (9.105)$$

which can be rewritten as

$$C_o = \frac{4P_c\pi^2}{f_r^2} \int_{-\infty}^{\infty} \frac{1}{\sqrt{2\pi\sigma_f^2}} \exp\left(-\frac{f^2}{2\sigma_f^2} \right) f^2 \, df \qquad (9.106)$$

The integral part in Eq. (9.106) is the second moment of a zero-mean Gaussian distribution with variance σ_f^2. Replacing the integral in Eq. (9.106) by σ_f^2 yields

$$C_o = \frac{4P_c\pi^2}{f_r^2} \sigma_f^2 \qquad (9.107)$$

Substituting Eq. (9.107) and Eq. (9.101) into Eq. (9.95) produces

$$CA = \frac{C_i}{C_o} = \left(\frac{f_r}{2\pi\sigma_f}\right)^2 \qquad (9.108)$$

It follows that the improvement factor for a single canceler is

$$I = \left(\frac{f_r}{2\pi\sigma_f}\right)^2 \frac{S_o}{S_i} \tag{9.109}$$

The power gain ratio for a single canceler is (remember that $|H(f)|$ is periodic with period f_r)

$$\frac{S_o}{S_i} = |H(f)|^2 = \frac{1}{f_r} \int_{-f_r/2}^{f_r/2} 4\left(\sin\frac{\pi f}{f_r}\right)^2 df \tag{9.110}$$

Using the trigonometric identity $(2 - 2\cos 2\vartheta) = 4(\sin\vartheta)^2$ yields

$$|H(f)|^2 = \frac{1}{f_r} \int_{-f_r/2}^{f_r/2} \left(2 - 2\cos\frac{2\pi f}{f_r}\right) df = 2 \tag{9.111}$$

It follows that

$$I = 2(f_r/2\pi\sigma_f)^2 \tag{9.112}$$

The expression given in Eq. (9.112) is an approximation valid only for $\sigma_f \ll f_r$. When the condition $\sigma_f \ll f_r$ is not true, then the autocorrelation function needs to be used in order to develop an exact expression for the improvement factor.

Example:

A certain radar has $f_r = 800\,Hz$. If the clutter rms is $\sigma_f = 6.4\,Hz$, find the improvement factor when a single delay line canceler is used.

Solution:

The clutter attenuation CA is

$$CA = \left(\frac{f_r}{2\pi\sigma_f}\right)^2 = \left(\frac{800}{(2\pi)(6.4)}\right)^2 = 395.771 = 25.974dB$$

and since $S_o/S_i = 2 = 3dB$ we get

$$I_{dB} = (CA + S_o/S_i)_{dB} = 3 + 25.97 = 28.974dB.$$

9.8.2. The General Case

A general expression for the improvement factor for the n-pulse MTI (shown for a 2-pulse MTI in Eq. (9.112)) is given by

$$I = \frac{1}{Q^2(2(n-1)-1)!!}\left(\frac{f_r}{2\pi\sigma_c}\right)^{2(n-1)} \tag{9.113}$$

where the double factorial notation is defined by

$$(2n-1)!! = 1 \times 3 \times 5 \times \ldots \times (2n-1) \tag{9.114}$$

$$(2n)!! = 2 \times 4 \times \ldots \times 2n \tag{9.115}$$

Of course $0!! = 1$; Q is defined by

$$Q^2 = \frac{1}{\sum_{n} A_i^2} \tag{9.116}$$

where A_i are the binomial coefficients for the MTI filter. It follows that Q^2 for a 2-pulse, 3-pulse, and 4-pulse MTI are, respectively,

$$\left\{\frac{1}{2}, \frac{1}{20}, \frac{1}{70}\right\} \tag{9.117}$$

Using this notation, then the improvement factor for a 3-pulse and 4-pulse MTI are, respectively, given by

$$I_{3-pulse} = 2\left(\frac{f_r}{2\pi\sigma_c}\right)^4 \tag{9.118}$$

$$I_{4-pulse} = \frac{4}{3}\left(\frac{f_r}{2\pi\sigma_c}\right)^6 \tag{9.119}$$

9.9. Subclutter Visibility (SCV)

Subclutter Visibility (SCV) describes the radar's ability to detect nonstationary targets embedded in a strong clutter background, for some probabilities of detection and false alarm. It is often used as a measure of MTI performance. For example, a radar with $10dB$ SCV will be able to detect moving targets whose returns are ten times smaller than those of clutter. A sketch illustrating the concept of SCV is shown in Fig. 9.24.

If a radar system can resolve the areas of strong and weak clutter within its field of view, then Interclutter Visibility (ICV) describes the radar's ability to detect nonstationary targets between strong clutter points. The subclutter visibility is expressed as the ratio of the improvement factor to the minimum MTI output SCR required for proper detection for a given probability of detection. More precisely,

$$SCV = I/(SCR)_o \qquad \text{(9.120)}$$

When comparing the performance of different radar systems on the basis of SCV, one should use caution since the amount of clutter power is dependent on the radar resolution cell (or volume), which may be different from one radar to another. Thus, only if the different radars have the same beamwidths and the same pulse widths can SCV be used as a basis of performance comparison.

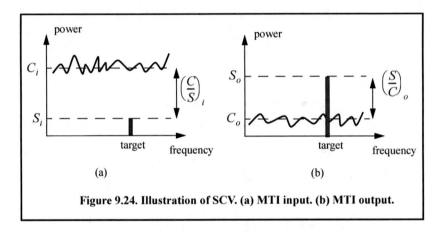

Figure 9.24. Illustration of SCV. (a) MTI input. (b) MTI output.

9.10. Delay Line Cancelers with Optimal Weights

The delay line cancelers discussed in this chapter belong to a family of transversal Finite Impulse Response (FIR) filters widely known as the "tapped delay line" filters. Figure 9.25 shows an N-stage tapped delay line implementation. When the weights are chosen such that they are the binomial coefficients (coefficients of the expansion $(1-x)^N$) with alternating signs, then the resultant MTI filter is equivalent to N-stage cascaded single line cancelers. This is illustrated in Fig. 9.26 for $N = 4$. In general, the binomial coefficients are given by

$$w_i = (-1)^{i-1} \frac{N!}{(N-i+1)!(i-1)!} \; ; \; i = 1, ..., N+1 \qquad \text{(9.121)}$$

Using the binomial coefficients with alternating signs produces an MTI filter that closely approximates the optimal filter in the sense that it maximizes the improvement factor, as well as the probability of detection. In fact, the difference between an optimal filter and one with binomial coefficients is so small that the latter one is considered to be optimal by most radar designers. However, being optimal in the sense of the improvement factor does not guarantee a deep notch or a flat pass-band in the MTI filter response. Consequently, many researchers have been investigating other weights that can produce a deeper notch around DC, as well as a better pass-band response.

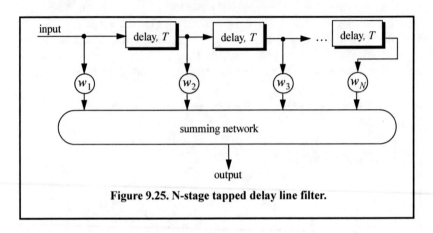

Figure 9.25. N-stage tapped delay line filter.

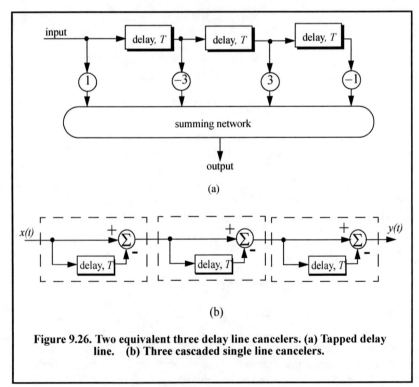

Figure 9.26. Two equivalent three delay line cancelers. (a) Tapped delay line.   (b) Three cascaded single line cancelers.

In general, the average power gain for an N-stage delay line canceler is

$$\frac{S_o}{S_i} = \prod_{i=1}^{N} |H_1(f)|^2 = \prod_{i=1}^{N} 4\left(\sin\left(\frac{\pi f}{f_r}\right)\right)^2 \qquad \text{(9.122)}$$

For example, $N = 2$ (double delay line canceler) gives

$$\frac{S_o}{S_i} = 16\left(\sin\left(\frac{\pi f}{f_r}\right)\right)^4 \qquad \text{(9.123)}$$

Equation (9.123) can be rewritten as

$$\frac{S_o}{S_i} = |H_1(f)|^{2N} = 2^{2N}\left(\sin\left(\frac{\pi f}{f_r}\right)\right)^{2N} \qquad \text{(9.124)}$$

As indicated by Eq. (9.124), blind speeds for an N-stage delay canceler are identical to those of a single canceler. It follows that blind speeds are independent from the number of cancelers used. It is possible to show that Eq. (9.124) can be written as

$$\frac{S_o}{S_i} = 1 + N^2 + \left(\frac{N(N-1)}{2!}\right)^2 + \left(\frac{N(N-1)(N-2)}{3!}\right)^2 + \dots \qquad \text{(9.125)}$$

A general expression for the improvement factor of an N-stage tapped delay line canceler is reported by Nathanson[1] to be

$$I = \frac{(S_o/S_i)}{\sum_{k=1}^{N}\sum_{j=1}^{N} w_k w_j^* \rho\left(\frac{(k-j)}{f_r}\right)} \qquad \text{(9.126)}$$

where the weights w_k and w_j are those of a tapped delay line canceler, and $\rho((k-j)/f_r)$ is the correlation coefficient between the kth and jth samples. For example, $N = 2$ produces

$$I = \frac{1}{1 - \frac{4}{3}\rho T + \frac{1}{3}\rho 2T} \qquad \text{(9.127)}$$

1. Nathanson, F. E., *Radar Design Principles*, 2nd edition, McGraw-Hill, Inc., NY, 1991.

9.11. MATLAB Program Listings

This section presents listings for all the MATLAB programs used to produce all of the MATLAB-generated figures in this chapter. They are listed in the same order they appear in the text.

9.11.1. MATLAB Function "clutter_rcs.m"

The function *"clutter_rcs.m"* implements Eq. (9.37). It generates plots of the clutter RCS versus the radar slant range. Its outputs include the clutter RCS in dBsm. The syntax is as follows:

function [sigmaC] = clutter_rcs(sigma0, thetaE, thetaA, SL, range, hr, ht, b,ant_id)

where

Symbol	Description	Units	Status
sigma0	clutter back scatterer coefficient	dB	input
thetaE	antenna 3dB elevation beamwidth	degrees	input
thetaA	antenna 3dB azimuth beamwidth	degrees	input
SL	antenna sidelobe level	dB	input
range	range; can be a vector or a single value	Km	input
hr	radar height	meters	input
ht	target height	meters	input
b	bandwidth	Hz	input
ant_id	1 for (sin(x)/x)^2 pattern 2 for Gaussian pattern	none	input
sigmac	clutter RCS; can be either vector or single value depending on "range"	dB	output

A GUI called *"clutter_rcs_gui"* was developed for this function. Executing this GUI generates plots of the σ_c versus range. Figure 9.26 shows the GUI workspace associated with this function.

MATLAB Function *"clutter_rcs.m"* Listing

function [sigmaC] = clutter_rcs(sigma0, thetaE, thetaA, SL, range, hr, ht, b,ant_id)
% This unction calculates the clutter RCS and the CNR for a ground based radar.
thetaA = thetaA * pi /180; % antenna azimuth beamwidth in radians
thetaE = thetaE * pi /180.; % antenna elevation beamwidth in radians
re = 6371000; % earth radius in meter
rh = sqrt(8.0*hr*re/3.); % range to horizon in meters

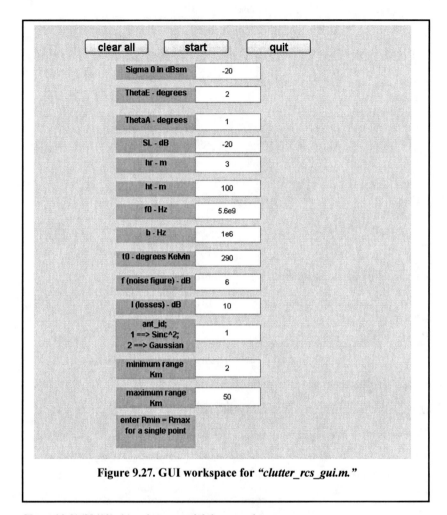

Figure 9.27. GUI workspace for *"clutter_rcs_gui.m."*

SLv = 10.0^(SL/10); % radar rms sidelobes in volts
sigma0v = 10.0^(sigma0/10); % clutter backscatter coefficient
deltar = 3e8 / 2 / b; % range resolution for unmodulated pulse
range_m = 1000 . range; % range in meters*
%%%%%%%%%%%%%%%%%%%%%%%%%%%%%%%%%%%%%%
thetar = asin(hr ./ range_m);
thetae = asin((ht-hr) ./ range_m);
% propagation attenuation due to round earth
propag_atten = 1. + ((range_m ./ rh).^4);
Rg = range_m . cos(thetar);*
deltaRg = deltar . cos(thetar);*
theta_sum = thetae + thetar;
% use sinc^2 antenna pattern when ant_id=1

```
% use Gaussian antenna pattern when ant_id=2
if(ant_id ==1) % use sinc^2 antenna pattern
   ant_arg = (theta_sum ) ./ (pi*thetaE);
   gain = (sinc(ant_arg)).^2;
else
   gain = exp(-2.776 .*(theta_sum./thetaE).^2);
end
% compute sigmac
sigmac = (sigma0v .* Rg .* deltaRg) .* ...
(pi * SLv * SLv + thetaA .* gain.^2) ./ propag_atten;
sigmaC = 10*log10(sigmac);
figure(1)
plot(range, sigmaC,'linewidth',1.5)
grid
xlabel('Slant Range in Km')
ylabel('Clutter RCS in dBsm')
%
```

9.11.2. MATLAB Function "single_canceler.m"

The function *"single_canceler.m"* computes and plots (as a function of f/f_r) the amplitude response for a single delay line canceler. The syntax is as follows:

$$[resp] = single\_canceler \ (fofr)$$

where *"fofr"* is the number of periods desired.

MATLAB Function *"single_canceler.m"* **Listing**

```
function [resp] = single_canceler (fofr1)
% single delay canceller
eps = 0.00001;
fofr = 0:0.01:fofr1;
arg1 = pi .* fofr;
resp = 4.0 .*((sin(arg1)).^2);
max1 = max(resp);
resp = resp ./ max1;
subplot(2,1,1)
plot(fofr,resp,'k')
xlabel ('Normalized frequency in f/fr')
ylabel( 'Amplitude response in Volts')
grid
subplot(2,1,2)
resp=10.*log10(resp+eps);
plot(fofr,resp,'k');
axis tight
grid
xlabel ('Normalized frequency in f/fr')
```

ylabel('Amplitude response in dB')

9.11.3. MATLAB Function "double_canceler.m"

The function *"double_canceler.m"* computes and plots (as a function of f/f_r) the amplitude response for a double delay line canceler. The syntax is as follows:

$$[resp] = double\_canceler\ (fofr)$$

where *"fofr"* is the number of periods desired.

MATLAB Function *"double_canceler.m"* Listing

```
function [resp] = double_canceler(fofr1)
eps = 0.00001;
fofr = 0:0.01:fofr1;
arg1 = pi .* fofr;
resp = 4.0 .* ((sin(arg1)).^2);
max1 = max(resp);
resp = resp ./ max1;
resp2 = resp .* resp;
subplot(2,1,1);
plot(fofr,resp,'k--',fofr, resp2,'k');
ylabel ('Amplitude response - Volts')
resp2 = 20. .* log10(resp2+eps);
resp1 = 20. .* log10(resp+eps);
subplot(2,1,2)
plot(fofr,resp1,'k--',fofr,resp2,'k');
legend ('single canceler','double canceler')
xlabel ('Normalized frequency f/fr')
ylabel ('Amplitude response in dB')
```

Problems

9.1. Compute the signal-to-clutter ratio (SCR) for the radar described in Section 9.2.1. In this case, assume antenna 3dB beam width $\theta_{3dB} = 0.03rad$, pulse width $\tau = 10\mu s$, range $R = 50Km$, grazing angle $\psi_g = 15°$, target RCS $\sigma_t = 0.1m^2$, and clutter reflection coefficient $\sigma^0 = 0.02(m^2/m^2)$.

9.2. Repeat the example in Section 9.3 for target RCS $\sigma_t = 0.15m^2$, pulse width $\tau = 0.1\mu s$, antenna beam width $\theta_a = \theta_e = 0.03radians$; the detection range is $R = 100Km$, and $\sum \sigma_i = 1.6 \times 10^{-9}(m^2/m^3)$.

9.3. The quadrature components of the clutter power spectrum are, respectively, given by

$$\overline{S}_I(f) = \delta(f) + \frac{C}{\sqrt{2\pi}\sigma_c} \exp(-f^2/2\sigma_c^2)$$

and

$$\overline{S}_Q(f) = \frac{C}{\sqrt{2\pi}\sigma_c} \exp(-f^2/2\sigma_c^2).$$

Compute the D.C. and A.C. power of the clutter. Let $\sigma_c = 10Hz$.

9.4. A certain radar has the following specifications: pulse width $\tau' = 1\mu s$, antenna beam width $\Omega = 1.5°$, and wavelength $\lambda = 3cm$. The radar antenna is $7.5m$ high. A certain target is simulated by two point targets (scatterers). The first scatterer is $4m$ high and has RCS $\sigma_1 = 20m^2$. The second scatterer is $12m$ high and has RCS $\sigma_2 = 1m^2$. If the target is detected at $10Km$, compute (a) SCR when both scatterers are observed by the radar, (b) SCR when only the first scatterer is observed by the radar. Assume a reflection coefficient of -1, and $\sigma^0 = -30dB$.

9.5. A certain radar has range resolution of $300m$ and is observing a target somewhere in a line of high towers each having RCS $\sigma_{tower} = 10^6 m^2$. If the target has RCS $\sigma_t = 1m^2$, (a) how much signal-to-clutter ratio should the radar have? (b) Repeat part (a) for range resolution of $30m$.

9.6. (a) Derive an expression for the impulse response of a single delay line canceler. (b) Repeat for a double delay line canceler.

9.7. (a) What is the transfer function, $H(z)$? (b) If the clutter power spectrum is $W(f) = w_0 \exp(-f^2/2\sigma_c^2)$, find an exact expression for the filter power gain. (c) Repeat part (b) for small values of frequency, f. (d) Compute the clutter attenuation and the improvement factor in terms of K and σ_c.

9.8. One implementation of a single delay line canceler with feedback is shown below

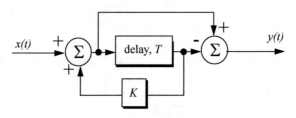

9.9. Plot the frequency response for the filter described in the previous problem for $K = -0.5, 0, and\ 0.5$.

9.10. An implementation of a double delay line canceler with feedback is shown below.

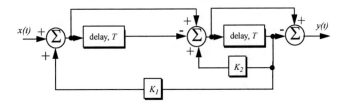

(a) What is the transfer function, $H(z)$? (b) Plot the frequency response for $K_1 = 0 = K_2$, and $K_1 = 0.2, K_2 = 0.5$.

9.11. Consider a single delay line canceler. Calculate the clutter attenuation and the improvement factor. Assume that $\sigma_c = 4Hz$ and PRF $f_r = 450Hz$.

9.12. Develop an expression for the improvement factor of a double delay line canceler.

9.13. Repeat Problem 9.10 for a double delay line canceler.

9.14. An experimental expression for the clutter power spectrum density is $W(f) = w_0 \exp(-f^2/2\sigma_c^2)$, where w_0 is a constant. Show that using this expression leads to the same result obtained for the improvement factor as developed in Section 9.8.

9.15. A certain radar uses two PRFs with stagger ratio 63/64. If the first PRF is $f_{r1} = 500Hz$, compute the blind speeds for both PRFs and for the resultant composite PRF. Assume $\lambda = 3cm$.

9.16. A certain filter used for clutter rejection has an impulse response $h(n) = \delta(n) - 3\delta(n-1) + 3\delta(n-2) - \delta(n-3)$. (a) Show an implementation of this filter using delay lines and adders. (b) What is the transfer function? (c) Plot the frequency response of this filter. (d) Calculate the output when the input is the unit step sequence.

9.17. The quadrature components of the clutter power spectrum are given in Problem 9.3. Let $\sigma_c = 10Hz$ and $f_r = 500Hz$. Compute the improvement of the signal-to-clutter ratio when a double delay line canceler is utilized.

9.18. Develop an expression for the clutter improvement factor for single and double line cancelers using the clutter autocorrelation function.

Chapter 10   *Doppler Processing*

In this chapter Doppler processing is analyzed in the context of continuous wave (CW) radars and pulsed Doppler radars. Continuous wave radars utilize CW waveforms, which may be considered to be a pure sinewave of the form $\cos 2\pi f_0 t$. Spectra of the radar echo from stationary targets and clutter will be concentrated at f_0. The center frequency for the echoes from moving targets will be shifted by f_d, the Doppler frequency. Thus, by measuring this frequency difference CW, radars can very accurately extract target radial velocity. Because of the continuous nature of CW emission, range measurement is not possible without some modifications to the radar operations and waveforms, which will be discussed later.

Alternatively, pulsed radars utilize a stream of pulses with a specific PRI (or PRF) to generate what is known as range-Doppler maps. Each map is divided into resolution cells. The dimensions of these resolution cells are range resolution along the time axis and Doppler resolution along the frequency axis.

10.1. CW Radar Functional Block Diagram

In order to avoid interruption of the continuous radar energy emission, two antennas are used in CW radars, one for transmission and one for reception. Figure 10.1 shows a simplified CW radar block diagram. The appropriate values of the signal frequency at different locations are noted on the diagram. The individual Narrow Band Filters (NBF) must be as narrow as possible in bandwidth in order to allow accurate Doppler measurements and minimize the amount of noise power. In theory, the operating bandwidth of a CW radar is infinitesimal (since it corresponds to an infinite duration continuous sinewave). However, systems with infinitesimal bandwidths cannot physically exist, and thus, the bandwidth of CW radars is assumed to correspond to that of a gated CW waveform.

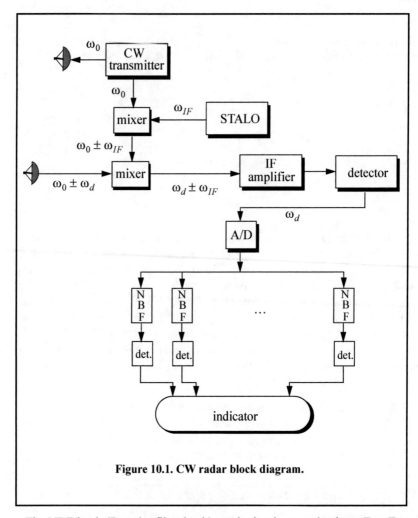

Figure 10.1. CW radar block diagram.

The NBF bank (Doppler filter bank) can be implemented using a Fast Fourier Transform (FFT). If the Doppler filter bank is implemented using an FFT of size N_{FFT}, and if the individual NBF bandwidth (FFT bin) is Δf, then the effective radar Doppler bandwidth is $N_{FFT}\Delta f/2$. The reason for the one-half factor is to account for both negative and positive Doppler shifts. The frequency resolution Δf is proportional to the inverse of the integration time.

Since range is computed from the radar echoes by measuring a two-way time delay, single frequency CW radars cannot measure target range. In order for CW radars to be able to measure target range, the transmit and receive waveforms must have some sort of timing marks. By comparing the timing marks at transmit and receive, CW radars can extract target range.

The timing mark can be implemented by modulating the transmit waveform, and one commonly used technique is Linear Frequency Modulation (LFM). Before we discuss LFM signals, we will first introduce the CW radar equation and briefly address the general Frequency Modulated (FM) waveforms using sinusoidal modulating signals.

10.1.1. CW Radar Equation

As indicated by Fig. 10.1, the CW radar receiver declares detection at the output of a particular Doppler bin if that output value passes the detection threshold within the detector box. Since the NBF bank is implemented by an FFT, only finite length data sets can be processed at a time. The length of such blocks is normally referred to as the dwell interval, integration time, or coherent processing interval. The dwell interval determines the frequency resolution or the bandwidth of the individual NBFs. More precisely,

$$\Delta f = 1/T_{Dwell} \tag{10.1}$$

T_{Dwell} is the dwell interval. Therefore, once the maximum resolvable frequency by the NBF bank is chosen the size of the NBF bank is computed as

$$N_{FFT} = 2B/\Delta f \tag{10.2}$$

B is the maximum resolvable frequency by the FFT. The factor 2 is needed to account for both positive and negative Doppler shifts. It follows that

$$T_{Dwell} = N_{FFT}/2B \tag{10.3}$$

The CW radar equation can now be derived. Consider the radar equation developed in Chapter 1. That is

$$SNR = \frac{P_{av}TG^2\lambda^2\sigma}{(4\pi)^3 R^4 kT_oFL} \tag{10.4}$$

where $P_{av} = (\tau/T)P_t$, τ/T, and P_t is the peak transmitted power. In CW radars the average transmitted power over the dwell interval P_{CW}, and T must be replaced by T_{Dwell}. Thus, the CW radar equation can be written as

$$SNR = \frac{P_{CW}T_{Dwell}G_tG_r\lambda^2\sigma}{(4\pi)^3 R^4 kT_oFLL_{win}} \tag{10.5}$$

where G_t and G_r are the transmit and receive antenna gains, respectively. The factor L_{win} is a loss term associated with the type of window (weighting) used in computing the FFT.

10.1.2. Linear Frequency Modulated CW Radar

CW radars may use LFM waveforms so that both range and Doppler information can be measured. In practical CW radars, the LFM waveform cannot be continually changed in one direction, and thus, periodicity in the modulation is normally utilized. Figure 10.2 shows a sketch of a triangular LFM waveform. The modulation does not need to be triangular; it may be sinusoidal, saw-tooth, or some other form. The dashed line in Fig. 10.2 represents the return waveform from a stationary target at range R. The beat frequency f_b is also sketched in Fig. 10.2. It is defined as the difference (due to heterodyning) between the transmitted and received signals. The time delay Δt is a measure of target range; that is,

$$\Delta t = \frac{2R}{c} \tag{10.6}$$

In practice, the modulating frequency f_m is selected such that

$$f_m = \frac{1}{2t_0} \tag{10.7}$$

The rate of frequency change, $\dot{f}$, is

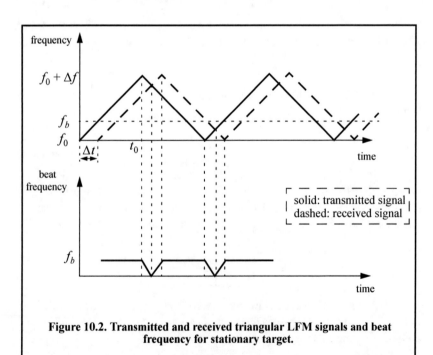

Figure 10.2. Transmitted and received triangular LFM signals and beat frequency for stationary target.

$$\dot{f} = \frac{\Delta f}{t_0} = \frac{\Delta f}{(1/2f_m)} = 2f_m\Delta f \qquad (10.8)$$

where Δf is the peak frequency deviation. The beat frequency f_b is given by

$$f_b = \Delta t\dot{f} = \frac{2R}{c}\dot{f} \qquad (10.9)$$

Equation (10.9) can be rearranged as

$$\dot{f} = \frac{c}{2R}f_b \qquad (10.10)$$

Equating Eqs. (10.8) and (10.10) and solving for f_b yield

$$f_b = \frac{4Rf_m\Delta f}{c} \qquad (10.11)$$

Now consider the case when Doppler is present (i.e., nonstationary target). The corresponding triangular LFM transmitted and received waveforms are sketched in Fig. 10.3, along with the corresponding beat frequency. As previously noted the beat frequency is defined as

$$f_b = f_{received} - f_{transmitted} \qquad (10.12)$$

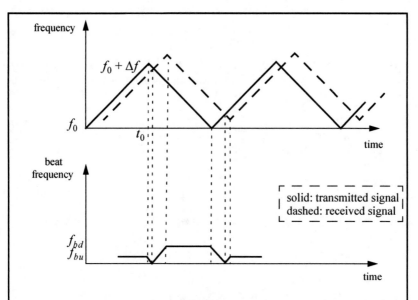

Figure 10.3. Transmitted and received LFM signals and beat frequency, for a moving target.

When the target is not stationary the received signal will contain a Doppler shift term in addition to the frequency shift due to the time delay Δt. In this case, the Doppler shift term subtracts from the beat frequency during the positive portion of the slope. Alternatively, the two terms add up during the negative portion of the slope. Denote the beat frequency during the positive (up) and negative (down) portions of the slope, respectively, as f_{bu} and f_{bd}. It follows that

$$f_{bu} = \frac{2R}{c}\dot{f} - \frac{2\dot{R}}{\lambda} \qquad (10.13)$$

where $\dot{R}$ is the range rate or the target radial velocity as seen by the radar. The first term of the right-hand side of Eq. (10.13) is due to the range delay defined by Eq. (10.6), while the second term is due to the target Doppler. Similarly,

$$f_{bd} = \frac{2R}{c}\dot{f} + \frac{2\dot{R}}{\lambda} \qquad (10.14)$$

Range is computed by adding Eq. (10.12) and Eq. (10.14). More precisely,

$$R = \frac{c}{4\dot{f}}(f_{bu} + f_{bd}) \qquad (10.15)$$

The range rate is computed by subtracting Eq. (10.14) from Eq. (10.13),

$$\dot{R} = \frac{\lambda}{4}(f_{bd} - f_{bu}) \qquad (10.16)$$

As indicated by Eq. (10.15) and Eq. (10.16), CW radars utilizing triangular LFM can extract both range and range rate information. In practice, the maximum time delay Δt_{max} is normally selected as

$$\Delta t_{max} = 0.1 t_0 \qquad (10.17)$$

Thus, the maximum range is given by

$$R_{max} = \frac{0.1 c t_0}{2} = \frac{0.1 c}{4 f_m} \qquad (10.18)$$

and the maximum unambiguous range will correspond to a shift equal to $2t_0$.

10.1.3. *Multiple Frequency CW Radar*

Continuous wave radars do not have to use LFM waveforms in order to obtain good range measurements. Multiple frequency schemes allow CW radars to compute very adequate range measurements without using frequency modulation. In order to illustrate this concept, first consider a CW radar with the following waveform

$$x(t) = A\sin 2\pi f_0 t \qquad (10.19)$$

The received signal from a target at range R is

$$x_r(t) = A_r \sin(2\pi f_0 t - \varphi) \qquad (10.20)$$

where the phase φ is equal to

$$\varphi = 2\pi f_0 (2R/c) \qquad (10.21)$$

Solving for R we obtain

$$R = \frac{c\varphi}{4\pi f_0} = \frac{\lambda}{4\pi}\varphi \qquad (10.22)$$

Clearly, the maximum unambiguous range occurs when φ is maximum, i.e., $\varphi = 2\pi$. Therefore, even for relatively large radar wavelengths, R is limited to impractical small values. Next, consider a radar with two CW signals, denoted by $s_1(t)$ and $s_2(t)$. More precisely,

$$x_1(t) = A_1 \sin 2\pi f_1 t \qquad (10.23)$$

$$x_2(t) = A_2 \sin 2\pi f_2 t \qquad (10.24)$$

The received signals from a moving target are

$$x_{1r}(t) = A_{r1} \sin(2\pi f_1 t - \varphi_1) \qquad (10.25)$$

and

$$x_{2r}(t) = A_{r2} \sin(2\pi f_2 t - \varphi_2) \qquad (10.26)$$

where $\varphi_1 = (4\pi f_1 R)/c$ and $\varphi_2 = (4\pi f_2 R)/c$. After heterodyning (mixing) with the carrier frequency, the phase difference between the two received signals is

$$\varphi_2 - \varphi_1 = \Delta\varphi = \frac{4\pi R}{c}(f_2 - f_1) = \frac{4\pi R}{c}\Delta f \qquad (10.27)$$

Again R is maximum when $\Delta\varphi = 2\pi$; it follows that the maximum unambiguous range is now

$$R = c/2\Delta f \qquad (10.28)$$

and since $\Delta f \ll c$, the range computed by Eq. (10.28) is much greater than that computed by Eq. (10.22).

10.2. Pulsed Radars

Pulsed radars transmit and receive a train of modulated pulses. Range is extracted from the two-way time delay between a transmitted and received pulse. Doppler measurements can be made in two ways. If accurate range measurements are available between consecutive pulses, then Doppler frequency can be extracted from the range rate $\dot{R} = \Delta R/\Delta t$. This approach works fine as long as the range is not changing drastically over the interval Δt. Otherwise, pulsed radars utilize a Doppler filter bank.

Pulsed radar waveforms can be completely defined by the following: (1) carrier frequency which may vary depending on the design requirements and radar mission; (2) pulse width, which is closely related to the bandwidth and defines the range resolution; (3) modulation; and finally (4) the pulse repetition frequency. Different modulation techniques are usually utilized to enhance the radar performance, or to add more capabilities to the radar that otherwise would not have been possible. The PRF must be chosen to avoid Doppler and range ambiguities as well as maximize the average transmitted power.

Radar systems employ low, medium, and high PRF schemes. Low PRF waveforms can provide accurate, long, unambiguous range measurements, but exert severe Doppler ambiguities. Medium PRF waveforms must resolve both range and Doppler ambiguities; however, they provide adequate average transmitted power as compared to low PRFs. High PRF waveforms can provide superior average transmitted power and excellent clutter rejection capabilities. Alternatively, high PRF waveforms are extremely ambiguous in range. Radar systems utilizing high PRFs are often called Pulsed Doppler Radars (PDR). Range and Doppler ambiguities for different PRFs are summarized in Table 10.1.

Distinction of a certain PRF as low, medium, or high PRF is almost arbitrary and depends on the radar mode of operations. For example, a $3KHz$ PRF is considered low if the maximum detection range is less than $30Km$. However, the same PRF would be considered medium if the maximum detection range is well beyond $30Km$.

Radars can utilize constant and varying (agile) PRFs. For example, Moving Target Indicator (MTI) radars use PRF agility to avoid blind speeds, as discussed in Chapter 9. This kind of agility is known as PRF staggering. PRF agility is also used to avoid range and Doppler ambiguities, as will be explained in the next three sections. Additionally, PRF agility is also used to prevent jammers from locking onto the radar's PRF. These two last forms of PRF agility are sometimes referred to as PRF jitter.

Figure 10.4 shows a simplified pulsed radar block diagram. The range gates can be implemented as filters that open and close at time intervals that correspond to the detection range. The width of such an interval corresponds to the desired range resolution. The radar receiver is often implemented as a series of contiguous (in time) range gates, where the width of each gate is achieved through pulse compression. The clutter rejection can be implemented using MTI or other forms of clutter rejection techniques. The NBF bank is normally implemented using an FFT, where bandwidth of the individual filters corresponds to the FFT frequency resolution.

TABLE 10.1. **PRF ambiguities.**

PRF	Range Ambiguous	Doppler Ambiguous
Low PRF	No	Yes
Medium PRF	Yes	Yes
High PRF	Yes	No

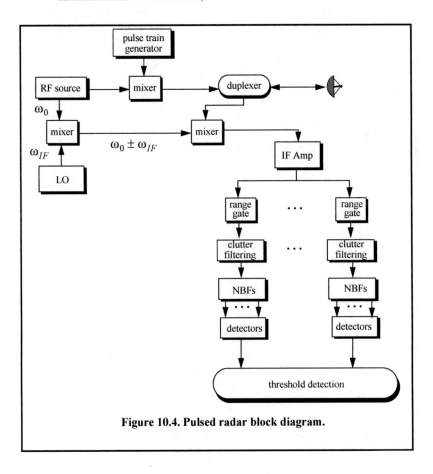

Figure 10.4. Pulsed radar block diagram.

10.2.1. Pulse Doppler Radars

In ground based radars, the amount of clutter in the radar receiver depends heavily on the radar-to-target geometry. The amount clutter is considerably higher when the radar beam has to face toward the ground. Furthermore, radars employing high PRFs have to deal with an increased amount of clutter due to folding in range. Clutter introduces additional difficulties for airborne radars when detecting ground targets and other targets flying at low altitudes. This is illustrated in Fig. 10.5. Returns from ground clutter emanate from ranges equal to the radar altitude to those which exceed the slant range along the mainbeam, with considerable clutter returns in the sidelobes and mainbeam. The presence of such large amounts of clutter interferes with radar detection capabilities and makes it extremely difficult to detect targets in the look-down mode. This difficulty in detecting ground or low altitude targets has led to the development of pulse Doppler radars where other targets, kinematics such as Doppler effects are exploited to enhance detection.

Pulse Doppler radars utilize high PRFs to increases the average transmitted power and rely on target's Doppler frequency for detection. The increase in the average transmitted power leads to an improved SNR which helps the detection process. However, using high PRFs compromise the radar's ability to detect long range target because of range ambiguities associated with high PRF applications.

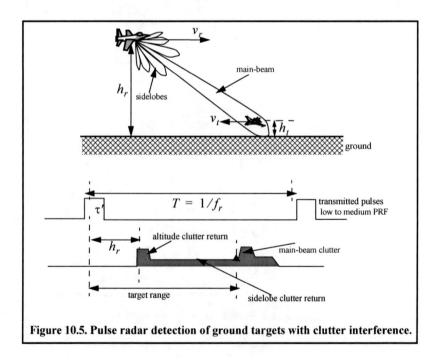

Figure 10.5. Pulse radar detection of ground targets with clutter interference.

As was explained in Chapter 9, pulse Doppler radars (or high PRF radars) have to deal with the additional increase in clutter power due to clutter folding. This has led to the development of a special class of airborne MTI filters, often referred to as AMTI. Techniques such as using specialized Doppler filters to reject clutter are very effective and are often employed by pulse Doppler radars. Pulse Doppler radars can measure target Doppler frequency (or its range rate) fairly accurately and use the fact that ground clutter typically possesses limited Doppler shift when compared with moving targets to separate the two returns. This is illustrated in Fig. 10.6. Clutter filtering (i.e., AMTI) is used to remove both main-beam and altitude clutter returns, and fast moving target detection is done effectively by exploiting its Doppler frequency. In many modern pulse Doppler radars the limiting factor in detecting slow moving targets is not clutter but rather another source of noise referred to as phase noise generated from the receiver local oscillator instabilities.

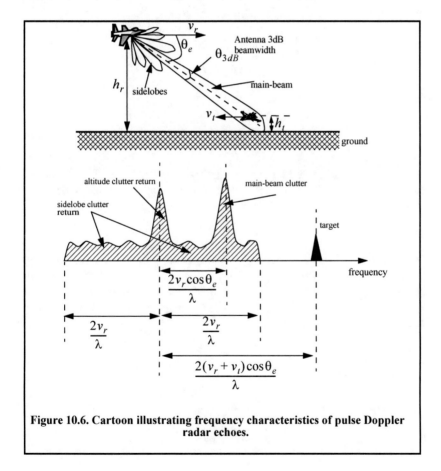

Figure 10.6. Cartoon illustrating frequency characteristics of pulse Doppler radar echoes.

10.2.2. High PRF Radar Equation

Consider a high PRF radar that uses a periodic train of very short pulses. The pulse width is τ and the period is T. This pulse train can be represented using an exponential Fourier series. The central power spectrum line (DC component) for this series contains most of the signal's power. Its value is $(\tau/T)^2$, and it is equal to the square of the transmit duty factor. Thus, the single pulse radar equation for a high PRF radar (in terms of the DC spectral power line) is

$$SNR = \frac{P_t G^2 \lambda^2 \sigma d_t^2}{(4\pi)^3 R^4 k T_o BFL d_r} \tag{10.29}$$

where, in this case, one can no longer ignore the receive duty factor since its value is comparable to the transmit duty factor. In fact, $d_r \approx d_t = \tau f_r$. Additionally, the operating radar bandwidth is now matched to the radar integration time (time on target), $B = 1/T_i$. It follows that

$$SNR = \frac{P_t \tau f_r T_i G^2 \lambda^2 \sigma}{(4\pi)^3 R^4 k T_o FL} \tag{10.30}$$

and finally,

$$SNR = \frac{P_{av} T_i G^2 \lambda^2 \sigma}{(4\pi)^3 R^4 k T_o FL} \tag{10.31}$$

where P_{av} was substituted for $P_t \tau f_r$. Note that the product $P_{av} T_i$ is a "kind of energy" product, which indicates that high PRF radars can enhance detection performance by using relatively low power and longer integration time.

Example:

Compute the single pulse SNR for a high PRF radar with the following parameters: peak power $P_t = 100KW$, antenna gain $G = 20dB$, operating frequency $f_0 = 5.6GHz$, losses $L = 8dB$, noise figure $F = 5dB$, dwell interval $T_i = 2s$, duty factor $d_t = 0.3$. The range of interest is $R = 50Km$. Assume target RCS $\sigma = 0.01m^2$.

Solution:

From Eq. (10.31) we have

$$(SNR)_{dB} = (P_{av} + G^2 + \lambda^2 + \sigma + T_i - (4\pi)^3 - R^4 - kT_o - F - L)_{dB}$$

The following table gives all parameters in dB:

P_{av}	λ^2	T_i	kT_0	$(4\pi)^3$	R^4	σ
44.771	−25.421	3.01	−23.977	32.976	187.959	−20

$$(SNR)_{dB} = 44.771 + 40 - 25.421 - 20 + 3.01 - 32.976 + 203.977$$
$$- 187.959 - 5 - 8 = 12.4dB$$

The same answer can be obtained by using the function "hprf_req.m" (see Section 10.3.2) with the following syntax:

$$hprf\_req \ (100e3, \ 2, \ 20, \ 5.6e9, \ 0.01, \ .3, \ 50e3, \ 5, \ 8)$$

10.2.3. Pulse Doppler Radar Signal Processing

The main idea behind pulse Doppler radar signal processing is to divide the footprint (the intersection of the antenna 3dB beamwidth with the ground) into resolution cells that constitute a range Doppler map, MAP. The sides of this map are range and Doppler, as illustrated in Fig. 10.7. Fine range resolution, ΔR, is accomplished in real time by utilizing range gating and pulse compression. Frequency (Doppler) resolution is obtained from the coherent processing interval.

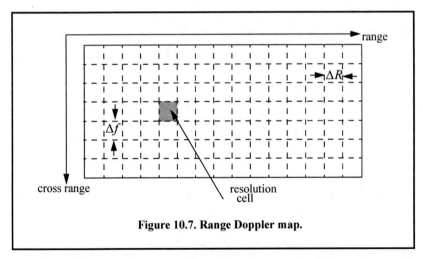

Figure 10.7. Range Doppler map.

To further illustrate this concept, consider the case where N_a is the number of azimuth (Doppler) cells, and N_r is the number of range bins. Hence, the MAP is of size $N_a \times N_r$, where the columns refer to range bins and the rows refer to azimuth cells. For each transmitted pulse within the dwell, the echoes from consecutive range bins are recorded sequentially in the first row of MAP. Once the first row is completely filled (i.e., returns from all range bins have been received), all data (in all rows) are shifted downward one row before the next pulse is transmitted. Thus, one row of MAP is generated for every transmitted pulse. Consequently, for the current observation interval, returns from the first transmitted pulse will be located in the bottom row of MAP, and returns from the last transmitted pulse will be in the top row of MAP.

Referring to Fig. 10.4, fine range resolution is achieved using the matched filter. Clutter rejection (filtering) is performed on each range bin (i.e, rows in the MAP). Then all samples from one dwell within each range bin are processed using an FFT to resolve targets in Doppler. It follows that a peak in a given resolution cell corresponds to a specific target detection at that range and Doppler frequency. Selection of the proper size FFT and its associated parameters were discussed in Chapter 2.

10.2.4. Resolving Range Ambiguities in Pulse Doppler Radars

Pulse Doppler radars exhibit serve range ambiguities because they use high PRF pulse streams. In order to resolve these ambiguities, pulse Doppler radars utilize multiple high PRFs (PRF staggering) within each processing interval (dwell). For this purpose, consider a pulse Doppler radar that uses two PRFs, f_{r1} and f_{r2}, on transmit to resolve range ambiguity, as shown in Fig. 10.8. Denote R_{u1} and R_{u2} as the unambiguous ranges for the two PRFs, respectively. Normally, these unambiguous ranges are relatively small and are short of the desired radar unambiguous range R_u (where $R_u \gg R_{u1}, R_{u2}$). Denote the radar desired PRF that corresponds to R_u as f_{rd}.

The choice of f_{r1} and f_{r2} is such that they are relatively prime with respect to one another. One choice is to select $f_{r1} = Nf_{rd}$ and $f_{r2} = (N+1)f_{rd}$ for some integer N. Within one period of the desired PRI ($T_d = 1/f_{rd}$) the two PRFs f_{r1} and f_{r2} coincide only at one location, which is the true unambiguous target position. The time delay T_d establishes the desired unambiguous range. The time delays t_1 and t_2 correspond to the time between the transmit of a pulse on each PRF and receipt of a target return due to the same pulse.

Let M_1 be the number of PRF1 intervals between transmit of a pulse and receipt of the true target return. The quantity M_2 is similar to M_1 except it is for PRF2. It follows that over the interval 0 to T_d, the only possible results are $M_1 = M_2 = M$ or $M_1 + 1 = M_2$. The radar needs only to measure t_1 and t_2. First, consider the case when $t_1 < t_2$. In this case,

$$t_1 + \frac{M}{f_{r1}} = t_2 + \frac{M}{f_{r2}} \qquad \text{(10.32)}$$

for which we get

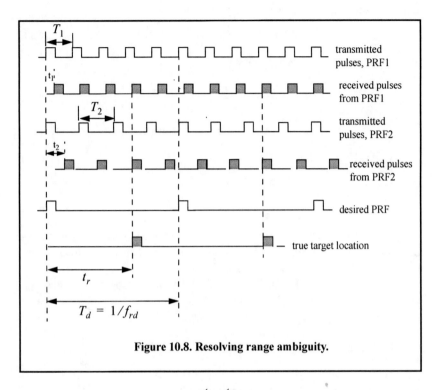

Figure 10.8. Resolving range ambiguity.

$$M = \frac{t_2 - t_1}{T_1 - T_2} \tag{10.33}$$

where $T_1 = 1/f_{r1}$ and $T_2 = 1/f_{r2}$. It follows that the round-trip time to the true target location is

$$t_r = MT_1 + t_1$$
$$t_r = MT_2 + t_2 \tag{10.34}$$

and the true target range is

$$R = ct_r/2 \tag{10.35}$$

Now, if $t_1 > t_2$, then

$$t_1 + \frac{M}{f_{r1}} = t_2 + \frac{M+1}{f_{r2}} \tag{10.36}$$

Solving for M we get

$$M = \frac{(t_2 - t_1) + T_2}{T_1 - T_2} \tag{10.37}$$

and the round-trip time to the true target location is

$$t_{r1} = MT_1 + t_1 \qquad (10.38)$$

and in this case, the true target range is

$$R = \frac{ct_{r1}}{2} \qquad (10.39)$$

Finally, if $t_1 = t_2$, then the target is in the first ambiguity. It follows that

$$t_{r2} = t_1 = t_2 \qquad (10.40)$$

and

$$R = ct_{r2}/2 \qquad (10.41)$$

Since a pulse cannot be received while the following pulse is being transmitted, these times correspond to blind ranges. This problem can be resolved by using a third PRF. In this case, once an integer N is selected, then in order to guarantee that the three PRFs are relatively prime with respect to one another, we may choose $f_{r1} = N(N+1)f_{rd}$, $f_{r2} = N(N+2)f_{rd}$, and $f_{r3} = (N+1)(N+2)f_{rd}$.

10.2.5. Resolving Doppler Ambiguity

In the case where the pulse Doppler radar is utilizing medium PRFs, it will be ambiguous in both range and Doppler. Resolving range ambiguities was discussed in the previous section. In this section Doppler ambiguity is addressed. Remember that the line spectrum of a train of pulses has $\sin x/x$ envelope (see Chapter 2), and the line spectra are separated by the PRF, f_r, as illustrated in Fig. 10.9. The Doppler filter bank is capable of resolving target Doppler as long as the anticipated Doppler shift is less than one half the bandwidth of the individual filters (i.e., one half the width of an FFT bin). Thus, pulsed radars are designed such that

$$f_r = 2f_{dmax} = (2v_{rmax})/\lambda \qquad (10.42)$$

where f_{dmax} is the maximum anticipated target Doppler frequency, v_{rmax} is the maximum anticipated target radial velocity, and λ is the radar wavelength.

If the Doppler frequency of the target is high enough to make an adjacent spectral line move inside the Doppler band of interest, the radar can be Doppler ambiguous. Therefore, in order to avoid Doppler ambiguities, radar systems require high PRF rates when detecting high speed targets. When a long-range radar is required to detect a high speed target, it may not be possible to be both range and Doppler unambiguous. This problem can be resolved by using multiple PRFs. Multiple PRF schemes can be incorporated sequentially within each dwell interval (scan or integration frame) or the radar can use a single PRF in one scan and resolve ambiguity in the next. The latter technique, however, may have problems due to changing target dynamics from one scan to the next.

The Doppler ambiguity problem is analogous to that of range ambiguity. Therefore, the same methodology can be used to resolve Doppler ambiguity. In this case, we measure the Doppler frequencies f_{d1} and f_{d2} instead of t_1 and t_2.

If $f_{d1} > f_{d2}$, then we have

$$M = \frac{(f_{d2} - f_{d1}) + f_{r2}}{f_{r1} - f_{r2}} \qquad (10.43)$$

And if $f_{d1} < f_{d2}$,

$$M = \frac{f_{d2} - f_{d1}}{f_{r1} - f_{r2}} \qquad (10.44)$$

and the true Doppler is

$$f_d = M f_{r1} + f_{d1} \qquad ; f_d = M f_{r2} + f_{d2} \qquad (10.45)$$

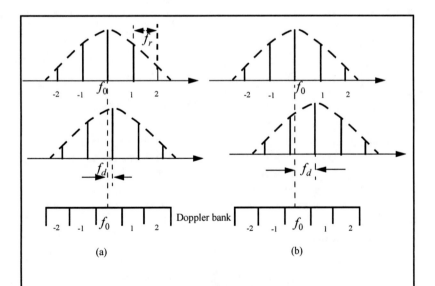

Figure 10.9. Spectra of transmitted and received waveforms, and Doppler bank. (a) Doppler is resolved. (b) Spectral lines have moved into the next Doppler filter. This results in an ambiguous Doppler measurement.

Finally, if $f_{d1} = f_{d2}$, then

$$f_d = f_{d1} = f_{d2} \tag{10.46}$$

Again, blind Dopplers can occur, which can be resolved using a third PRF.

Example:

A certain radar uses two PRFs to resolve range ambiguities. The desired unambiguous range is $R_u = 100Km$. Choose $N = 59$. Compute f_{r1}, f_{r2}, R_{u1}, and R_{u2}.

Solution:

First let us compute the desired PRF, f_{rd}

$$f_{rd} = \frac{c}{2R_u} = \frac{3 \times 10^8}{200 \times 10^3} = 1.5 KHz$$

It follows that

$$f_{r1} = N f_{rd} = (59)(1500) = 88.5 KHz$$

$$f_{r2} = (N+1)f_{rd} = (59+1)(1500) = 90 KHz$$

$$R_{u1} = \frac{c}{2f_{r1}} = \frac{3 \times 10^8}{2 \times 88.5 \times 10^3} = 1.695 Km$$

$$R_{u2} = \frac{c}{2f_{r2}} = \frac{3 \times 10^8}{2 \times 90 \times 10^3} = 1.667 Km.$$

Example:

Consider a radar with three PRFs; $f_{r1} = 15 KHz$, $f_{r2} = 18 KHz$, and $f_{r3} = 21 KHz$. Assume $f_0 = 9 GHz$. Calculate the frequency position of each PRF for a target whose velocity is $550 m/s$. Calculate f_d (Doppler frequency) for another target appearing at $8 KHz$, $2 KHz$, and $17 KHz$ for each PRF.

Solution:

The Doppler frequency is

$$f_d = 2\frac{v f_0}{c} = \frac{2 \times 550 \times 9 \times 10^9}{3 \times 10^8} = 33 KHz$$

Then by using Eq. (10.42) $n_i f_{ri} + f_{di} = f_d$ where $i = 1, 2, 3$, we can write

$$n_1 f_{r1} + f_{d1} = 15 n_1 + f_{d1} = 33$$

$$n_2 f_{r2} + f_{d2} = 18n_2 + f_{d2} = 33$$

$$n_3 f_{r3} + f_{d3} = 21n_3 + f_{d3} = 33$$

We will show here how to compute n_1, and leave the computations of n_2 and n_3 to the reader. First, if we choose $n_1 = 0$, that means $f_{d1} = 33\,KHz$, which cannot be true since f_{d1} cannot be greater than f_{r1}. Choosing $n_1 = 1$ is also invalid since $f_{d1} = 18KHz$ cannot be true either. Finally, if we choose $n_1 = 2$ we get $f_{d1} = 3KHz$, which is an acceptable value. It follows that the minimum n_1, n_2, n_3 that may satisfy the above three relations are $n_1 = 2$, $n_2 = 1$, and $n_3 = 1$. Thus, the apparent Doppler frequencies are $f_{d1} = 3KHz$, $f_{d2} = 15KHz$, and $f_{d3} = 12KHz$, as seen below.

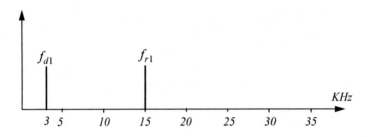

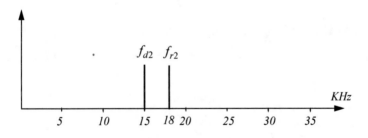

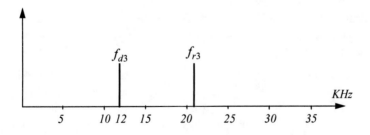

Now for the second part of the problem. Again by using Eq. (10.61) we have

$$n_1 f_{r1} + f_{d1} = f_d = 15n_1 + 8$$

$$n_2 f_{r2} + f_{d2} = f_d = 18n_2 + 2$$

$$n_3 f_{r3} + f_{d3} = f_d = 21n_3 + 17$$

We can now solve for the smallest integers n_1, n_2, n_3 that satisfy the above three relations. See the table below.

n	0	1	2	3	4
f_d from f_{r1}	8	23	<u>38</u>	53	68
f_d from f_{r2}	2	20	<u>38</u>	56	
f_d from f_{r3}	17	<u>38</u>	39		

Thus, $n_1 = 2 = n_2$, and $n_3 = 1$, and the true target Doppler is $f_d = 38KHz$. It follows that

$$v_r = 38000 \times \frac{0.0333}{2} = 632.7 \frac{m}{\sec}$$

10.3. MATLAB Programs and Routines

10.3.1. MATLAB Program "range_calc.m"

The program *"range_calc.m"* solves the radar range equation of the form

$$R = \left(\frac{P_t \tau f_r T_i G_t G_r \lambda^2 \sigma}{(4\pi)^3 k T_0 F L (SNR)_o} \right)^{\frac{1}{4}} \tag{10.47}$$

where P_t is peak transmitted power, τ is pulse width, f_r is PRF, G_t and G_r are respectively the transmitting and receiving antenna gain, λ is wavelength, σ is target cross section, k is Boltzman's constant, T_0 is 290 kelvin, F is system noise figure, L is total system losses, and $(SNR)_o$ is the minimum SNR required for detection.

One can choose either CW or pulsed radars. In the case of CW radars, the terms $P_t \tau f_r$ is replaced within the code by the average CW power P_{CW}. Additionally, the term T_i refers to the dwell interval. Alternatively, in the case of pulse radars T_i denotes the time on target. The plot inside Fig. 10.10 shows an example of the SNR versus the detection range for a pulse radar using the parameters shown in the figure. A MATLAB-based Graphical User Interface

(GUI) (see Fig. 10.10) is utilized in inputting and editing all input parameters. The outputs include the maximum detection range versus minimum SNR plots. The following MATLAB function is used by this GUI to generate the desired outputs.

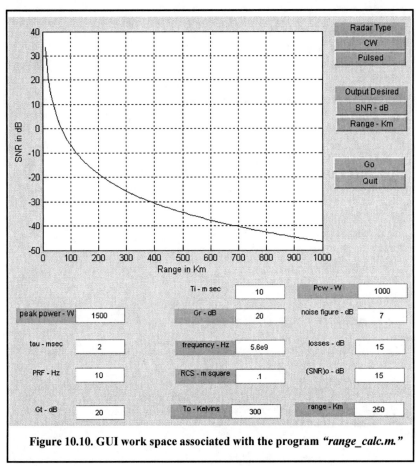

Figure 10.10. GUI work space associated with the program *"range_calc.m."*

```
function [output_par] = range_calc (pt, tau, fr, time_ti, gt, gr, freq, ...
    sigma, te, nf, loss, snro, pcw, range, radar_type, out_option)
c = 3.0e+8;
lambda = c / freq;
if (radar_type == 0)
    pav = pcw;
else
    % Compute the duty cycle
    dt = tau * 0.001 * fr;
    pav = pt * dt;
```

```
end
pav_db = 10.0 * log10(pav);
  lambda_sqdb = 10.0 * log10(lambda^2);
  sigmadb = 10.0 * log10(sigma);
  for_pi_cub = 10.0 * log10((4.0 * pi)^3);
  k_db = 10.0 * log10(1.38e-23);
  te_db = 10.0 * log10(te);
  ti_db = 10.0 * log10(time_ti);
  range_db = 10.0 * log10(range * 1000.0);
if (out_option == 0)
  %compute SNR
  snr_out = pav_db + gt + gr + lambda_sqdb + sigmadb + ti_db - ...
  for_pi_cub - k_db - te_db - nf - loss - 4.0 * range_db
  index = 0;
  for range_var = 10:10:1000
    index = index + 1;
    rangevar_db = 10.0 * log10(range_var * 1000.0);
    snr(index) = pav_db + gt + gr + lambda_sqdb + sigmadb + ti_db - ...
      for_pi_cub - k_db - te_db - nf - loss - 4.0 * rangevar_db;
  end
  var = 10:10:1000;
  plot(var,snr,'k')
  xlabel ('Range in Km');
  ylabel ('SNR in dB');
  grid
else
  range4 = pav_db + gt + gr + lambda_sqdb + sigmadb + ti_db - ...
    for_pi_cub - k_db - te_db - nf - loss - snro;
  range = 10.0^(range4/40.) / 1000.0
  index = 0;
  for snr_var = -20:1:60
    index = index + 1;
    rangedb = pav_db + gt + gr + lambda_sqdb + sigmadb + ti_db - ...
      for_pi_cub - k_db - te_db - nf - loss - snr_var;
    range(index) = 10.0^(rangedb/40.) / 1000.0;
  end
  var = -20:1:60;
  plot(var,range,'k')
  xlabel ('Minimum SNR required for detection in dB');
  ylabel ('Maximum detection range in Km');
  grid
end
return
```

10.3.2. MATLAB Function "hprf_req.m"

The function "*hprf_req.m*" implements the high PRF radar equation. Its syntax is as follows:

$$[snr] = hprf\_req (pt, Ti, g, freq, sigma, dt, range, nf, loss)$$

where

Symbol	Description	Units	Status
pt	peak power	W	input
Ti	time on target	seconds	input
g	antenna gain	dB	input
freq	frequency	Hz	input
sigma	target RCS	m^2	input
dt	duty cycle	none	input
range	target range (can be a single value or a vector)	m	input
nf	noise figure	dB	input
loss	radar losses	dB	input
snr	SNR (can be a single value or a vector)	dB	output

MATLAB Function "*hprf_req.m*" **Listing**

```
function [snr] = hprf_req (pt, Ti, g, freq, sigma, dt, range, nf, loss)
% This program implements Eq. (10.31)
c = 3.0e+8; % speed of light
lambda = c / freq; % wavelength
pav = 10*log10(pt*dt); % compute average power in dB
Ti_db = 10*log10(Ti); % time on target in dB
lambda_sqdb = 10*log10(lambda^2); % compute wavelength square in dB
sigmadb = 10*log10(sigma); % convert sigma to dB
four_pi_cub = 10*log10((4.0 * pi)^3); % (4pi)^3 in dB
k_db = 10*log10(1.38e-23); % Boltzman's constant in dB
to_db = 10*log10(290); % noise temp. in dB
range_pwr4_db = 10*log10(range.^4); % vector of target range^4 in dB
% Implement Equation (1.72)
num = pav + Ti_db + g + g + lambda_sqdb + sigmadb;
den = four_pi_cub + k_db + to_db + nf + loss + range_pwr4_db;
snr = num - den;
return
```

Problems

10.1. In a multiple frequency CW radar, the transmitted waveform consists of two continuous sinewaves of frequencies $f_1 = 105KHz$ and $f_2 = 115KHz$. Compute the maximum unambiguous detection range.

10.2. Consider a radar system using linear frequency modulation. Compute the range that corresponds to $\dot{f} = 20, 10MHz$. Assume a beat frequency $f_b = 1200Hz$.

10.3. A certain radar using linear frequency modulation has a modulation frequency $f_m = 300Hz$ and frequency sweep $\Delta f = 50MHz$. Calculate the average beat frequency differences that correspond to range increments of 10 and 15 meters.

10.4. A CW radar uses linear frequency modulation to determine both range and range rate. The radar wavelength is $\lambda = 3cm$, and the frequency sweep is $\Delta f = 200KHz$. Let $t_0 = 20ms$. (a) Calculate the mean Doppler shift; (b) compute f_{bu} and f_{bd} corresponding to a target at range $R = 350Km$, which is approaching the radar with radial velocity of $250m/s$.

10.5. Consider a medium PRF radar on board an aircraft moving at a speed of 350 m/s with PRFs $f_{r1} = 10KHz$, $f_{r2} = 15KHz$, and $f_{r3} = 20KHz$; the radar operating frequency is $9.5GHz$. Calculate the frequency position of a nose-on target with a speed of 300 m/s. Also calculate the closing rate of a target appearing at 6, 5, and $18KHz$ away from the center line of PRF 10, 15, and $20KHz$, respectively.

10.6. A certain radar operates at two PRFs, f_{r1} and f_{r2}, where $T_{r1} = (1/f_{r1}) = T/5$ and $T_{r2} = (1/f_{r2}) = T/6$. Show that this multiple PRF scheme will give the same range ambiguity as that of a single PRF with PRI T.

10.7. Consider an X-band radar with wavelength $\lambda = 3cm$ and bandwidth $B = 10MHz$. The radar uses two PRFs, $f_{r1} = 50KHz$ and $f_{r2} = 55.55KHz$. A target is detected at range bin 46 for f_{r1} and at bin 12 for f_{r2}. Determine the actual target range.

10.8. A certain radar uses two PRFs to resolve range ambiguities. The desired unambiguous range is $R_u = 150Km$. Select a reasonable value for N. Compute the corresponding f_{r1}, f_{r2}, R_{u1}, and R_{u2}.

10.9. A certain radar uses three PRFs to resolve range ambiguities. The desired unambiguous range is $R_u = 250Km$. Select $N = 43$. Compute the corresponding $f_{r1}, f_{r2}, f_{r3}, R_{u1}, R_{u2}$, and R_{u3}.

10.10.   In Chapter 1 we developed an expression for the Doppler shift associated with a CW radar (i.e., $f_d = \pm 2v/\lambda$, where the plus sign is used for closing targets and the negative sign is used for receding targets). CW radars can use the system shown below to determine whether the target is closing or receding. Assuming that the emitted signal is $A\cos\omega_0 t$ and the received signal is $kA\cos((\omega_0 \pm \omega_d)t + \varphi)$, show that the direction of the target can be determined by checking the phase shift difference in the outputs $y_1(t)$ and $y_2(t)$.

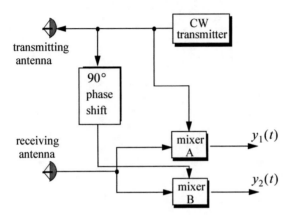

Chapter 11   *Adaptive Array*
                Processing

11.1. Introduction

The emphasis in this chapter is on adaptive array processing. For this purpose, a top level overview of phased array antennas is first introduced. Phased array antennas are capable of forming multiple beams at the transmitting or receiving modes. Beamforming can be carried out at the Radio frequency (RF), Intermediate Frequency (IF), base band, or digital levels. RF beamforming is the simplest and most common technique. In this case, multiple narrow beams are formed through the use of phase shifters. IF and base band beamforming require complex coherent hardware. However, the system is operated at lower frequencies where tolerance is not as critical. Digital beamforming is more flexible than RF, IF, or base band techniques, but it requires a demanding level of processing hardware.

Adaptive arrays mostly employ phased arrays to automatically sense and eliminate unwanted signals entering the radar's Field of View (FOV) while enhancing reception about the desired target returns. For this purpose, adaptive arrays utilize a rather complicated combination of hardware and require demanding levels of software implementation. Through feedback networks, a proper set of complex weights is computed and applied to each channel of the array. A successful implementation of adaptive arrays depends heavily on two factors: first, a proper choice of the reference signal, which is used for comparison against the received target/jammer returns. A good estimate of the reference signal makes the computation of the weights systematic and effective. On the other hand, a bad estimate of the reference signal increases the array's adapting time and limits the system to impractical (non-real time) situations. Second, a fast (real time) computation of the optimum weights is essential. There have been many algorithms developed for this purpose. Nevertheless, they all share a common problem, that is, the computation of the inverse of a complex matrix. This drawback has limited the implementation of adaptive arrays to experimental systems or small arrays.

11.2. General Arrays

An array is a composite antenna formed from two or more basic radiators. Each radiator is denoted as an element. The elements forming an array could be dipoles, dish reflectors, slots in a wave guide, or any other type of radiator. Array antennas synthesize narrow directive beams that may be steered, mechanically or electronically, in many directions. Electronic steering is achieved by controlling the phase of the current feeding the array elements. Arrays with electronic beam steering capability are called phased arrays. Phased array antennas, when compared with other simple antennas such as dish reflectors, are costly and complicated to design. However, the inherent flexibility of phased array antennas to steer the beam electronically and also the need for specialized multifunction radar systems have made phased array antennas attractive for radar applications.

Figure 11.1 shows the geometrical fundamentals associated with this problem. Consider the radiation source located at (x_1, y_1, z_1) with respect to a phase reference at $(0, 0, 0)$. The electric field measured at far field point P is

$$E(\theta, \phi) = I_0 \frac{e^{-jkR_1}}{R_1} f(\theta, \phi) \tag{11.1}$$

where I_0 is the complex amplitude, $k = 2\pi/\lambda$ is the wave number, and $f(\theta, \phi)$ is the radiation pattern.

Now, consider the case where the radiation source is an array made of many elements, as shown in Fig. 11.2. The coordinates of each radiator with respect to the phase reference are (x_i, y_i, z_i), and the vector from the origin to the *ith* element is given by

$$\vec{r}_i = \hat{a}_x x_i + \hat{a}_y y_i + \hat{a}_z z_i \tag{11.2}$$

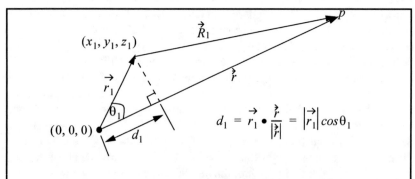

Figure 11.1. Geometry for an array antenna. Single element.

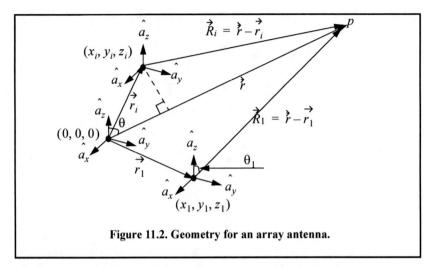

Figure 11.2. Geometry for an array antenna.

The far field components that constitute the total electric field are

$$E_i(\theta, \phi) = I_i \frac{e^{-jkR_i}}{R_i} f(\theta_i, \phi_i)$$ (11.3)

where

$$R_i = |\vec{R_i}| = |\vec{r} - \vec{r_i}| = \sqrt{(x - x_i)^2 + (y - y_i)^2 + (z - z_i)^2}$$

$$= r\sqrt{1 + (x_i^2 + y_i^2 + z_i^2)/r^2 - 2(xx_i + yy_i + zz_i)/r^2}$$ (11.4)

Using spherical coordinates, where $x = r\sin\theta\cos\varphi$, $y = r\sin\theta\sin\varphi$, and $z = r\cos\theta$, yields

$$\frac{(x_i^2 + y_i^2 + z_i^2)}{r^2} = \frac{|\vec{r_i}|^2}{r^2} \ll 1$$ (11.5)

Thus, a good approximation (using binomial expansion) for Eq. (11.4) is

$$R_i = r - r(x_i\sin\theta\cos\phi + y_i\sin\theta\sin\phi + z_i\cos\theta)$$ (11.6)

It follows that the phase contribution at the far field point from the *ith* radiator with respect to the phase reference is

$$e^{-jkR_i} = e^{-jkr} e^{jk(x_i\sin\theta\cos\phi + y_i\sin\theta\sin\phi + z_i\cos\theta)}$$ (11.7)

Remember, however, that the unit vector $\vec{r_0}$ along the vector $\vec{r}$ is

$$\vec{r}_0 = \frac{\vec{r}}{|\vec{r}|} = \hat{a}_x sin\theta cos\phi + \hat{a}_y sin\theta sin\phi + \hat{a}_z cos\theta \tag{11.8}$$

Hence, we can rewrite Eq. (11.7) as

$$e^{-jkR_i} = e^{-jkr} e^{jk(\hat{r}_i \bullet \vec{r}_0)} = e^{-jkr} e^{j\Psi_i(\theta, \phi)} \tag{11.9}$$

Finally, by virtue of superposition, the total electric field is

$$E(\theta, \phi) = \sum_{i=1}^{N} I_i e^{j\Psi_i(\theta, \phi)} \tag{11.10}$$

which is known as the array factor for an array antenna where the complex current for the *ith* element is I_i.

In general, an array can be fully characterized by its array factor. This is true since knowing the array factor provides the designer with knowledge of the array's (1) 3-dB beamwidth, (2) null-to-null beamwidth, (3) distance from the main peak to the first side-lobe, (4) height of the first side-lobe as compared to the main beam, (5) location of the nulls, (6) rate of decrease of the side-lobes, and (7) grating lobes' locations.

11.3. Linear Arrays

Figure 11.3 shows a linear array antenna consisting of N identical elements. The element spacing is d (normally measured in wavelength units). Let element #1 serve as a phase reference for the array. From the geometry, it is clear that an outgoing wave at the *nth* element leads the phase at the $(n+1)th$ element by $kdsin\theta$, where $k = 2\pi/\lambda$. The combined phase at the far field observation point P is independent of ϕ and can be written as

$$\Psi(\theta, \phi) = k(\vec{r}_n \bullet \vec{r}_0) = (n-1)kdsin\theta \tag{11.11}$$

Thus, from Eq. (11.10), the electric field at a far field observation point with direction-sine equal to $sin\theta$ (assuming isotropic elements) is

$$E(sin\theta) = \sum_{n=1}^{N} e^{j(n-1)(kdsin\theta)} \tag{11.12}$$

Expanding the summation in Eq. (11.12) yields

$$E(sin\theta) = 1 + e^{jkdsin\theta} + ... + e^{j(N-1)(kdsin\theta)} \tag{11.13}$$

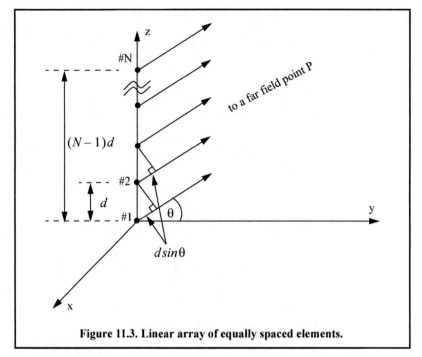

Figure 11.3. Linear array of equally spaced elements.

The right-hand side of Eq. (11.13) is a geometric series, which can be expressed in the form

$$1 + a + a^2 + a^3 + \ldots + a^{(N-1)} = \frac{1-a^N}{1-a} \tag{11.14}$$

Replacing a by $e^{jkd\sin\theta}$ yields

$$E(\sin\theta) = \frac{1 - e^{jNkd\sin\theta}}{1 - e^{jkd\sin\theta}} = \frac{1 - (\cos Nkd\sin\theta) - j(\sin Nkd\sin\theta)}{1 - (\cos kd\sin\theta) - j(\sin kd\sin\theta)} \tag{11.15}$$

The far field array intensity pattern is then given by

$$|E(\sin\theta)| = \sqrt{E(\sin\theta)E^*(\sin\theta)} \tag{11.16}$$

Substituting Eq. (11.15) into Eq. (11.16) and collecting terms yield

$$|E(\sin\theta)| = \sqrt{\frac{(1 - \cos Nkd\sin\theta)^2 + (\sin Nkd\sin\theta)^2}{(1 - \cos kd\sin\theta)^2 + (\sin kd\sin\theta)^2}} \tag{11.17}$$

which can be written as

$$|E(sin\theta)| = \sqrt{\frac{1 - cos\,Nkd\,sin\theta}{1 - cos\,kd\,sin\theta}} \qquad (11.18)$$

and using the trigonometric identity $1 - cos\theta = 2(sin\theta/2)^2$ yields

$$|E(sin\theta)| = \left|\frac{sin(Nkd\,sin\theta/2)}{sin(kd\,sin\theta/2)}\right| \qquad (11.19)$$

which is a periodic function of $kd\,sin\theta$, with a period equal to 2π.

The maximum value of $|E(sin\theta)|$, which occurs at $\theta = 0$, is equal to N. It follows that the normalized intensity pattern is equal to

$$|E_n(sin\theta)| = \frac{1}{N}\left|\frac{sin((Nkd\,sin\theta)/2)}{sin((kd\,sin\theta)/2)}\right| \qquad (11.20)$$

The normalized two-way array pattern (radiation pattern) is given by

$$G(sin\theta) = |E_n(sin\theta)|^2 = \frac{1}{N^2}\left(\frac{sin((Nkd\,sin\theta)/2)}{sin((kd\,sin\theta)/2)}\right)^2 \qquad (11.21)$$

Figure 11.4 shows a plot of Eq. (11.21) versus $sin\theta$ for $N = 8$. This plot can be reproduced using the following MATLAB code.

```
% Use this code to produce figure 11.4a and 11.4b
clear all; close all;
eps = 0.00001;
k = 2*pi;
theta = -pi : pi / 10791 : pi;
var = sin(theta);
nelements = 8;
d = 1;      % d = 1;
num = sin((nelements * k * d * 0.5) .* var);
if(abs(num) <= eps)
   num = eps;
end
den = sin((k* d * 0.5) .* var);
if(abs(den) <= eps)
   den = eps;
end
pattern = abs(num ./ den);
maxval = max(pattern);
pattern = pattern ./ maxval;
figure(1)
plot(var,pattern)
xlabel('sine angle - dimensionless')
ylabel('Array pattern')
grid
```

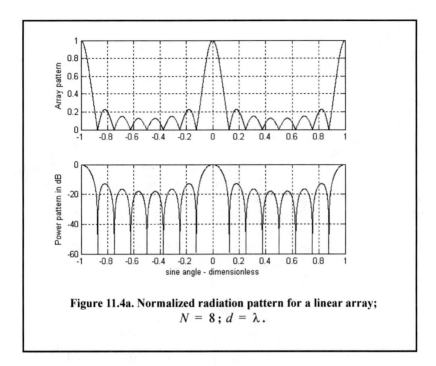

Figure 11.4a. Normalized radiation pattern for a linear array;
$$N = 8 \,; d = \lambda.$$

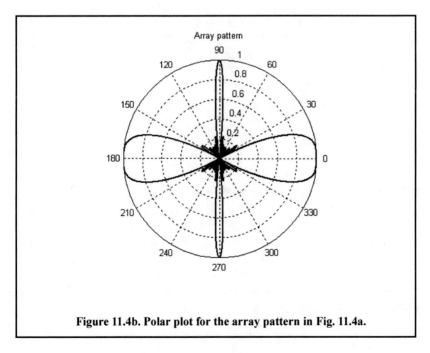

Figure 11.4b. Polar plot for the array pattern in Fig. 11.4a.

figure(2)
plot(var,20*log10(pattern))
axis ([-1 1 -60 0])
xlabel('sine angle - dimensionless')
ylabel('Power pattern in dB')
grid;
figure(3)
theta = theta +pi/2;
polar(theta,pattern)
title ('Array pattern')

   The radiation pattern $G(sin\theta)$ has cylindrical symmetry about its axis ($sin\theta = 0$) and is independent of the azimuth angle. Thus, it is completely determined by its values within the interval $(0 < \theta < \pi)$. The main beam of an array can be steered electronically by varying the phase of the current applied to each array element. Steering the main beam into the direction-sine $sin\theta_0$ is accomplished by making the phase difference between any two adjacent elements equal to $kdsin\theta_0$. In this case, the normalized radiation pattern can be written as

$$G(sin\theta) = \frac{1}{N^2} \left(\frac{sin[(Nkd/2)(sin\theta - sin\theta_0)]}{sin[(kd/2)(sin\theta - sin\theta_0)]} \right)^2 \tag{11.22}$$

If $\theta_0 = 0$, then the main beam is perpendicular to the array axis, and the array is said to be a broadside array. Alternatively, the array is called an endfire array when the main beam points along the array axis. The radiation pattern maxima are computed using L'Hopital's rule when both the denominator and numerator of Eq. (11.22) are zeros. More precisely,

$$\left(\frac{kdsin\theta}{2} = \pm m\pi \right); \quad m = 0, 1, 2, ... \tag{11.23}$$

Solving for θ yields

$$\theta_m = asin\left(\pm \frac{\lambda m}{d} \right); \quad m = 0, 1, 2, ... \tag{11.24}$$

where the subscript m is used as a maxima indicator. The first maximum occurs at $\theta_0 = 0$ and is denoted as the main beam (lobe). Other maxima occurring at $|m| \geq 1$ are called grating lobes. Grating lobes are undesirable and must be suppressed. The grating lobes occur at non-real angles when the absolute value of the arc-sine argument in Eq. (11.24) is greater than unity; it follows that $d < \lambda$. Under this condition, the main lobe is assumed to be at $\theta = 0$ (broadside array). Alternatively, when electronic beam steering is considered, the grating lobes occur at

$$\left|\sin\theta - \sin\theta_0\right| = \pm\frac{\lambda n}{d}; \quad n = 1, 2, \dots \tag{11.25}$$

Thus, in order to prevent the grating lobes from occurring between $\pm 90°$, the element spacing should be $d < \lambda/2$.

The radiation pattern attains secondary maxima (side-lobes) when the numerator of Eq. (11.24) is maximum, or equivalently

$$\frac{Nkd\sin\theta}{2} = \pm(2l+1)\frac{\pi}{2}; \quad l = 1, 2, \dots \tag{11.26}$$

Solving for θ yields

$$\theta_l = a\sin\left(\pm\frac{\lambda}{2d}\frac{2l+1}{N}\right); \quad l = 1, 2, \dots \tag{11.27}$$

where the subscript l is used as an indication of side-lobe maxima. The nulls of the radiation pattern occur when only the numerator of Eq. (11.24) is zero. More precisely,

$$\frac{N}{2}kd\sin\theta = \pm n\pi; \quad \begin{matrix} n = 1, 2, \dots \\ n \neq N, 2N, \dots \end{matrix} \tag{11.28}$$

Again solving for θ yields

$$\theta_n = a\sin\left(\pm\frac{\lambda n}{dN}\right); \quad \begin{matrix} n = 1, 2, \dots \\ n \neq N, 2N, \dots \end{matrix} \tag{11.29}$$

where the subscript n is used as a null indicator. Define the angle that corresponds to the half power point as θ_h. It follows that the half power (3-dB) beamwidth is $2\left|\theta_m - \theta_h\right|$. This occurs when

$$\frac{N}{2}kd\sin\theta_h = 1.391 \ radians \Rightarrow \theta_h = a\sin\left(\frac{\lambda}{2\pi d}\frac{2.782}{N}\right) \tag{11.30}$$

In order to reduce the side-lobe levels, the array must be designed to radiate more power toward the center and much less at the edges. This can be achieved through tapering (windowing) the current distribution over the face of the array. There are many possible tapering sequences that can be used for this purpose. However, as known from spectral analysis, windowing reduces side-lobe levels at the expense of widening the main beam. Thus, for a given radar application, the choice of the tapering sequence must be based on the trade-off between side-lobe reduction and main-beam widening.

Figures 11.5 through Fig. 11.13 show plots of the array gain pattern versus steering angle for a few. These plots can be reproduced using the following MATLAB code

```
% produce figures 11.5 through 11.13
clear all; close all; clc
win = hamming(19);
[theta,patternr,patterng] = linear_array(19, 0.5, 0, -1, -1, -3);
figure(5)
plot(theta, patterng,'linewidth',1.5)
xlabel('Steering angle in degrees'); ylabel('Antenna gain pattern in dB')
title('N = 19; d = 0.5\lambda; \theta = 0 degrees; Perfect phase shifters')
grid on; axis tight
[theta, patternr, patterng] = linear_array(19, 0.5, 0, 1, win, -3);
figure(6)
plot(theta, patterng,'linewidth',1.5)
xlabel('Steering angle - degrees')
ylabel('Antenna gain pattern - dB')
title('N = 19; d = 0.5\lambda; \theta = 0 degrees; Perfect phase shifters; Hamming window')
grid on; axis tight
[theta, patternr, patterng] = linear_array(19, 0.5, -15, -1, -1, 3);
figure(7)
plot(theta, patterng,'linewidth',1.5)
xlabel('Steering angle in degrees'); ylabel('Antenna gain pattern in dB')
title('N = 19; d = 0.5\lambda; \theta = -15 degrees; 3-bits phase shifters')
grid on; axis tight
[theta, patternr, patterng] = linear_array(19, 0.5, 5, 1, win, 3);
figure(8)
plot(theta, patterng,'linewidth',1.5)
xlabel('Steering angle - degrees')
ylabel('Antenna gain pattern - dB')
title('N = 19; d = 0.5\lambda; \theta = 5 degrees; 3-bits phase shifters; Hamming window')
grid on; axis tight
[theta, patternr, patterng] = linear_array(19, 0.5, 25, 1, win, 3);
figure(9)
plot(theta, patterng,'linewidth',1.5)
xlabel('Steering angle in degrees')
ylabel('Antenna gain pattern - dB')
title('N = 19; d = 0.5\lambda; \theta = 25 degrees; 3-bits phase shifters; Hamming window')
grid on; axis tight
[theta, patternr, patterng] = linear_array(19, 1.5, 48, -1, -1, -3);
figure(10)
plot(theta, patterng,'linewidth',1.5)
xlabel('Steering angle in degrees'); ylabel('Antenna gain pattern in dB')
title('N = 19; d = 1.5\lambda; \theta = 48 degrees; Perfect phase shifters')
```

grid on; axis tight
[theta, patternr, patterng] = linear_array(19, 1.5, 48, 1, win, -3);
figure(11)
plot(theta, patterng,'linewidth',1.5)
xlabel('Steering angle in degrees'); ylabel('Antenna gain pattern in dB')
title('N = 19; d = 1.5\lambda; \theta = 48 degrees; Perfect phase shifters; Hamming
window')
grid on; axis tight
[theta, patternr, patterng] = linear_array(19, 1.5, -53, -1, -1, 3);
figure(12)
plot(theta, patterng,'linewidth',1.5)
xlabel('Steering angle in degrees'); ylabel('Antenna gain pattern in dB')
title('N = 19; d = 1.5\lambda; \theta = -53 degrees; 3-bits phase shifters')
grid on; axis tight
[theta, patternr, patterng] = linear_array(19, 1.5, -33, 1, win, 3);
figure(13)
plot(theta, patterng,'linewidth',1.5)
xlabel('Steering angle in degrees')
ylabel('Antenna gain pattern - dB')
title('N = 19; d = 1.5\lambda; \theta = -33 degrees; 3-bits phase shifters; ...
Hamming window')
grid on;
axis tight

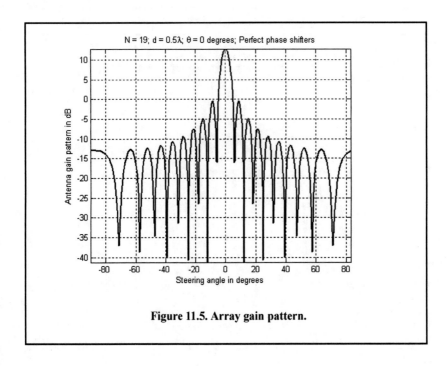

Figure 11.5. Array gain pattern.

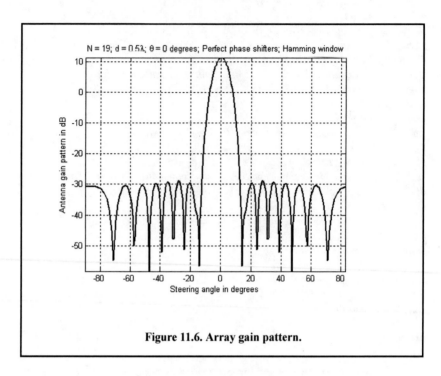

Figure 11.6. Array gain pattern.

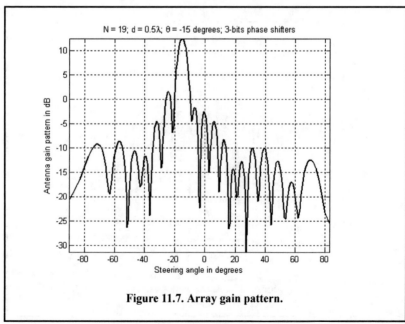

Figure 11.7. Array gain pattern.

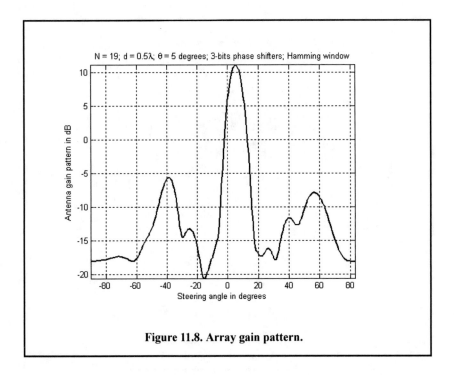

Figure 11.8. Array gain pattern.

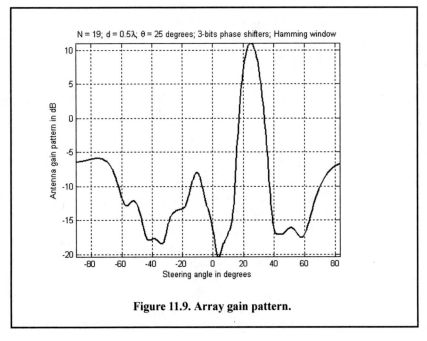

Figure 11.9. Array gain pattern.

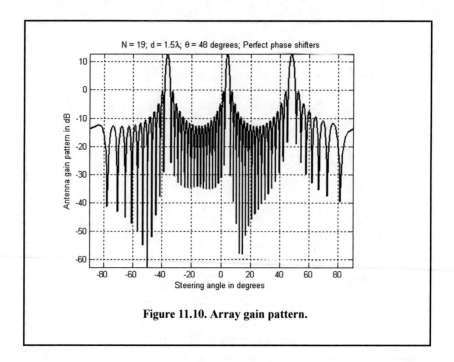

Figure 11.10. Array gain pattern.

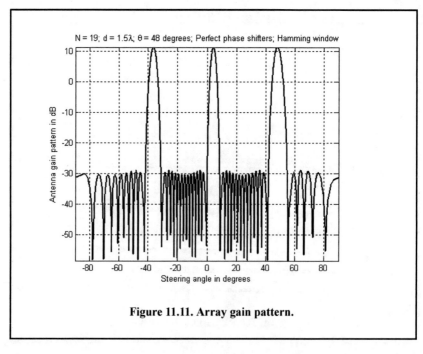

Figure 11.11. Array gain pattern.

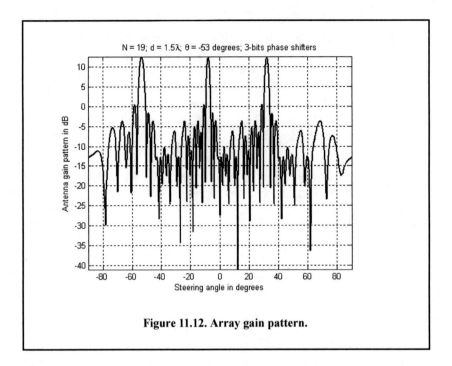

Figure 11.12. Array gain pattern.

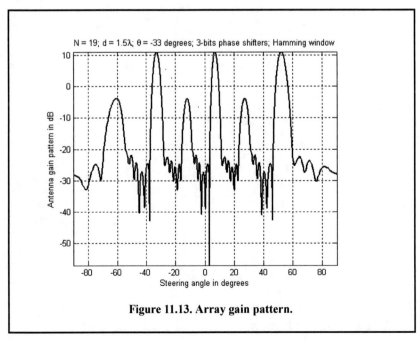

Figure 11.13. Array gain pattern.

11.4. Nonadaptive Beamforming

In adaptive beamforming the beam of interest is formed (generated) by continuously changing a set of weights through feedback circuits to minimize an output error signal. Nonadaptive or conventional beamformers do the same thing in the sense that the beam of interest is generated using a set of unique weights. Except in this case, these weights are determined a priori so that interference from a specific angle of arrival is minimized or eliminated. Different sets of weights will produce nulls in different directions in the array's field of view.

Consider a linear array of N equally spaced elements, and a plane wave $exp(j2\pi f_0 t))$ incident on the aperture with direction-sine $sin\theta$, as shown in Fig. 11.14. The weights w_i, $i = 0, 1, ...N-1$ are, in general, complex constants. The output of the beamformer is

$$y(t) = \sum_{n=0}^{N-1} w_n x_n(t - \tau_n) \tag{11.31}$$

$$\tau_n = n\frac{d}{c}sin\theta; \quad n = 0, 1, ..., (N-1) \tag{11.32}$$

where d is the element spacing and c is the speed of light. Fourier transformation of Eq. (11.31) yields

$$Y(\omega) = \sum_{n=0}^{N-1} w_n X_n(\omega)exp(-j\omega\tau_n) = \sum_{n=0}^{N-1} w_n X_n(\omega)e^{-jn\Delta\theta} \tag{11.33}$$

The phase term $\Delta\theta$ is defined as

$$\Delta\theta = 2\pi f_0\frac{d}{c}sin\theta = \frac{2\pi}{\lambda}d sin\theta \tag{11.34}$$

$\omega = 2\pi f_0$ and $f_0/c = 1/\lambda$. Eq. (11.33) can be written in vector form as

$$\mathbf{Y} = \mathbf{s}^\dagger\mathbf{x} \tag{11.35}$$

$$\mathbf{s}^\dagger = \begin{bmatrix} 1 & e^{j\Delta\theta} & ... & e^{j(N-1)\Delta\theta} \end{bmatrix} \tag{11.36}$$

$$\mathbf{x}^\dagger = \begin{bmatrix} w_0 X_o & w_1 X_1 & ... & ...w_{N-1}X_{N-1} \end{bmatrix}^* \tag{11.37}$$

where the superscripts $*$ and $\dagger$, respectively, indicate complex conjugate and complex conjugate transpose.

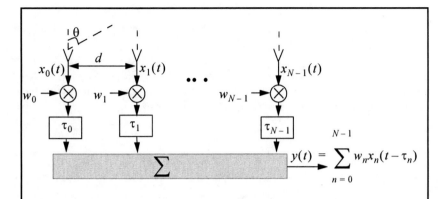

Figure 11.14. A linear array of size N, element spacing d, and an incident plane wave defined by $\sin\theta$.

Let A_1 be the amplitude of the wavefront defined by $\sin\theta_1$; it follows that the vector $\mathbf{x}$ is given by

$$\mathbf{x} = A_1 \mathbf{s}_1{}^* \tag{11.38}$$

where $\mathbf{s}_1$ is a steering vector can be written as,

$$\mathbf{s}^\dagger{}_1 = \begin{bmatrix} w_0 & w_1 e^{-j\Delta\theta_1} & \dots w_{N-1} e^{-j(N-1)\Delta\theta_1} \end{bmatrix} ; \quad \Delta\theta_1 = \frac{2\pi d}{\lambda} \cdot \sin\theta_1 \tag{11.39}$$

Using this notation, Eq. (11.35) can be expressed in the form

$$\mathbf{Y} = \mathbf{s}^\dagger \mathbf{x} = A_1 \mathbf{s}^\dagger \mathbf{s}^*{}_1 \tag{11.40}$$

The array pattern of the beam steered at θ_1 is computed as the expected value of $\mathbf{Y}$. In other words, the power spectrum density for the beamformer output is given by

$$S(k) = E[\mathbf{Y}\mathbf{Y}^\dagger] = P_1 \mathbf{s}^\dagger \Re \mathbf{s} \tag{11.41}$$

where $P_1 = E[|A_1|^2]$ and $\Re$ is the correlation matrix given by

$$\Re = E\{\mathbf{s}_1 \mathbf{s}^\dagger{}_1\} \tag{11.42}$$

Consider L incident plane waves with directions of arrival defined by

$$\Delta\theta_i = \frac{2\pi d}{\lambda} \sin\theta_i; \quad i = 1, L \tag{11.43}$$

The *nth* sample at the output of the *mth* sensor is

$$y_m(n) = \upsilon(n) + \sum_{i=1}^{L} A_i(n)exp(-jm\Delta\theta_i); \quad m = 0, N-1 \qquad \textbf{(11.44)}$$

where $A_i(n)$ is the amplitude of the *ith* plane wave and $\upsilon(n)$ is white, zero-mean noise with variance σ_υ^2, and it is assumed to be uncorrelated with the signals. Equation (11.44) can be written in vector notation as

$$\mathbf{y}(n) = \upsilon(n) + \sum_{i=1}^{L} A_i(n)\mathbf{s}_i^* \qquad \textbf{(11.45)}$$

A set of L steering vectors is needed to simultaneously form L beams. Define the steering matrix $\aleph$ as

$$\aleph = \begin{bmatrix} \mathbf{s}_1 & \mathbf{s}_2 & \dots & \mathbf{s}_L \end{bmatrix} \qquad \textbf{(11.46)}$$

Then the autocorrelation matrix of the field measured by the array is

$$\Re = E\{\mathbf{y}_m(n)\mathbf{y}_m^\dagger(n)\} = \sigma_\upsilon^2 I + \aleph \, \mathbf{C} \aleph^\dagger \qquad \textbf{(11.47)}$$

where $\mathbf{C} = dig\begin{bmatrix} P_1 & P_2 & \dots & P_L \end{bmatrix}$, and I is the identity matrix.

For example, consider the case depicted in Fig. 11.15, where an interfering signal located at angle $\theta_i = \pi/6$ off the antenna boresight. The desired signal is at $\theta_t = 0°$. The desired output should contain only the signal $s(t)$. From Eq. (11.33) and Eq. (11.34) the desired output is

$$y_d(t) = \sum_{n=0}^{1} w_n x_n(t - \tau_{n_t}) = w_0 x_0 + w_1 x_1 e^{-j\frac{2\pi}{\lambda}d\sin\theta_t} \qquad \textbf{(11.48)}$$

Since the angle $\theta_t = 0°$, it follows that

$$y_d(t) = \{Ae^{j2\pi f_0 t}\}\{(w_{0R} + jw_{0I}) + (w_{1R} + jw_{1I})\} \qquad \textbf{(11.49)}$$

$$\begin{aligned} w_0 &= w_{0R} + jw_{0I} \\ w_1 &= w_{1R} + jw_{1I} \end{aligned} \qquad \textbf{(11.50)}$$

Thus, in order to produce the desired signal, $s(t)$, at the output of the beamformer, it is required that

$$\begin{aligned} w_{0R} + w_{1R} &= 1 \Rightarrow w_{0R} = 1 - w_{1R} \\ w_{0I} + w_{1I} &= 0 \Rightarrow w_{0I} = -w_{1I} \end{aligned} \qquad \textbf{(11.51)}$$

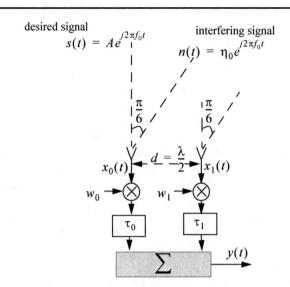

Figure 11.15. Two element array with an interfering signal at $\theta_i = \pi/6$.

Next, the output due to the interfering signal is

$$y_i(t) = \sum_{n=0}^{1} w_n x_n(t - \tau_{n_i}) = w_0 x_0 + w_1 x_1 e^{-j\frac{2\pi}{\lambda} d \sin\theta_i} \qquad (11.52)$$

Since the angle $\theta_i = \pi/6$, it follows that

$$y_d(t) = \{\eta_0 e^{j2\pi f_0 t}\} \{(w_{0R} + jw_{0I}) - j(w_{1R} + jw_{1I})\} \qquad (11.53)$$

and in order to eliminate the interference signal from the output of the beam-former, it is required that

$$\begin{aligned} w_{0R} + w_{1I} = 0 &\Rightarrow w_{0R} = -w_{1I} \\ w_{0I} - w_{1R} = 0 &\Rightarrow w_{0I} = w_{1R} \end{aligned} \qquad (11.54)$$

Solving Eq. (11.51) and Eq. (11.54) yields

$$w_{0R} = \frac{1}{2}; w_{0I} = \frac{1}{2}; w_{1R} = \frac{1}{2}; w_{1I} = \frac{-1}{2} \qquad (11.55)$$

Using the weights given in Eq. (11.55) will allow the desired signal to get through the beamformer unaffected; however, the interference signal will be completely eliminated from the output.

11.5. Adaptive Array Processing

11.5.1. Adaptive Signal Processing Using Least Mean Squares (LMS)

Adaptive signal processing evolved as a natural evolution from adaptive control techniques of time varying systems. Advances in digital processing computation techniques and associated hardware have facilitated maturing adaptive processing techniques and algorithms. Consider the basic adaptive digital system shown in Fig. 11.16. The system input is the sequence $x[k]$ and its output is the sequence $y[k]$. What differentiates adaptive from nonadaptive systems is that in adaptive systems the transfer function $H_k(z)$ is now time varying. The arrow through the transfer function box is used to indicate adaptive processing (or time varying transfer function). The sequence $d[k]$ is referred to as the *desired* response sequence. The error sequence is the difference between the desired response and the actual response. Remember that the desired sequence is not completely known; otherwise, if it were completely known, one would not need any adaptive processing to compute it. The definition of this desired response is dependent on the system specific requirements.

Many different techniques and algorithms have been developed to minimize the error sequence. Using one technique over another depends heavily on the operating environment under consideration. For example, if the input sequence is a stationary random process, then minimizing the error signal is nothing more than solving the least mean squares problem. However, in most adaptive processing systems the input signal is a nonstationary process. In this section the least mean squares technique is examined.

The least mean squares (LMS) algorithm is the most commonly utilized algorithm in adaptive processing, primary because of its simplicity. The time varying transfer function of order L can be written as a Finite Impulse Response (FIR) filter defined by

$$H_k(z) = b_0 + b_1 z^{-1} + \ldots + b_L z^{-L} \qquad (11.56)$$

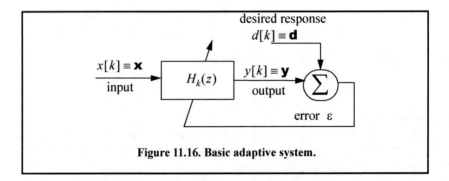

Figure 11.16. Basic adaptive system.

The input output relationship is given by the discrete convolution

$$y(k) = \sum_{n=0}^{L} b_n(k)x(k-n) \qquad (11.57)$$

The goal of the adaptive LMS process is to adjust the filter coefficients toward an optimum minimum mean square error (MMSE). The most common approach to achieving this MMSE utilizes the method of steepest descent. For this purpose, define the filter coefficients in vector notation as

$$\mathbf{b}_k = \begin{bmatrix} b_o(k) & b_1(k) \dots & b_L(k) \end{bmatrix}^{\dagger} \qquad (11.58)$$

then

$$\mathbf{b}_{k+1} = \mathbf{b}_k - \mu\nabla_k \qquad (11.59)$$

where μ is a parameter that controls how fast the error converges to the desired MMSE value, and the gradient vector ∇_k is defined by

$$\nabla_k = \frac{\partial}{\partial \mathbf{b}_k}E[\varepsilon_k^2] = \begin{bmatrix} \frac{\partial}{\partial \mathbf{b}_0(k)}E[\varepsilon_k^2] & \dots & \frac{\partial}{\partial \mathbf{b}_L(k)}E[\varepsilon_k^2] \end{bmatrix}^{\dagger} \qquad (11.60)$$

As clearly indicated by Eq. (11.59) the adaptive filter coefficients update rate is proportional to the negative gradient; thus, if the gradient is known at each step of the adaptive process, then better computation of the coefficient is obtained. In other words, the MMSE decreases from step k to step $k+1$. Of course, once the solution is found the gradient becomes zero and the coefficient will not change any more.

When the gradient is not known, estimates of the gradient are used based only on the instantaneous squared error. These estimates are defined by

$$\hat{\nabla}_k = \frac{\partial}{\partial \mathbf{b}_k}[\varepsilon_k^2] = 2\varepsilon_k\frac{\partial}{\partial \mathbf{b}_k}(d_k - y_k) \qquad (11.61)$$

Since the desired sequence $d[k]$ is independent from the output $y[k]$, Eq. (11.61) can be written as

$$\hat{\nabla}_k = -2\varepsilon_k\mathbf{x}_k \qquad (11.62)$$

where the vector $\mathbf{x}_k$ is the input signal sequence. Substituting Eq. (11.62) into Eq. (11.59) yields

$$\mathbf{b}_{k+1} = \mathbf{b}_k + 2\varepsilon_k\mu\mathbf{x}_k \qquad (11.63)$$

The choice of the convergence parameter μ plays a significant role in determining the system performance. This is clear because as indicated by Eq. (11.63), a successful implementation of the LMS algorithm depends on the input signal, the choice of the desired signal, and the convergence parameter. Much research and effort has been devoted toward selecting the optimal value for μ. Nonetheless, no universal value has been found. However, a range for this parameter has been determined to be $0 < \mu < 1$.

Often, a normalized value for the convergence parameter μ_N can be used instead of its absolute value. That is,

$$\mu_N = \frac{\mu}{(L+1)\sigma^2} \qquad (11.64)$$

where L is the order of the adaptive FIR filter and σ^2 is the variance (power) of the input signal. When the input signal is not stationary and its variance is varying with time, a time varying estimate of σ^2 is used. That is

$$\hat{\sigma}_k^2 = \alpha x_k^2 + (1 - \alpha)\hat{\sigma}_{k-1}^2 \qquad (11.65)$$

where α is a factor selected such that $0 < \alpha < 1$. Finally, Eq. (11.63) can be written as

$$\mathbf{b}_{k+1} = \mathbf{b}_k + \frac{2\varepsilon_k \mu \mathbf{x}_k}{(L+1)\hat{\sigma}_k^2} \qquad (11.66)$$

As an example and in reference to Fig. 11.15, let the input and desired signals be defined as

$$x[k] = \sqrt{2}\ sin\left(\frac{2\pi k}{20}\right) + n[k] \qquad ; \ k = 0, 1, ..., 500 \qquad (11.67)$$

$$d[k] = \sqrt{2}\ sin\left(\frac{2\pi k}{20}\right) \qquad ; \ k = 0, 1, ..., 500 \qquad (11.68)$$

where $n[k]$ is additive white noise with zero mean and variance $\sigma_n^2 = 2$. Figure 11.17 shows the output of the LMS algorithm defined in Eq. (11.66) when $\mu = 0.1$ and $\alpha = 0$. Figure 11.18 is similar to Fig. 11.17 except in this case, $\mu = 0.01$ and $\alpha = 0.1$. Note that in Fig. 11.18 the rate of convergences is reduced since μ is smaller than that used in Fig. 11.17; however, the filter's output is less noise because α is greater than zero which allows for more accurate updates of the noise variance as defined in Eq. (11.65). These plots can be reproduced using the following MATLAB code which utilizes the function "LMS.m" (see Section 11.6.2).

```
% Figures 11.17 and 11.18
close all; clear all
N = 501;
mu = 0.1; % convergence parameter
L = 20; % FIR filter order
B = zeros(1,L+1); % FIR coefficients
sigma = 2; %Initial estimate for noise power
alpha = .00; % forgetting factor
 k = 1:N;
noise = rand(1, length(k)) - .5; % Random noise
D = sqrt(2)*sin(2*pi*k/20);
X = D + sqrt(7)*noise;
Y = LMS(X, D, B, mu, sigma, alpha);
subplot(3,1,1)
plot(D,'linewidth',1); xlim([0 501]); grid on;
ylabel('Desired response'); title('\mu = 0.1; \alpha = 0.')
subplot(3,1,2)
plot(X,'linewidth',1); xlim([0 501]); grid on;
ylabel('Corrupted signal')
subplot(3,1,3)
plot(Y,'linewidth',1); xlim([0 501]); grid on;
xlabel('time in sec');
ylabel('LMS output')
```

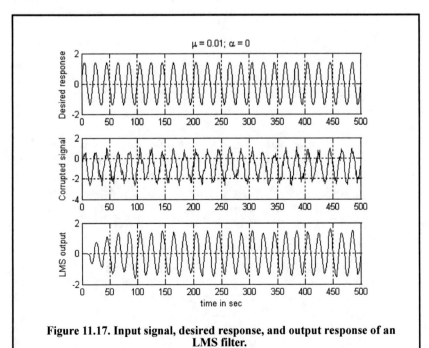

Figure 11.17. Input signal, desired response, and output response of an LMS filter.

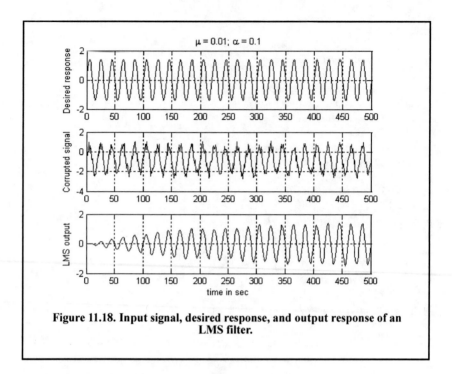

Figure 11.18. Input signal, desired response, and output response of an LMS filter.

11.5.2. The LMS Adaptive Array Processing

Consider the LMS adaptive array shown in Fig. 11.19. The difference between the reference signal and the array output constitutes an error signal. The error signal is then used to adaptively calculate the complex weights, using a predetermined convergence algorithm. The reference signal is assumed to be an accurate approximation of the desired signal (or desired array response). This reference signal can be computed using a training sequence or spreading code which is supposed to be known at the radar receiver. The format of this reference signal will vary from one application to another. But in all cases, the reference signal is assumed to be correlated with the desired signal. An increased amount of this correlation significantly enhances the accuracy and speed of the convergence algorithm being used. In this section, the LMS algorithm is assumed.

In general, the complex envelope of a bandpass signal and its corresponding analytical (pre-envelope) signal can be written using the quadrature components pair $(x_I(t), x_Q(t))$. Recall that the quadrature components are related using the Hilbert transform as follows:

$$x_Q(t) = \hat{x}_I(t) \qquad (11.69)$$

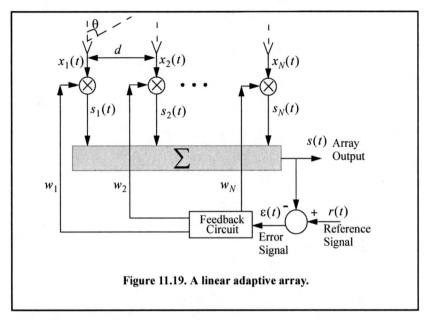

Figure 11.19. A linear adaptive array.

where $\hat{x}_I$ is the Hilbert transform of x_I. A bandpass signal $x(t)$ can be expressed as follows (see Chapter 2):

$$x(t) = x_I(t)\cos 2\pi f_0 t - x_Q(t)\sin 2\pi f_0 t \tag{11.70}$$

$$\psi(t) = x(t) + j\hat{x}(t) \equiv \tilde{x}(t)e^{j2\pi f_0 t} \tag{11.71}$$

$$\tilde{x}(t) = x_I(t) + jx_Q(t) \tag{11.72}$$

where $\psi(t)$ is the pre-envelope and $\tilde{x}(t)$ is the complex envelope. Equation (11.72) can be written using Eq. (11.69) as

$$\tilde{x}(t) = x_I(t) + jx_Q(t) = x_I(t) + j\hat{x}_I(t) \tag{11.73}$$

Using this notation, the adaptive array output signal, its reference signal, and the error signal can also be written using the same notation as

$$\tilde{s}(t) = s(t) + j\hat{s}(t) \tag{11.74}$$

$$\tilde{r}(t) = r(t) + j\hat{r}(t) \tag{11.75}$$

$$\tilde{\varepsilon}(t) = \varepsilon(t) + j\hat{\varepsilon}(t) \tag{11.76}$$

Referencing Fig. 11.19, denote the output of the nth array input signal as $s_n(t)$ and assume complex weights given by

$$w_n = w_{nI} + jw'_{nQ} = w_{nI} - jw_{nQ} \tag{11.77}$$

It follows that

$$s_n(t) = w_{nI} \, x_{nI}(t) + w_{nQ} \, x_{nQ}(t) \tag{11.78}$$

Taking the Hilbert transform of Eq. (11.78) yields

$$\hat{s}_n(t) = w_{nI} \, \hat{x}_{nI}(t) + w_{nQ} \, \hat{x}_{nQ}(t) \tag{11.79}$$

By using Eq. (11.67) into Eq. (11.79), one gets

$$\hat{x}_n(t) = w_{nI} \, \hat{x}_{nQ}(t) - w_{nQ} \, \hat{x}_{nI}(t) \tag{11.80}$$

The nth channel analytic signal is

$$\psi_n(t) = s_n(t) + j\hat{s}_n(t) \tag{11.81}$$

Substituting Eq. (11.78) and Eq. (11.79) into Eq. (11.80) gives

$$\psi_n(t) = w_{nI} \, x_{nI}(t) + w_{nQ} \, x_{nQ}(t) + j[w_{nI} \, \hat{x}_{nQ}(t) - w_{nQ} \, \hat{x}_{nI}(t)] \tag{11.82}$$

Collecting terms yields, using complex notation,

$$\psi_n(t) = w_n \tilde{x}_n(t) \tag{11.83}$$

Therefore, the output of the entire adaptive array is

$$\tilde{s}(t) = \sum_{n=1}^{N} \tilde{s}_n(t) = \sum_{n=1}^{N} w_n \tilde{x}_n(t) \tag{11.84}$$

which can be written using vector notation as

$$\tilde{s} = \mathbf{w}^t \tilde{\mathbf{x}} = \tilde{\mathbf{x}}^t \mathbf{w} \tag{11.85}$$

where the vectors $\tilde{\mathbf{x}}$ and $\mathbf{w}$ are given by

$$\tilde{\mathbf{x}} = \left[\tilde{x}_1(t) \, \tilde{x}_2(t) \, ... \tilde{x}_N(t) \right]^t \tag{11.86}$$

$$\mathbf{w} = \left[w_1 \, w_2 \, ... w_N \right]^t \tag{11.87}$$

The superscript $\{ ^t \}$ indicates the transpose operation.

As discussed earlier, one common technique to achieving the MMSE of an LMS algorithm is to use steepest descent. Thus, the complex weights in the LMS adaptive array are related as defined in Eq. (11.59). That is,

$$\mathbf{w}_{k+1} = \mathbf{w}_k - \mu \nabla_k \tag{11.88}$$

where again μ is the convergence parameter. The subscript k indicates time samples. In this case, the gradient vector ∇_k is defined by

$$\nabla_k = \frac{\partial}{\partial \mathbf{w}_k} E[\tilde{\varepsilon}_k^2] = \left[\frac{\partial}{\partial \mathbf{w}_0(k)} E[\tilde{\varepsilon}_k^2] \cdots \frac{\partial}{\partial \mathbf{w}_N(k)} E[\tilde{\varepsilon}_k^2] \right]^t \tag{11.89}$$

Rearranging Eq. (11.88) so that the rate of change between consecutive estimates of the complex weights is on one side of the equation yields

$$\mathbf{w}_{k+1} - \mathbf{w}_k = -\mu \frac{\partial}{\partial \mathbf{w}_k} (E[\tilde{\varepsilon}_k^2]) \tag{11.90}$$

where the middle portion of Eq. (11.89) was also substituted for the gradient vector. In this format, the left hand side of Eq. (11.90) represents the rate of change of the complex weights with respect to time (i.e., the derivative of the weights with respect to time). It follows that

$$\frac{d}{dt} \mathbf{w} = -\mu \frac{\partial}{\partial \mathbf{w}} (E[\tilde{\varepsilon}^2(t)]) \tag{11.91}$$

However, see from Fig. 11.18, that the error signal complex envelope is

$$\tilde{\varepsilon}(t) = \tilde{r}(t) - \sum_{n=1}^{N} w_n \tilde{x}_n(t) \Rightarrow \tilde{\varepsilon}(t) = \tilde{r}(t) - \tilde{\mathbf{x}}^t \mathbf{w} \tag{11.92}$$

It can be shown (see Problem 11.6) that

$$\frac{\partial}{\partial \mathbf{w}} E[\tilde{\varepsilon}^2(t)] = -E[\tilde{\mathbf{x}}^* \tilde{\varepsilon}(t)] \tag{11.93}$$

Therefore, Eq. (11.91) can be written as

$$\frac{d}{dt} \mathbf{w} = \mu E[\tilde{\mathbf{x}}^* \tilde{\varepsilon}(t)] \tag{11.94}$$

substituting Eq. (11.92) into Eq. (11.94) gives

$$\frac{d}{dt} \mathbf{w} = \mu E[\tilde{\mathbf{x}}^* (\tilde{r}(t) - \tilde{\mathbf{x}}^t \mathbf{w})] \tag{11.95}$$

Equivalently,

$$\frac{d}{dt}\mathbf{w} + \mu E[\tilde{\mathbf{x}}^*\tilde{\mathbf{x}}^t]\mathbf{w} = \mu E[\tilde{\mathbf{x}}^*\tilde{r}(t)] \tag{11.96}$$

The covariance matrix is by definition

$$\boldsymbol{C} = E[\tilde{\mathbf{x}}^*\tilde{\mathbf{x}}^t] = \begin{bmatrix} \tilde{x}_1^*\tilde{x}_1 & \tilde{x}_1^*\tilde{x}_2 & \cdots \\ \tilde{x}_2^*\tilde{x}_1 & \tilde{x}_2^*\tilde{x}_2 & \cdots \\ \cdots & \cdots & \end{bmatrix} \tag{11.97}$$

and the reference signal correlation vector $\mathbf{s}$ is

$$\mathbf{s} = E[\tilde{\mathbf{x}}^*\tilde{d}(t)] = E\begin{bmatrix} \tilde{x}_1^*\tilde{r} & \tilde{x}_2^*\tilde{r} & \cdots \end{bmatrix}^t \tag{11.98}$$

Using Eq. (11.98) and Eq. (11.97), one can rewrite the differential equation (DE) given Eq. (11.96) as

$$\frac{d}{dt}\mathbf{w} + \mu \boldsymbol{C}\mathbf{w} = \mu\mathbf{s} \tag{11.99}$$

The steady state solution for the DE defined in Eq. (11.99) (provided that the covariance matrix is not singular) is

$$\mathbf{w} = \boldsymbol{C}^{-1}\mathbf{s} \tag{11.100}$$

As the size of the covariance matrix increase (i.e., number of channels in the adaptive array) so does the complexity associated with computing the adaptive weights in real time. This is true because computing the inverse of large matrices in real time can be extremely challenging and demands significant amount of computing power. Consequently, the effectiveness of adaptive arrays has been limited to small-sized arrays, where only a few interfering signals can be eliminated (cancelled). Additionally, computing of a good estimate of the covariance matrix in real time is also difficult in practical applications. In order to mitigate that effect, a reasonable estimate for $E\{x_i x_j^*\}$ (the i,j element of the covariance matrix) is derived by averaging m independent samples of data from the same distribution. This approach can be extended to the entire covariance matrix by collecting M independent "snapshots" of data from N channels. Thus, the estimate of the covariance matrix can be given as,

$$\tilde{\boldsymbol{C}} \approx (\tilde{\mathbf{x}}^\dagger\tilde{\mathbf{x}})/M \tag{11.101}$$

The transient solution of Eq. (11.99) (see Problem 11.7) is

$$\mathbf{w}(t) = \sum_{n=1}^{N} \mathbf{p}_i e^{-\mu\lambda_i t} \tag{11.102}$$

where the vectors $\mathbf{p}_i$ are constants that depend on the initial value of $\mathbf{w}(t)$, and λ_i are the eigenvalues of the matrix $\boldsymbol{C}$. It follows that the complete solution of Eq. (11.99) is

$$\mathbf{w}(t) = \sum_{n=1}^{N} \mathbf{p}_i e^{-\mu\lambda_i t} + \boldsymbol{C}^{-1} \mathbf{s} \tag{11.103}$$

A very common measure of effectiveness of an adaptive array is the ratio of the total output interference power, S_o to the internal noise power, S_n.

Example:

Consider the two-element array in Section 11.4. Assume the desired signal is at directional-sine $\sin(\theta_t)$ and the interference signal is at $\sin(\theta_i)$. Calculate the adaptive weights so that the interference signal is cancelled.

Solution:

From Fig. 11.19

$$\tilde{x}_1(t) = \tilde{d}_1(t) + \tilde{n}_1(t) + \tilde{I}_1(t)$$

$$\tilde{x}_2(t) = \tilde{d}_2(t) + \tilde{n}_2(t) + \tilde{I}_2(t)$$

where d is the desired response, n is the noise, signal, and I is the interference signal. The noise signal is spatially incoherent, more specifically

$$E[\tilde{n}_i{}^*(t)\tilde{n}_j(t)] = \begin{cases} 0 & i \neq j \\ \sigma_n^2 & i = j \end{cases}$$

Also

$$E[\tilde{d}_i{}^*(t)\tilde{n}_i(t)] = 0 \qquad for \ \ all \ \ (i,j)$$

The desired signal is

$$\tilde{d}(t) = \tilde{d}_1(t) + \tilde{d}_2(t) = A_d e^{j2\pi f_0 t} e^{j\Theta_d} + A_d e^{j2\pi f_0 t} e^{j\Theta_d} e^{-j\pi \sin\theta_d}$$

where Θ_d is a uniform random variable. The interference signal is

$$\tilde{I}(t) = \tilde{I}_1(t) + \tilde{I}_2(t) = A_i e^{j2\pi f_0 t} e^{j\Theta_i} + A_i e^{j2\pi f_0 t} e^{j\Theta_i} e^{-j\pi \sin\theta_i}$$

where Θ_i *is a uniform random variable. Of course the random variables* Θ_d *and* Θ_i *are assumed to be statistically independent. In vector format*

$$\tilde{\mathbf{x}}_d = A_d e^{j2\pi f_0 t} e^{j\Theta_d} \begin{bmatrix} 1 \\ e^{-j\pi \sin\theta_d} \end{bmatrix}$$

$$\tilde{\mathbf{x}}_i = A_i e^{j2\pi f_0 t} e^{j\Theta_i} \begin{bmatrix} 1 \\ e^{-j\pi \sin\theta_i} \end{bmatrix}$$

Of course the noise vector is

$$\tilde{\mathbf{x}}_n = \begin{bmatrix} \tilde{n}_1(t) \\ \tilde{n}_2(t) \end{bmatrix}$$

and the reference signal is (this is an assumption so that the desired and reference signal are correlated)

$$\tilde{r}(t) = A_r e^{j2\pi f_0 t} e^{j\Theta_d}$$

Note that the input SNR is

$$SNR_d = A_d^2 / \sigma_n^2$$

and the interference to noise ratio is

$$SNR_i = A_i^2 / \sigma_n^2$$

The input signal can be written using vector notation as

$$\tilde{\mathbf{x}} = \tilde{\mathbf{x}}_d + \tilde{\mathbf{x}}_i + \tilde{\mathbf{x}}_n$$

The covariance matrix is computed from Eq. (11.97) as

$$\mathbf{C} = E[\tilde{\mathbf{x}}_d^* \tilde{\mathbf{x}}_d^t] = \begin{bmatrix} A_d^2 + A_i^2 + \sigma_n^2 & A_d^2 e^{-j\pi \sin\theta_d} + A_i^2 e^{-j\pi \sin\theta_i} \\ A_d^2 e^{j\pi \sin\theta_d} + A_i^2 e^{j\pi \sin\theta_i} & A_d^2 + A_i^2 + \sigma_n^2 \end{bmatrix}$$

In order to compute the covariance matrix eigenvalue, one needs to compute the determinant first

$$|\mathbf{C}| = 4A_d^2 A_i^2 \left(sin\left(\frac{\theta_d + \theta_i}{s}\right)\right)^2 + 2A_d^2 \sigma_n^2 + 2A_i^2 \sigma_n^2 + \sigma_n^4$$

Thus,

$$\mathbf{C}^{-1} = \frac{1}{|\mathbf{C}|}\begin{bmatrix} A_d^2 + A_i^2 + \sigma_n^2 & -A_d^2 e^{-j\pi\,sin\theta_d}-A_i^2 e^{-j\pi\,sin\theta_i} \\ -A_d^2 e^{j\pi\,sin\theta_d}-A_i^2 e^{j\pi\,sin\theta_i} & A_d^2 + A_i^2 + \sigma_n^2 \end{bmatrix}$$

The reference correlation vector is

$$\mathbf{s} = E[\tilde{\mathbf{x}}^* \tilde{r}(t)] = A_d A_r \begin{bmatrix} 1 \\ e^{j\pi\,sin\theta_d} \end{bmatrix}$$

It follows that the weights are

$$\mathbf{w} = \frac{A_d A_r}{|\mathbf{C}|}\begin{bmatrix} A_i^2 + \sigma_n^2 - A_i^2 e^{j\pi(sin\theta_d - sin\theta_i)} \\ e^{j\pi\,sin\theta_d}\{A_i^2 + \sigma_n^2 - A_i^2 e^{j\pi(sin\theta_i - sin\theta_d)}\} \end{bmatrix}$$

11.5.3. Sidelobe Cancelers (SLC)

Sidelobe cancelers typically consist of a main antenna (which can be a phased array or a single element) and one or more auxiliary antennas. The main antenna is referred to as the main channel; it is assumed to be highly directional and is pointed toward the desired signal angular location. The interfering signal is assumed to be located somewhere off the main antenna boresight (in the sidelobes). Because of this configuration the main channel receives returns from both the desired and the interfering signals. However, returns from the interfering signal in the main channel are weak because of the low main antenna sidelobe gain in the direction of the interfering signal. Also the auxiliary antenna returns are primarily from the interfering signal. This is illustrated in Fig. 11.20.

Referring to Fig. 11.20, $\tilde{s}(t)$ is the desired signal, $\tilde{n}(t)$ is the main channel noise signal which is primarily from the interfering signal, while $\tilde{n}'(t)$ is the interfering signal in the auxiliary array. It is assumed that the signals $\tilde{s}(t)$ and $\tilde{n}(t)$ are uncorrelated. It is also assumed that the interfering signal is highly correlated with the noise signal in the main channel. The basic idea behind SLC is to have the adaptive auxiliary channel produce an accurate estimate of the noise signal first, then to subtract that estimate from the main channel signal so that the output signal is mainly the desired signal.

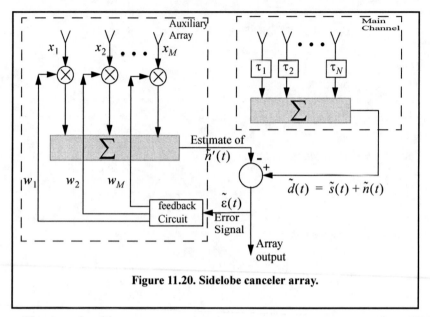

Figure 11.20. Sidelobe canceler array.

The error signal is

$$\tilde{\varepsilon} = \tilde{\mathbf{d}} - \mathbf{w}'\tilde{\mathbf{x}} \tag{11.104}$$

where $\tilde{\mathbf{x}}$ is the vector of auxiliary array signal, $\mathbf{w}$ is the adapted weights. The vector $\mathbf{d}$ of size M. The residual power is

$$P_{res} = E[\tilde{\varepsilon}\tilde{\varepsilon}^\dagger] \tag{11.105}$$

$$P_{res} = E[(\tilde{\mathbf{d}} - \mathbf{w}'\tilde{\mathbf{x}})(\tilde{\mathbf{d}}^* - \tilde{\mathbf{x}}^\dagger\mathbf{w}^*)] \tag{11.106}$$

It follows that

$$P_{res} = E[|\tilde{\mathbf{d}}|^2] - E[\tilde{\mathbf{d}}\tilde{\mathbf{x}}^\dagger\mathbf{w}^*] - E[\tilde{\mathbf{d}}^*\mathbf{w}'\tilde{\mathbf{x}}] - \mathbf{w}'E[\tilde{\mathbf{x}}\tilde{\mathbf{x}}^\dagger]\mathbf{w}^* \tag{11.107}$$

Differentiate the residual power with respect to $\mathbf{w}$ and setting the answer equal to zero (to compute the optimal weights that minimize the power residual) yields

$$\frac{\partial P_{res}}{\partial \mathbf{w}} = \mathbf{0} = -\tilde{\mathbf{x}}\tilde{\mathbf{d}} + \boldsymbol{C}_a\mathbf{w} \tag{11.108}$$

where $\boldsymbol{C}_a$ is the covariance matrix of the auxiliary channel. Finally, the optimal weights are given by

$$\mathbf{w} = \mathbf{C}_a^{-1} \; \tilde{\mathbf{x}}\tilde{\mathbf{d}} \qquad\qquad (11.109)$$

Note that the vector $\tilde{\mathbf{x}}\tilde{\mathbf{d}}$ represents the components that are common to both main and auxiliary channels. Note that Eq. (11.109) makes intuitive sense where the objective is to isolate the components in the data which are common to the main and auxiliary channels and we then wish to give them some heavy attenuation (which comes from inverting $\mathbf{C}_a$).

11.6. MATLAB Program Listings

This section presents listings for all the MATLAB programs used in this chapter. They are listed in the same order they appear in the text.

11.6.1. MATLAB Function "linear_array.m"

The function *"linear_array.m"* computes and plots the linear array gain pattern as a function of real sine-space. The syntax is as follows:

[theta, patternr, patterng] = linear_array(Nr, dolr, theta0, winid, win, nbits)

where

Symbol	Description	Units	Status
Nr	*number of elements in array*	*none*	*input*
dolr	*element spacing in lambda units*	*wavelengths*	*input*
theta0	*steering angle*	*degrees*	*input*
winid	*-1: No weighting is used* *1: Use weighting defined in win*	*none*	*input*
win	*window for side-lobe control*	*none*	*input*
nbits	*negative #: perfect quantization* *positive #: use 2^{nbits} quantization levels*	*none*	*input*
theta	*real angle available for steering*	*degrees*	*output*
patternr	*array pattern*	*dB*	*output*
patterng	*gain pattern*	*dB*	*output*

MATLAB Function *"linear_array.m"* Listing

```
function [theta,patternr,patterng] = linear_array(Nr,dolr,theta0,winid,win,nbits);
% This function computes and returns the gain radiation pattern for a linear array
% It uses the FFT to computes the pattern
%%%% *INPUTS ********** %%%%%%%%%%%%%%%%
```

```
% Nr ==> number of elements; dolr ==> element spacing (d) in lambda units divided
by lambda
% theta0 ==> steering angle in degrees; winid ==> use winid negative for no window,
winid positive to enter your window of size(Nr)
% win is input window, NOTE that win must be an NrX1 row vector; nbits ==> number
of bits used in the pahse shifters
% negative nbits mean no quantization is used
%%%% *OUTPUTS ********** %%%%%%%%%%%%%%%%%
% theta ==> real-space angle; patternr ==> array radiation pattern in dBs
% patterng ==> array directive gain pattern in dBs
%%%%%%%% ******************* %%%%%%%%%%%%%
eps = 0.00001;
n = 0:Nr-1;
i = sqrt(-1);
%if dolr is > 0.5 then; choose dol = 0.25 and compute new N
if(dolr <=0.5)
   dol = dolr;
   N = Nr;
else
   ratio = ceil(dolr/.25);
   N = Nr * ratio;
   dol = 0.25;
end
% choose proper size fft, for minimum value choose 256
Nrx = 10 * N;
nfft = 2^(ceil(log(Nrx)/log(2)));
if nfft < 256
   nfft = 256;
end
% convert steering angle into radians; and compute the sine of angle
theta0 = theta0 *pi /180.;
sintheta0 = sin(theta0);
% detrmine and comput quantized steering angle
if nbits < 0
   phase0 = exp(i*2.0*pi .* n * dolr * sintheta0);
else
   % compute and add the phase shift terms (WITH nbits quantization)
   % Use formula thetal = (2*pi*n*dol) * sin(theta0) divided into 2^nbits
   % and rounded to the nearest quantization level
   levels = 2^nbits;
   qlevels = 2.0 * pi / levels; % compute quantization levels
% compute the phase level and round it to the closest quantization level
   angleq = round(dolr .* n * sintheta0 * levels) .* qlevels; % vector of possible angles
   phase0 = exp(i*angleq);
end
% generate array of elements with or without window
if winid < 0
   wr(1:Nr) = 1;
```

```
else
   wr = win';
end
% add the phase shift terms
 wr = wr .* phase0;
 % determine if interpolation is needed (i.e N > Nr)
if N > Nr
   w(1:N) = 0;
   w(1:ratio:N) = wr(1:Nr);
else
   w = wr;
end
% compute the sine(theta) in real space that correspond to the FFT index
arg = [-nfft/2:(nfft/2)-1] ./ (nfft*dol);
idx = find(abs(arg) <= 1);
sinetheta = arg(idx);
theta = asin(sinetheta);
% convert angle into degrees
theta = theta .* (180.0 / pi);
% Compute fft of w (radiation pattern)
patternv = (abs(fftshift(fft(w,nfft)))).^2;
% convert radiation pattern to dBs
patternr = 10*log10(patternv(idx) ./Nr + eps);
% Compute directive gain pattern
rbarr  = 0.5 *sum(patternv(idx)) ./ (nfft * dol);
patterng = 10*log10(patternv(idx) + eps) - 10*log10(rbarr + eps);
return
```

11.6.2. MATLAB Function "LMS.m"

The function *"LMS.m"* implements Eq. (11.66). Its syntax is as follows

$$Y = LMS(X, D, B, mu, sigma, alpha)$$

where X is the corrupted sequence, D is the desired response, B is a vector containing the FIR filter coefficients (its initial value can be set to zero), *mu* is the convergence parameter, *sigma* is the SNR, and *alpha* is the forgetting factor.

MATLAB Function *"LMS.m"* Listing

```
function X = LMS(X, D, B, mu, sigma, alpha)
%   This program was written by Stephen Robinson a senior radar
%   engineer at deciBel Research, Inc. in Huntsville, AL
%   X = data vector ; size = 1 x N
%   D = desired signal vector; size = 1 x N
%   N = number of data samples and of adaptive iterations
%   B = adaptive coefficients of Lht order fFIRfilter; size = 1 x L
%   L = order of adaptive system
%   mu = convergence parameter
```

```
%  sigma = input signal power estimate
%  alpha = exponential forgetting factor
N = size(X,2)
L = size(B,2)-1
px = B;
for k = 1:N
   px(1) = X(k);
   X(k) = sum(B.*px);
   E = D(k) - X(k);
   sigma = alpha*(px(1)^2) + (1 - alpha)*sigma;
   tmp = 2*mu/((L+1)*sigma);
   B = B + tmp*E*px;
   px(L+1:-1:2) = px(L:-1:1);
end
return
```

Problems

11.1.   Consider an antenna whose diameter is $d = 3m$. What is the far field requirement for an X-band or an L-band radar that is using this antenna?

11.2.   Consider an antenna with electric field intensity in the xy-plane $E(\varsigma)$. This electric field is generated by a current distribution $D(y)$ in the yz-plane. The electric field intensity is computed using the integral

$$E(\varsigma) = \int_{-r/2}^{r/2} D(y)\exp\left(2\pi j\frac{y}{\lambda}\sin\varsigma\right)dy$$

where λ is the wavelength and r is the aperture. (a) Write an expression for $E(\varsigma)$ when $D(Y) = d_0$ (a constant). (b) Write an expression for the normalized power radiation pattern and plot it in dB.

11.3.   A linear phased array consists of 50 elements with $\lambda/2$ element spacing. (a) Compute the 3dB beam width when the main-beam steering angle is $0°$ and $45°$. (b) Compute the electronic phase difference for any two consecutive elements for steering angle $60°$.

11.4.   A linear phased array antenna consists of eight elements spaced with $d = \lambda$ element spacing. (a) Give an expression for the antenna gain pattern (assume no steering and uniform aperture weighting). (b) Sketch the gain pattern versus sine of the off-boresight angle β. What problems do you see is using $d = \lambda$ rather than $d = \lambda/2$?

11.5.   In Section 10.4.2 we showed how a DFT can be used to compute the radiation pattern of a linear phased array. Consider a linear of 64 elements at half wavelength spacing, where an FFT of size 512 is used to compute the pattern. What are the FFT bins that correspond to steering angles $\beta = 30°, 45°$?

11.6.   Derive Eq. (11.93).

11.7. Compute the transient solution of the DE defined in Eq. (11.99).

11.8. Compute the interference power to the intput power ratio of the example in Section 11.5.3.

11.9. To generate the sum and difference patterns for a linear array of size N follow this algorithm: To form the difference pattern, multiply the first $N/2$ elements by -1 and the second $N/2$ elements by +1. Plot the sum and difference patterns for a linear array of size 60.

11.10. Generate the delta/sum patterns for a 21-element linear array using the form $\frac{\Delta}{\Sigma} = j\dfrac{V_\Delta}{\sqrt{|V_\Delta|^2 + |V_\Sigma|^2}}$ where V_Δ is the difference voltage pattern and V_Σ is the sum voltage pattern.

Bibliography

Abramowitz, M. and Stegun, I. A., eds., *Handbook of Mathematical Functions, with Formulas, Graphs, and Mathematical Tables*, Dover Publications, New York, NY, 1970.

Balanis, C. A., *Antenna Theory, Analysis and Design*, Harper & Row, New York, 1982.

Barkat, M., *Signal Detection and Estimation*, Artech House, Norwood, MA, 1991.

Barton, D. K., *Modern Radar System Analysis*, Artech House, Norwood, MA, 1988.

Benedict, T. and Bordner, G., Synthesis of an Optimal Set of Radar Track-While-Scan Smoothing Equations, *IRE Transaction on Automatic Control, Ac-7*, July 1962, pp. 27-32.

Berkowitz, R. S., *Modern Radar - Analysis, Evaluation, and System Design*, John Wiley & Sons, Inc, New York, 1965.

Beyer, W. H., *CRC Standard Mathematical Tables*, 26th edition, CRC Press, Boca Raton, FL, 1981.

Billetter, D. R., *Multifunction Array Radar*, Artech House, Norwood, MA, 1989.

Blackman, S. S., *Multiple-Target Tracking with Radar Application*, Artech House, Norwood, MA, 1986.

Blake, L. V., *A Guide to Basic Pulse-Radar Maximum Range Calculation. Part-I: Equations, Definitions, and Aids to Calculation*, Naval Res. Lab. Report 5868, 1969.

Blake, L. V., *Radar-Range Performance Analysis*, Lexington Books, Lexington, MA, 1980.

Boothe, R. R., *A Digital Computer Program for Determining the Performance of an Acquisition Radar Through Application of Radar Detection Probability Theory*, U.S. Army Missile Command, Report No. RD-TR-64-2, Redstone Arsenal, Alabama, 1964.

Bowman, J. J., Piergiorgio, L. U., and Senior, T. B., *Electromagnetic and Acoustic Scattering by Simple Shapes*, North-Holland Pub. Co, Amsterdam, 1969.

Brookner, E., ed., *Aspects of Modern Radar*, Artech House, Norwood, MA, 1988.

Brookner, E., ed., *Practical Phased Array Antenna System*, Artech House, Norwood, MA, 1991.

Brookner, E., *Radar Technology*, Lexington Books, Lexington, MA, 1996.

Burdic, W. S., *Radar Signal Analysis*, Prentice Hall, Englewood Cliffs, NJ, 1968.

Brookner, E., *Tracking and Kalman Filtering Made Easy*, John Wiley & Sons, New York, 1998.

Cadzow, J. A., *Discrete-Time Systems, An Introduction with Interdisciplinary Applications*, Prentice Hall, Englewood Cliffs, NJ, 1973.

Carlson, A. B., *Communication Systems, An Introduction to Signals and Noise in Electrical Communication*, 3rd edition, McGraw-Hill, New York, 1986.

Carpentier, M. H., *Principles of Modern Radar Systems*, Artech House, Norwood, MA, 1988.

Compton, R. T., *Adaptive Antennas*, Prentice Hall, Englewood Cliffs, NJ, 1988.

Cook, E. C. and Bernfeld, M., *Radar Signals An Introduction to Theory and Application*, Artech House, Norwood, MA, 1993.

Costas, J. P., A Study of a Class of Detection Waveforms Having Nearly Ideal Range-Doppler Ambiguity Properties, *Proc. IEEE 72*, 1984, pp. 996-1009.

Curry, G. R., *Radar System Performance Modeling*, Artech House, Norwood, 2001.

DiFranco, J. V. and Rubin, W. L., *Radar Detection*. Artech House, Norwood, MA, 1980.

Dillard, R. A. and Dillard, G. M., *Detectability of Spread-Spectrum Signals*, Artech House, Norwood, MA, 1989.

Edde, B., *Radar Principles, Technology, Applications*, Prentice Hall, Englewood Cliffs, NJ, 1993.

Elsherbeni, A., Inman, M. J., and Riley, C., Antenna Design and Radiation Pattern Visualization, *The 19th Annual Review of Progress in Applied Computational Electromagnetics*, ACES'03, Monterey, CA, March 2003.

Fehlner, L. F., *Marcum's and Swerling's Data on Target Detection by a Pulsed Radar*, Johns Hopkins University, Applied Physics Lab. Rpt. # TG451, July 2, 1962, and Rpt. # TG451A, September 1964.

Fielding, J. E. and Reynolds, G. D., *VCCALC: Vertical Coverage Calculation Software and Users Manual*, Artech House, Norwood, MA, 1988.

Gabriel, W. F., Spectral Analysis and Adaptive Array Superresolution Techniques, *Proc. IEEE*, Vol. 68, June 1980, pp. 654-666.

Gelb, A., ed., *Applied Optimal Estimation*, MIT Press, Cambridge, MA, 1974.

Goldman, S. J., *Phase Noise Analysis in Radar Systems, Using Personal Compurters*, John Wiley & Sons, New York, NY, 1989.

Grewal, M. S. and Andrews, A. P., *Kalman Filtering - Theory and Practice Using MATLAB*, 2nd edition, Wiley & Sons Inc., New York, 2001.

Hamming, R. W., *Digital Filters*, 2nd edition, Prentice Hall, Englewood Cliffs, NJ, 1983.

Hanselman, D. and Littlefield, B., *Mastering MATLAB 5, A Complete Tutorial and Reference,* MATLAB Curriculum Series, Prentice Hall, Englewood Cliffs, NJ, 1998.

Hirsch, H. L. and Grove, D. C., *Practical Simulation of Radar Antennas and Radomes,* Artech House, Norwood, MA, 1987.

Hovanessian, S. A., *Radar System Design and Analysis,* Artech House, Norwood, MA, 1984.

James, D. A., *Radar Homing Guidance for Tactical Missiles,* John Wiley & Sons, New York, 1986.

Jin, J., *The Finite Element Method in Electromagnetics,* John Wiley & Sons, New York, 2002.

Kanter, I., Exact Detection Probability for Partially Correlated Rayleigh Targets, *IEEE Trans, AES-22,* March 1986, pp. 184-196.

Kay, S. M., *Fundamentals of Statistical Signal Processing - Estimation Theory,* Volume I, Prentice Hall Signal Processing Series, Englewood Cliffs, NJ, 1993.

Kay, S. M., *Fundamentals of Statistical Signal Processing - Detection Theory,* Volume II, Prentice Hall Signal Processing Series, Englewood Cliffs, NJ, 1993.

Keller, J. B., Geometrical Theory of Diffraction, *Journal Opt. Soc. Amer.,* Vol. 52, February 1962, pp. 116-130.

Klauder, J. R., Price, A. C., Darlington, S., and Albershiem, W. J., The Theory and Design of Chirp Radars, *The Bell System Technical Journal,* Vol. 39, No. 4, 1960.

Klemm, R., *Principles of Space-Time Adaptive Processing,* 3rd Ed, IET, London UK, 2006.

Knott, E. F., Shaeffer, J. F., and Tuley, M. T., *Radar Cross Section,* 2nd edition, Artech House, Norwood, MA, 1993.

Lativa, J., Low-Angle Tracking Using Multifrequency Sampled Aperture Radar, *IEEE-AES Trans.,* Vol. 27, No. 5, September 1991, pp.797-805.

Lee, S. W. and Mittra, R., Fourier Transform of a Polygonal Shape Function and Its Application in Electromagnetics, *IEEE Trans. Antennas and Propagation,* Vol. 31, January 1983, pp. 99-103.

Levanon, N., *Radar Principles,* John Wiley & Sons, New York, 1988.

Levanon, N. and Mozeson, E., Nullifying ACF Grating Lobes in Stepped-frequency Train of LFM Pulses, *IEEE-AES Trans.,* Vol. 39, No. 2, April 2003, pp. 694-703.

Levanon, N. and Mozeson, E., *Radar Signals,* John Wiley-Interscience, Hoboken, NJ, 2004.

Lewis, B. L., Kretschmer, Jr., F. F., and Shelton, W. W., *Aspects of Radar Signal Processing,* Artech House, Norwood, MA, 1986.

Long, M. W., *Radar Reflectivity of Land and Sea,* Artech House, Norwood, MA, 1983.

Lothes, R. N., Szymanski, M. B., and Wiley, R. G., *Radar Vulnerability to Jamming*, Artech House, Norwood, MA, 1990.

Mahafza, B. R., *Introduction to Radar Analysis*, CRC Press, Boca Raton, FL, 1998.

Mahafza, B. R., *Radar Systems Analysis and Design Using MATLAB,* 2nd Ed, Taylor & Francis, Boca Raton, FL, 2005.

Mahafza, B. R. and Polge, R. J., Multiple Target Detection Through DFT Processing in a Sequential Mode Operation of Real Two-Dimensional Arrays, *Proc. of the IEEE Southeast Conf. '90*, New Orleans, LA, April 1990, pp. 168-170.

Mahafza, B. R., Heifner, L.A., and Gracchi, V. C., Multitarget Detection Using Synthetic Sampled Aperture Radars (SSAMAR), *IEEE-AES Trans.*, Vol. 31, No. 3, July 1995, pp. 1127-1132.

Mahafza, B. R. and Sajjadi, M., Three-Dimensional SAR Imaging Using a Linear Array in Transverse Motion, *IEEE-AES Trans.*, Vol. 32, No. 1, January 1996, pp. 499-510.

Marchand, P., *Graphics and GUIs with MATLAB*, 2nd edition, CRC Press, Boca Raton, FL, 1999.

Marcum, J. I., A Statistical Theory of Target Detection by Pulsed Radar, Mathematical Appendix, *IRE Trans.*, Vol. IT-6, April 1960, pp. 259-267.

Medgyesi-Mitschang, L. N. and Putnam, J. M., Electromagnetic Scattering from Axially Inhomogenous Bodies of Revolution, *IEEE Trans. Antennas and Propagation.*, Vol. 32, August 1984, pp. 797-806.

Meeks, M. L., *Radar Propagation at Low Altitudes*, Artech House, Norwood, MA, 1982.

Melsa, J. L. and Cohn, D. L., *Decision and Estimation Theory*, McGraw-Hill, New York, 1978.

Mensa, D. L., *High Resolution Radar Imaging*, Artech House, Norwood, MA, 1984.

Meyer, D. P. and Mayer, H. A., *Radar Target Detection: Handbook of Theory and Practice*, Academic Press, New York, 1973.

Monzingo, R. A. and Miller, T. W., *Introduction to Adaptive Arrays,* John Wiley & Sons, New York, 1980.

Morchin, W., *Radar Engineer's Sourcebook*, Artech House, Norwood, MA, 1993.

Morris, G. V., *Airborne Pulsed Doppler Radar*, Artech House, Norwood, MA, 1988.

Nathanson, F. E., *Radar Design Principles*, 2nd edition, McGraw-Hill, New York, 1991.

Navarro, Jr., A. M., *General Properties of Alpha Beta and Alpha Beta Gamma Tracking Filters*, Physics Laboratory of the National Defense Research Organization TNO, Report PHL 1977-92, January 1977.

North, D. O., An Analysis of the Factors Which Determine Signal/Noise Discrimination in Pulsed Carrier Systems, *Proc. IEEE 51*, No. 7, July 1963, pp. 1015-1027.

Oppenheim, A. V. and Schafer, R. W., *Discrete-Time Signal Processing*, Prentice Hall, Englewood Cliffs, NJ, 1989.

Oppenheim, A. V., Willsky, A. S., and Young, I. T., *Signals and Systems*, Prentice Hall, Englewood Cliffs, NJ, 1983.

Orfanidis, S. J., *Optimum Signal Processing, an Introduction*, 2nd edition, McGraw-Hill, New York, 1988.

Papoulis, A., *Probability, Random Variables, and Stochastic Processes*, 2nd edition, McGraw-Hill, New York, 1984.

Parl, S. A., New Method of Calculating the Generalized Q Function, *IEEE Trans. Information Theory*, Vol. IT-26, No. 1, January 1980, pp. 121-124.

Peebles, Jr., P. Z., *Probability, Random Variables, and Random Signal Principles*, McGraw-Hill, New York, 1987.

Peebles, Jr., P. Z., *Radar Principles*, John Wiley & Sons, New York, 1998.

Pettit, R. H., *ECM and ECCM Techniques for Digital Communication Systems*, Lifetime Learning Publications, New York, 1982.

Polge, R. J., Mahafza, B. R., and Kim, J. G., *Extension and Updating of the Computer Simulation of Range Relative Doppler Processing for MM Wave Seekers*, Interim Technical Report, Vol. I, prepared for the U.S. Army Missile Command, Redstone Arsenal, Alabama, January 1989.

Polge, R. J., Mahafza, B. R., and Kim, J. G., Multiple Target Detection Through DFT Processing in a Sequential Mode Operation of Real or Synthetic Arrays, *IEEE 21st Southeastern Symposium on System Theory*, Tallahassee, FL, 1989, pp. 264-267.

Poularikas, A., *Signals and Systems Primer with MATLAB*, Taylor & Francis, Boca Raton, FL, 2007.

Poularikas, A. and Ramadan, Z. M., *Adaptive Filtering Primer with MATLAB*, Taylor & Francis, Boca Raton, FL, 2006.

Poularikas, A. and Seely, S., *Signals and Systems*, PWS Publishers, Boston, MA, 1984.

Putnam, J. N. and Gerdera, M. B., CARLOS TM: A General-Purpose Three-Dimensional Method of Moments Scattering Code, *IEEE Trans. Antennas and Propagation*, Vol. 35, April 1993, pp. 69-71

Reed, H. R. and Russell, C. M., *Ultra High Frequency Propagation*, Boston Technical Publishers, Inc., Lexington, MA, 1964.

Resnick, J. B., *High Resolution Waveforms Suitable for a Multiple Target Environment*, MS Thesis, MIT, Cambridge, MA, June 1962.

Richards, M. A., *Fundamentals of Radar Signal Processing*, McGraw-Hill, New York, 2005.

Rihaczek, A. W., *Principles of High Resolution Radars*, McGraw-Hill, New York, 1969.

Robertson, G. H., Operating Characteristics for a Linear Detector of CW Signals in Narrow-band Gaussian Noise, *Bell Sys. Tech. Journal*, Vol. 46 April 1967, pp. 755-774.

Ross, R. A., Radar Cross Section of Rectangular Flat Plate as a Function of Aspect Angle, *IEEE Trans.* AP-14, 1966, p. 320.

Ruck, G. T., Barrick, D. E., Stuart, W. D., and Krichbaum, C. K., *Radar Cross Section Handbook*, Volume 1, Plenum Press, New York, 1970.

Ruck, G. T., Barrick, D. E., Stuart, W. D., and Krichbaum, C. K., *Radar Cross Section Handbook*, Volume 2, Plenum Press, New York, 1970.

Rulf, B. and Robertshaw, G. A., *Understanding Antennas for Radar, Communications, and Avionics*, Van Nostrand Reinhold, 1987.

Scanlan, M.J., ed., *Modern Radar Techniques*, Macmillan, New York, 1987.

Scheer, J. A. and Kurtz, J. L., ed., *Coherent Radar Performance Estimation*, Artech House, Norwood, MA, 1993.

Shanmugan, K. S. and Breipohl, A. M., *Random Signals: Detection, Estimation and Data Analysis*, John Wiley & Sons, New York, 1988.

Shatz, M. P. and Polychronopoulos, G. H., *An Algorithm for Evaluation of Radar Propagation in the Spherical Earth Diffraction Region.* IEEE Transactions on Antenna and Propagation, VOL. 38, NO.8, August 1990, pp. 1249-1252.

Sherman, S. M., *Monopulse Principles and Techniques*, Artech House, Norwood, MA.

Singer, R. A., Estimating Optimal Tracking Filter Performance for Manned Maneuvering Targets, *IEEE Transaction on Aerospace and Electronics, AES-5*, July 1970, pp. 473-483.

Skillman, W. A., *DETPROB: Probability of Detection Calculation Software and User's Manual*, Artech House, Norwood, MA, 1991.

Skolnik, M. I., *Introduction to Radar Systems*, McGraw-Hill, New York, 1982.

Skolnik, M. I., ed., *Radar Handbook*, 2nd edition, McGraw-Hill, New York, 1990.

Song, J. M., Lu, C. C., Chew, W. C., and Lee, S. W., Fast Illinios SolverCode (FSIC), *IEEE Trans. Antennas and Propagation*, Vol. 40, June 1998, pp. 27-34.

Stearns, S. D. and David, R. A., *Signal Processing Algorithms*, Prentice Hall, Englewood Cliffs, NJ, 1988.

Stimson, G. W., *Introduction to Airborne Radar*, Hughes Aircaft Company, El Segundo, CA, 1983.

Stratton, J. A., *Electromagnetic Theory*, McGraw-Hill, New York, 1941.

Stremler, F. G., *Introduction to Communication Systems*, 3rd edition, Addison-Wesley, New York, 1990.

Stutzman, G. E., Estimating Directivity and Gain of Antennas, *IEEE Antennas and Propagation Magazine 40*, August 1998, pp. 7-11.

Swerling, P., Probability of Detection for Fluctuating Targets, *IRE Transaction on Information Theory,* Vol. IT-6, April 1960, pp. 269-308.

Taflove, A., *Computational Electromagnetics: The Finite-Difference Time-Domain Method,* Artech House, Norwood, MA, 1995.

Van Trees, H. L., *Detection, Estimation, and Modeling Theory,* Part I, Wiley & Sons, Inc., New York, 2001.

Van Trees, H. L., *Detection, Estimation, and Modeling Theory,* Part III, Wiley & Sons, New York, 2001.

Van Trees, H. L., *Optimum Array Processing,* Part IV of *Detection, Estimation, and Modeling Theory,* Wiley & Sons, New York, 2002.

Tzannes, N. S., *Communication and Radar Systems,* Prentice Hall, Englewood Cliffs, NJ, 1985.

Urkowitz, H., *Decision and Detection Theory,* Unpublished Lecture Notes, Lockheed Martin Co., Moorestown, NJ.

Urkowtiz, H., *Signal Theory and Random Processes,* Artech House, Norwood, MA, 1983.

Vaughn, C. R., Birds and Insects as Radar Targets: A Review, *Proc. IEEE,* Vol. 73, No. 2, February 1985, pp. 205-227.

Wehner, D. R., *High Resolution Radar,* Artech House, Norwood, MA, 1987.

Weiner, M. M., ed., *Adaptive Antennas and Recivers,* Taylor & Francis, Boca Raton, FL, 2006.

White, J. E., Mueller, D. D., and Bate, R. R., *Fundamentals of Astrodynamics,* Dover Publications, New York, NY, 1971.

Ziemer, R. E. and Tranter, W. H., *Principles of Communications, Systems, Modulation, and Noise,* 2nd edition, Houghton Mifflin, Boston, MA, 1985.

Zierler, N., *Several Binary-Sequence Generators,* MIT Technical Report No. 95, Sept. 1955.

Index

A

Active correlation, 326
  also see Pulse compression
Adaptive arrays
  adaptive weights, 454, 456, 461
  convergence parameter, 449, 450
  covariance matrix, 456, 460
  beamforming *see* nonadaptive beamforming
  LMS, 448, 452
  reference correlation vector, 456
  SLC, 459-461
Ambiguity function, 171, 172, 187, 188
  Barker code, 233-241
  contour diagrams, 216
  ideal, 189
  LFM, 192-197
  NLFM, 208
  PRN, 241-249
  properties, 188
  pulse train, 197-201
  pulse train with LFM, 202-206
  single pulse, 189-192
  SFW, 206-208
Amplitude estimate, 183
Analytic signal *see* Signals
Arrays
  general array, 430-432
  linear, 432-4443
Atmosphere, 41, 42
  stratified, 44-47
Atmospheric attenuation, 65-66

B

Bandpass signal *see* Signals
Bandwidth *see* Effective bandwidth
Barker code, 233-241
Bessel-Jacobi equation, 101
Binary phase codes
  see Barker code
  see Codes, PRN
Blind speeds, 377, 384
Boltzmann's constant, 12

C

Cancelers *see* Moving Target Indicator
  (MTI)
Chirp waveforms
  down-chirp, 110
  up-chirp, 110
Clutter
  CNR, 364
  components, 374-375
  definition, 353
  density, 353, 354
  main beam, 361
  RCS, 361-373
  sidelode clutter, 361
  spectrum, 373-374, 376
  statistical models, 373, 374
  subclutter visibility, 392-393
  surface clutter, 354-356
  surface height irregularity, 355
  volume, 358-361
Codes
  Barker, 233-241
  binary phase codes, 232
  Costas, 252-255
  definition, 225, 226
  Franks, 249
  frequency, 252
  phase codes, 232
  Polyphase code, 249
  PRN, 241-249
  pulse-train codes, 226-231
Coherent integration *see* Pulse integration
Coherence, 10
Complementary error function, 265
Complex envelope, 96
Compressed pulse width, 196
Compression gain, 196
Compression ratio, 196
Constant false alarm rate (CFAR)
  cell averaging (single pulse), 293-295
  cell-averaging CFAR (noncoherent
  integration), 295-296
Convolution integral, 89

Windowing techniques, 128-133
   Hamming, 131
   Hanning, 131
   Kaiser, 131

Z

Z-transform, 124, 125